Microbial Biostimulants for Sustainable Agriculture and Environmental Bioremediation

Microbial Biostimulants for Sustainable Agriculture and Environmental Bioremediation

Edited by
Inamuddin
Charles Oluwaseun Adetunji
Mohd Imran Ahamed
Tariq Altalhi

CRC Press
Taylor & Francis Group
Boca Raton London New York

CRC Press is an imprint of the
Taylor & Francis Group, an **informa** business

First edition published 2022
by CRC Press
6000 Broken Sound Parkway NW, Suite 300, Boca Raton, FL 33487-2742

and by CRC Press
4 Park Square, Milton Park, Abingdon, Oxon, OX14 4RN

CRC Press is an imprint of Taylor & Francis Group, LLC

ISBN: 978-1-032-03575-8 (hbk)
ISBN: 978-1-032-03580-2 (pbk)
ISBN: 978-1-003-18803-2 (ebk)

DOI: 10.1201/9781003188032

Typeset in Times
by codeMantra

Contents

Preface

Today, the agriculture industry is confronted with simultaneous issues in terms of how to fully embrace mass production of safer food in terms of both quality and quantity, in order to fulfil the demands of the ever-increasing population. Other difficulties include how to improve resource use efficacy while reducing environmental impacts, particularly on human and environmental health. Most industries have been concerned to avoid significant levels of soil pollution and environmental threats as a result of the excessive and harmful use of synthetic products on crops. Therefore, there is a need to adopt sustainable technological innovations that could ensure the sustainability of agricultural production systems mainly through an effective decrease in the level of synthetic agrochemicals such as fertilizers and pesticides. Hence, the application of eco-friendly, innovative, and biocompatible-biotechnological techniques through the application of biostimulants derived from a beneficial microorganism is highly accepted. This could ensure an increase in crop production, improvement in the nutritional attributes, enhanced fertility and maintenance of the soil health, high level of tolerance to different abiotic stressors, the release of required nutrients to the soil, resistance to plant pathogens/pests, efficient use of nutrients, and tolerance to abiotic/biotic stress rejuvenation of the polluted environment.

As per industry estimates, the worldwide microbial biostimulants market was valued at USD 1.74 billion in 2016 and is expected to grow at a rate of 10.2% from 2017 to 2025. As a result, the daily growth in demand for increased productivity, along with rapid soil deterioration, is likely to shift the market in the time allotted.

This book is intended to provide extensive information on the use of microbial biostimulants in the bioremediation of contaminated environments as well as their efficacy in improving agricultural output to satisfy the demands of the ever-increasing population. Under a multitude of organizational and environmental stratagems, relevant information on the physiological, cellular, and molecular mechanisms of action underpinning microbial-biostimulant interactions has been studied. It puts emphasis on the agricultural role in improving nutrient utilization efficiency, their tolerance to biotic/abiotic stress, and the specific synergetic effect of microbial biostimulants derived from humic substances, complex organic materials produced from urban waste products, and agroindustrial products such as manure, compost, and sludge extract, as well as N-containing substances such as betaines, peptides, and polyamines together. It also emphasizes the importance of microbial biostimulants for crop–microbe interactions and nutrient utilization efficiency, while timing, location, soil condition, and rate all play a role in the potential success of microbial biostimulants. Nanomaterials generated from physiologically active metabolites obtained from a new biostimulant strain are studied for their bioremediation and crop enhancement properties. It is an invaluable guide to planners, governmental and non-governmental organizations, environmentalists, biotechnologists, engineers, students, professors, scientists, and R&D industrial experts working in the field of sustainability through microbial biostimulants. The summaries of the work reported in the following 14 chapters are as follows.

Chapter 1 discusses binary bacterial–macromycete cultures. Mushrooms are characterized as a precious source of nutritive food ingredients, additionally capable of producing a lot of physiologically active compounds. Attention is paid to the great biotechnological interest of wide-scale cultivated mushrooms that encourages research into microbial biostimulants as the factors of biological origin that promote their growth and assist in tolerating different stresses.

Chapter 2 focuses on the effectiveness of microbial biostimulants against insects and diseases to minimize crop losses. These biostimulants are of fungal, bacterial, and viral origins having huge biocontrol potential. These are environmentally friendly, are compatible with other biocontrol agents and insecticides, and make good replacements for chemical pesticides.

Chapter 3 details the role of plant microbial biostimulants in plant–insect interactions. Plants have strong symbiotic relationships with mycorrhizal fungi, plant growth-promoting rhizobacteria (PGPR), and other endophytes. They improve the plant immune system and suppress the herbivory injuries by changing the behaviour and fitness of insect herbivores. Microbial biostimulants and their secondary metabolic structures can be effective insecticidal tools.

Chapter 4 provides an overview of the efficacy of microbial biostimulants against soilborne diseases associated with roots or root-borne diseases.

Chapter 5 emphasizes the use of microbial biostimulants to treat plants because biotic/abiotic constraints limit plant development and production, and microbial biostimulants increase the plant growth under unfavourable situations. Plants undergo several changes in morphological, physiological, and biochemical mechanisms in response to microbial pretreatment. Consequently, plants manifest a notable tolerance to environmental constraints.

Chapter 6 gives a note on crop losses due to plant nematodes and their management through bacterial and fungal biostimulants. These biostimulants trigger several activities suppressing the nematode population, stimulating ISR, and changing biochemical processes in plants. These biostimulants are available commercially, and their application is eco-friendly and sustainable in the environment.

Chapter 7 reviews the classification of nitrogenous biofertilizers of bacterial, fungal, and algal origins and nitrogen-fixing biostimulants along with their role in promoting plant growth, development, and resistance towards stress conditions in detail. It also puts emphasis on need-based research in order to better comprehend the role of biofertilizers and biostimulants.

Chapter 8 elaborates on the role of microbe-based biostimulants as a sustainable resource for bioremediation of polluted environments. It further details various types of microorganisms used as potential biostimulants and their general mechanism of action employed for pollutant stress alleviation.

Chapter 9 summarizes several applications of microbial biostimulants in improving bioremediation technology to manage, treat, and recover municipal solid waste. Additionally, it presents an update of the principle, types, and benefits of the bioremediation process and microbial biostimulant concepts. A new ex situ bioremediation technique for this type of waste based on microbiological and enzymatic processes generates a lot of interest.

Chapter 10 addresses the use of bioremediation of organic and inorganic compounds as a substitute for conventional methods in treating the environment. Several highly effective and environmentally friendly techniques are also detailed. Additionally, the role of microbes in the remediation of water and soil in various industries is presented.

Chapter 11 outlines a summary of biostimulation requirements for their beneficial use under both aerobic and anaerobic conditions. This is supported by numerous studies and biostimulation examples.

Chapter 12 briefly focuses on commercially available bacterial and fungal biostimulants, which serve as biofertilizers and biocontrol agents for agriculture. The aspects of biofertilizer production and spectrum of antagonistic activity of biocontrol agents are discussed. The indigenous guidelines and regulations governing their production and marketing are covered.

Chapter 13 presents the concepts, definitions, terminologies, and overview of microbial biostimulants. It further examines the formulation and commercialization of these products and outlines the bottlenecks in the industry. Finally, it provides the future perspectives that can advance the industry and promote the development of sustainable agricultural systems globally.

Editors

Inamuddin, PhD, is an Assistant Professor in the Department of Applied Chemistry, Aligarh Muslim University, Aligarh, India. He earned an M.Sc. in organic chemistry at Chaudhary Charan Singh (CCS) University, Meerut, India, in 2002. He earned a M.Phil. and a PhD in applied chemistry at Aligarh Muslim University (AMU), India, in 2004 and 2007, respectively. He has extensive research experience in the multidisciplinary fields of analytical chemistry, materials chemistry, electrochemistry, and, more specifically, renewable energy and environment. He has worked on different research projects as project fellow and senior research fellow funded by the University Grants Commission (UGC), Government of India, and the Council of Scientific and Industrial Research (CSIR), Government of India. He has received the Fast Track Young Scientist Award from the Department of Science and Technology, India, for his work in the area of bending actuators and artificial muscles. He has completed four major research projects sanctioned by the UGC, Department of Science and Technology, the CSIR, and the Council of Science and Technology, India. He has published 199 research articles in international journals of repute and 19 book chapters in knowledge-based book editions published by renowned international publishers. He has published 150 edited books with Springer (UK), Elsevier, Nova Science Publishers, Inc. (USA), CRC Press – Taylor & Francis Asia Pacific, Trans Tech Publications Ltd. (Switzerland), IntechOpen Limited (UK), Wiley-Scrivener (USA), and Materials Research Forum LLC (USA). He is a member of various journals' editorial boards. He is an Associate Editor (*Environmental Chemistry Letter, Applied Water Science* and *Euro-Mediterranean Journal for Environmental Integration*, Springer-Nature), Frontiers Section Editor (*Current Analytical Chemistry*, Bentham Science Publishers), Editorial Board Member (*Scientific Reports*, Nature), Editor (*Eurasian Journal of Analytical Chemistry*), and Review Editor (*Frontiers in Chemistry*, Frontiers, the UK) for several journals. He has also guest-edited various thematic special issues to the journals of Elsevier, Bentham Science Publishers, and John Wiley & Sons, Inc. He has attended as well as chaired sessions at various international and national conferences. He has worked as a Postdoctoral Fellow, leading a research team at the Creative Research Initiative Center for Bio-Artificial Muscle, Hanyang University, South Korea, in the field of renewable energy, especially biofuel cells. He has also worked as a Postdoctoral Fellow at the Center of Research Excellence in Renewable Energy, King Fahd University of Petroleum and Minerals, Saudi Arabia, in the field of polymer electrolyte membrane fuel cells and computational fluid dynamics of polymer electrolyte membrane fuel cells. He is a life member of the *Journal of the Indian Chemical Society*. His research interests include ion exchange materials, sensors for heavy metal ions, biofuel cells, supercapacitors, and bending actuators.

Charles Oluwaseun Adetunji, PhD, is a Faculty member at the Microbiology Department, Faculty of Science, Edo University Iyamho (EUI), Edo State, Nigeria, where he utilized the application of biological techniques and microbial bioprocesses for the actualization of sustainable development goals and Agrarian revolution, through quality teaching, research, and community development. He is presently the Ag. Director of Intellectual Properties and Technology Transfer; Chairman Committee on Research Grant and Associate Professor of Microbiology and Biotechnology at EUI. He has won several scientific awards and grants from renowned academic bodies like Council of Scientific and Industrial Research (CSIR) India, and Department of Biotechnology (DBT) India, The World Academy of Science (TWAS) Italy, Netherlands Fellowship Programme (NPF) Netherlands, The Agency for International Development Cooperation; Israel, Royal Academy of Engineering, UK among many others. He has filled several scientific patents on Bioherbicides, Biopesticides, Nanobiosurfactants, Nanobiopesticdes and many more. He has published over 150 scientific journals and conference proceedings in international and local refereed journals. He is currently editing several Biotechnology text book with Elsevier. The breadth of his scholarly contributions to research are evident from his contributions, which cover topics relating to food security, Agriculture and Environmental sustainability. His research interest includes Microbiology, Biotechnology, Post-harvest management, and Nanotechnology. He is an editorial board member of many international journals and serves as a reviewer to many double-blind peer review journals like Elsevier, Springer, Francis and Taylor, Wiley, PLOS One, Nature, American Chemistry Society, Bentham Science Publishers etc. He is a member of many scientific and professional including bodies like American Society for Microbiology, Nigerian Young Academy, and, Biotechnology Society of Nigeria, Nigerian Society for Microbiology. He has won a lot of international recognition and also acted as a keynote speaker delivering invited talk/ position paper at various Universities, research institutes and several centers of excellence which span across several continent of the globe. He has over the last fifteen years built strong working collaborations with reputable research groups in numerous and leading Universities across the globe. He is the convener for Recent Advances in Biotechnology, which is an annual international conference where renown Microbiologist and Biotechnologist come together to share their latest discoveries. He is the Founder & CEO of BECTIK Biotechnology and Nanotechnology Company. His Biotechnology Company consist a team of leading academics in their fields and thrive to deliver solutions all around Bio & Nanotechnology. The company also deliver solutions that are tailored to your business needs and for anyscale.

Mohd Imran Ahamed, PhD, is working as Research Associate at the Department of Chemistry, Aligarh Muslim University (AMU), Aligarh, India. He received his B.Sc. (Hons) Chemistry and Ph.D. (Chemistry) degrees from AMU. He has completed his M.Sc. (Organic Chemistry) from Dr. Bhimrao Ambedkar University, Agra, India. He has published several research and review articles in various international scientific journals. He has co-edited 57 books with Springer (UK), Elsevier, CRC Press – Taylor & Francis Asia Pacific, Materials Research Forum LLC (USA) and Wiley-Scrivener (USA). His research work includes ion-exchange chromatography, wastewater treatment, and analysis, bending actuator and electrospinning.

Tariq Altalhi, PhD, joined the Department of Chemistry at Taif University, Saudi Arabia as Assistant Professor in 2014. He received his doctorate degree from University of Adelaide, Australia in the year 2014 with Dean's Commendation for Doctoral Thesis Excellence. He was promoted to the position of the head of Chemistry Department at Taif university in 2017 and Vice Dean of Science college in 2019 till now. In 2015, one of his works was nominated for Green Tech awards from Germany, Europe's largest environmental and business prize, amongst top 10 entries. He has co-edited various scientific books. His group is involved in fundamental multidisciplinary research in nanomaterial synthesis and engineering, characterization, and their application in molecular separation, desalination, membrane systems, drug delivery, and biosensing. In addition, he has established key contacts with major industries in Kingdom of Saudi Arabia.

Contributors

Hafiz Muhammad Aatif
Department of Plant Pathology
Bahauddin Zakariya University
Bahadur sub-campus, Layyah

Maqshoof Ahmad
Department of Soil Science
The Islamia University of Bahawalpur
Bahawalpur, Pakistan

Saqib Ajmal
Department of Entomology
The Islamia University of Bahawalpur
Bahawalpur, Pakistan

Ali Akbar
Department of Botany
Government College University Faisalabad
Faisalabad, Pakistan

Sajjad Ali
Department of Entomology
Faculty of Agriculture and Environment
The Islamia University of Bahawalpur
Bahawalpur, Pakistan

Shafaqat Ali
Department of Environmental Sciences and
   Engineering
Government College University Faisalabad
Faisalabad, Pakistan

B.N. Aloo
Department of Biological Sciences
University of Eldoret
Eldoret, Kenya

M. Anjum Aqueel
Department of Entomology
The Islamia University of Bahawalpur
Bahawalpur, Pakistan

Muhammad Arslan Ashraf
Department of Botany
Government College University Faisalabad
Faisalabad, Pakistan

Waqas Ashraf
Department of Plant Pathology
Faculty of Agriculture and Environment
The Islamia University of Bahawalpur
Bahawalpur, Pakistan

Muhammad Naveed Aslam
Department of Plant Pathology
Faculty of Agriculture and Environment
The Islamia University of Bahawalpur
Bahawalpur, Pakistan

Rabia Tahir Bajwa
Department of Plant Pathology
Faculty of Agriculture and Environment
The Islamia University of Bahawalpur
Bahawalpur, Pakistan

Fatima Boubrik
Plant Technology Laboratory
Ibn Zohr University
Agadir, Morocco

Emilio Bucio
Department of Radiation Chemistry and
   Radiochemistry
Institute of Nuclear Sciences
National Autonomous University of Mexico
Mexico City, Mexico

Moises Bustamante-Torres
Biomedical Engineering Department
School of Biological and Engineering
Yachay Tech University
Urcuqui City, Ecuador
and
Department of Radiation Chemistry and
   Radiochemistry
Institute of Nuclear Sciences
National Autonomous University of Mexico
Mexico City, Mexico

Peter Dart
Plant-Microbe Interactions Laboratory
School of Agriculture and Food Sciences
The University of Queensland
Brisbane, Queensland, Australia

Iram Gul
Department of Earth and Environmental Sciences
Hazara University
Mansehra, Pakistan

Iqbal Hussain
Department of Botany
Government College University Faisalabad
Faisalabad, Pakistan

Yasir Iftikhar
Department of Plant Pathology
College of Agriculture
University of Sargodha
Sargodha, Pakistan

Muhammad Iqbal
Department of Botany
Government College University Faisalabad
Faisalabad, Pakistan

Irum Iqrar
Department of Biotechnology
Quaid-i-Azam University
Islamabad, Pakistan

Aruna Jyothi Kora
National Centre for Compositional
    Characterisation of Materials (NCCCM)
Bhabha Atomic Research Centre (BARC)
Hyderabad, India
and
Homi Bhabha National Institute (HBNI)
Anushakti Nagar
Mumbai, India

B.A. Makumba
Department of Biological Sciences
Moi University
Eldoret, Kenya

Muhammad Zeeshan Mansha
Department of Plant Pathology
Bahauddin Zakariya University
Bahadur sub-campus, Layyah

E.R. Mbega
Department of Sustainable Agriculture and
    Biodiversity Conservation
Nelson Mandela African Institution of
    Science and Technology
Arusha, Tanzania

Jalal Mouadi
Plant Technology Laboratory
Ibn Zohr University
Agadir, Morocco

David Morales-Pérez
Environmental Engineering Department
Faculty of Life Science
Salesian Polytechnic University
Quito City, Ecuador

Mustansar Mubeen
Department of Plant Pathology
College of Agriculture
University of Sargodha
Sargodha, Pakistan

Fahim Nawaz
Department of Agronomy
MNS University of Agriculture
Multan

El Asri Ouahid
Plant Technology Laboratory
Ibn Zohr University
Agadir, Morocco

Samantha Pardo
Environmental Engineering Department
Faculty of Life Science
Salesian Polytechnic University
Quito City, Ecuador

Abida Parveen
Department of Botany
Government College University Faisalabad
Faisalabad, Pakistan

Mauricio Pérez-Albornoz
Environmental Engineering Department
Faculty of Life Science
Salesian Polytechnic University
Quito City, Ecuador

Muhammad Raheel
Department of Plant Pathology
Faculty of Agriculture and Environment
The Islamia University of Bahawalpur
Bahawalpur, Pakistan

Rizwan Rasheed
Department of Botany
Government College University Faisalabad
Faisalabad, Pakistan

Ifrah Rashid
Department of Plant Pathology
Faculty of Agriculture and Environment
The Islamia University of Bahawalpur
Bahawalpur, Pakistan

Amir Riaz
Department of Agricultural Extension
Faculty of Agriculture and Environment
The Islamia University of Bahawalpur
Bahawalpur, Pakistan

Muhammad Riaz
Department of Environmental Sciences and
 Engineering
Government College University Faisalabad
Faisalabad, Pakistan

Gulab Khan Rohela
Central Sericultural Research and Training
 Institute
Central Silk Board
Pampore, India

Pawan Saini
Central Sericultural Research and Training
 Institute
Central Silk Board
Pampore, India

Asif Sajjad
Department of Entomology
The Islamia University of Bahawalpur
Bahawalpur, Pakistan

Peer M. Schenk
Plant-Microbe Interactions Laboratory
School of Agriculture and Food Sciences
The University of Queensland
Brisbane, Queensland, Australia

Mudassir Iqbal Shad
Department of Botany
Government College University Faisalabad
Faisalabad, Pakistan

Qaiser Shakeel
Department of Plant Pathology
Faculty of Agriculture and Environment
The Islamia University of Bahawalpur
Bahawalpur, Pakistan

Ziyu Shao
Plant-Microbe Interactions Laboratory
School of Agriculture and Food Sciences
The University of Queensland
Brisbane, Queensland, Australia

Zabta Khan Shinwari
Department of Plant Sciences
Quaid-i-Azam University
Islamabad, Pakistan

Faouzia Tanveer
Department of Biotechnology
Quaid-i-Azam University
Islamabad, Pakistan

Kaleem Tariq
Department of Agriculture
Abdul Wali Khan University Mardan
Mardan, Pakistan

Olga Tsivileva
Institute of Biochemistry and
 Physiology of Plants and Microorganisms
Saratov Scientific Centre of the Russian
 Academy of Sciences

J.B. Tumuhairwe
Department of Agricultural Production
College of Agricultural and
 Environmental Sciences
Makerere University
Kampala, Uganda

Brian A. Wartell
Department of Physical Sciences
Community College of Baltimore County
Catonsville, Maryland, USA

1 Microbial Biostimulants for Tolerance to Abiotic and Biotic Stresses in Mushrooms

Olga Tsivileva

Institute of Biochemistry and Physiology of Plants and Microorganisms, Saratov Scientific Centre of the Russian Academy of Sciences

CONTENTS

1.1 INTRODUCTION

Mushrooms, the macromycete flora representatives, play a significant role as food and biological subjects capable of producing many physiologically active compounds. Numerous scientific works are dedicated to the investigation of mushrooms' physiology, ecology, biochemistry and biotechnology. Owing to their commercialization, many edible mushrooms are readily available [1] and provide a precious source of nutritive food ingredients. Broad spectrum and diversity of mushrooms have led to an exciting pharmacological potentiality of bioactive preparations from their cultivation. The discovery that a lot of mushrooms produce a broad range of medicinally useful metabolites [2] was followed by the intense development of biomedicinal area of fungi exploitation [3]. For instance, various chemical agents entering the composition of mycelia exhibit antiviral activity against many viruses pathogenic for humans [4], which currently presents a great pharmaceutical challenge. Thus, in view of the great biotechnological interest in wide-scale cultivation of basidiomycetes, the factors of biological origin that promote their growth, including microbial biostimulants, are of major importance.

One of the preferential lines in the current mushroom science is decoding the ways of inducement of biochemical processes of the basidiomycetes' ontogenesis. Complex physiological and biochemical processes that occur at growth and development of fungus organisms are determined largely by the biotic factors of environment, which must be studied in detail. That provides the necessity of studying the joint cultivation with bacteria in relation to the processes of cytodifferentiation and morphogenesis of mushrooms, and of revealing the role of these factors in the vital activity of fungal cultures. The research of such kind is useful for the directive selection of effective inducers enhancing the stress resistivity of edible cultivable mushrooms, as it is true for plants. Understanding the mechanisms and peculiarities of associative interactions at the bacterial–fungal communities could assist in developing more efficient and productive methods of growing the cultivable edible and medicinal mushrooms.

DOI: 10.1201/9781003188032-1

Hence, the challenging problem is the development of ecologically safe ways of applying the growth-promoting bacteria and bioactive substances of bacterial origin to the technologies of enhancing the resistivity of cultivable basidiomycetes against the pathogenic bacterial and lower-fungal contamination, and to the technologies of controlling the abundance of the contaminating microflora representatives in the industrial mushroom growing.

Analyzing the current literature in respect of works on xylotrophic basidiomycetes and bacteria co-cultivation shows the absolute insufficiency of these problems under consideration, in spite of the high potentiality of fundamental and applied research in this area. Bacteria from the genus *Azospirillum* are rhizospheric microorganisms that enhance plant growth and development via nitrogen fixation and hormone production [5]. Worthy to mention is the bactericidal and fungicidal activity of azospirilla against several bacteria and lower micromycetes [6,7]. Systemic research into the basidiomycetes joint culture with bacteria from the genus *Azospirillum* under the artificial conditions was scarce in the literature until studies were initiated at the Lab. of Microbiology of IBPPM RAS in relation to the double culture of *Lentinula edodes* (shiitake mushroom) strain F-249 and *Azospirillum brasilense* Sp7 [8,9]. The effect of *A. brasilense* Sp7 on growth and morphological peculiarities of *L. edodes* F-249 was studied [10,11]; therewith, the reasons for enhanced development of mycelium remain unclear. Currently, the search for appropriate, effective strains of azospirilla in respect of mushrooms from different systematic groups becomes a necessity.

1.2 PGPR INTERACTIONS WITH MYCORRHIZAL FUNGI

Bacteria involved in the associative and symbiotic relations with plant roots are known to contribute positively to plant metabolism and growth parameters. Such bacteria are called "plant growth-promoting rhizobacteria" (PGPR), i.e., rhizobacteria facilitating plant growth [12–15]. Phytostimulating properties of such PGPR representatives as *Azospirillum* are well studied [16].

PGPR capable of influencing positively the development of mycorrhizal structures or the mycorrhizal fungi, on the formation of mycorrhizal symbiosis, constitute the group "mycorrhization helper bacteria" (MHB) [17–19]. The intense citation of works dealt with mycorrhizal fungi is out of the scope of this review. That is, follow only few representative examples. *Pseudomonas fluorescens* could exert a positive effect on the mycelium development of the ectomycorrhizal fungus *Laccaria bicolor in vitro* by means of production of thiamine assimilated by the fungus. At the same time, *L. bicolor* is capable of accumulating trehalose in hyphae, which promotes the growth of colony of *P. fluorescens*. Both organisms demonstrate synergetic interactions important for their vital activity [20]. Endophytic bacteria *Pseudomonas aeruginosa* and *Burkholderia cepacia* form associations with the mycorrhizal fungi *Glomus intraradices* and *Glomus clarum* and are capable of influencing positively the mycelium of these fungi [21]. Even in the absence of host plant, the mutual positive effects of mycorrhizal fungi and MHB are manifested [22]. MHB could be non-specific and promote the mycorrhiza formation by various fungi [23].

Reasonability of studying different aspects of favorable MHB action on the mycorrhizal basidiomycetes is commonly recognized; however, the mechanisms responsible for the favorable effect of soil bacteria on the mycorrhizal fungi are investigated insufficiently [24,25].

1.3 PGPR INTERACTIONS WITH CULTIVABLE MUSHROOMS

Putting into practice the biological techniques of mycelium growth stimulation, and its protection from the interfering microflora, would allow one to improve the growing technology via the less prolonged cultivation period and concurrent suppression of the contaminating microflora. The research into the microbial populations isolated from the surface of mycelium during the fruit body formation of *Pleurotus ostreatus* was performed. Dominating in bacterial community, *Pseudomonas* spp. appear to facilitate fruiting when being used as inoculum for pure cultures of the basidiomycete *in vitro* [26]. In double culture of the same basidiomycete with *P. fluorescens*, the formation of a bacterial biofilm was observed. The biofilm was successfully applied by the authors [27] in their experiments with tomato plants. In the joint cultivation of *P. ostreatus* with the strain of diazotroph *Bradyrhizobium elkanii*, not only the dense colonization of mycelium by bacteria, but also the active process of nitrogen fixation within the biofilm on hyphae was observed. More recently, the same mushroom has attracted the attention of Febriansyah and co-authors [28]. They presented a work on the impact of the bacterial cell-free extract on *P. ostreatus* mycelium development. Nineteen bacteria strains were involved, leading to an increase in mycelial growth; nevertheless, there was no significant influence statistically. Indole-3-acetic acid content was assayed with three chosen bacterial isolates: *Bacillus cereus*, *Bacillus aryabhattai* and *Acinetobacter pittii* [28], the highest content being produced by *Bacillus aryabhattai* strain B8W22.

Agaricus bisporus (champignon) is an edible mushroom cultivated for food worldwide. *A. bisporus* vegetative growth should be done at compost, and the presence of bacteria in the casing layer greatly influences the fruit body initiation of this mushroom [29]. In the latter work, greenhouse experiments were conducted using numerous bacterial isolates; 60 rhizobia and 30 pseudomonades strains were tested to estimate the effect of these plant growth-promoting bacteria (PGPB) on mushroom fruit body formation. Rhizobial bacteria were *Sinorhizobium meliloti* and *Rhizobium leguminosarum* biovar *phaseoli*, and pseudomonades belonged to *Pseudomonas fluorescens* species. The results obtained testified to the increased yield and quality of *A. bisporus* fruit bodies provided that the growth-promoting bacteria were implemented in this edible mushroom cultivation at compost. Thus, PGPR occurred beneficial in the production of healthy food [29].

Zarenejad and co-workers [30] evaluated the prospective bacteria with mushroom growth-promoting features from soils sampled at the bisporic champignon farms. Casing layers used for *A. bisporus* fruit body production contained two strains as potent MGPB, which were isolated and both assigned to genus *Pseudomonas* [30]. It seems obvious that the bacteria inhabiting the casing soil and compost should move onto fungal mycelium and subsequently occur within the basidiomes of *Agaricus* species mushrooms. That is why the microbes in the mushroom fruit bodies have received a little attention. Xiang and co-authors reported [31] on the diversity of bacterial isolates sampled in basidiomes of bisporic champignon. Among 55 strains of bacteria belonging to seven biological families, *Bacillus* spp. was a dominant genus in this mushroom. But only further studies could clarify whether these isolates are capable of serving as a growth-promoting factor for *A. bisporus*.

Fifty-six specimens were obtained from the culture substrate of another species of *Agaricus*, *A. blazei*, in order to estimate the impact of the isolated bacteria on the development of fungal mycelium and the yield of fruit bodies [32]. Few isolates were selected and shortlisted on the grounds of the effect on this mushroom productivity: *Agaricicola taiwanensis*, *Advenella incenata*, *Curtobacterium citreum*, *Gordonia hydrophobica*, *Streptomyces violaceorubidus* and *Microbacterium humi*. Several of the tested species promoted *A. blazei* fruiting, reduced efficiently the required time for harvesting and increased the total fresh matter yield to different extents. On sterilized casing soil, inoculation by *Exiguobacterium* sp. appeared to be capable of shortening a period of time till harvest and increasing both the fresh fungal biomass and total polysaccharide content in this mushroom compared to the non-inoculated control [33].

Taking into consideration the conducted studies on the PGPR interactions with cultivable mushrooms, we should recognize that the exact role of PGPR in increasing the growth and yield of cultivable mushrooms is not clearly elucidated. The amount of mushrooms under research is very limited, and the same is for presumably beneficial bacteria, being in fact restricted by the most common *Agaricus bisporus*, *A. blazei* and *Pleurotus ostreatus*, and as for bacteria, by rhizobia and pseudomonades. Such an approach is partially substantiated by previous studies; e.g., it is known that fluorescent pseudomonades account for 14–41% of the overall bacteria revealed in solid substrate for *A. bisporus* growing [30]. Frequently in the research, a very substantial number of bacterial isolates have been studied as potent MGPB. But the researchers start from a total analogy with plant benefits gained from PGPR, when the phytohormone production, phosphate solubilization, siderophore secretion, antipathogenic properties, nitrogen fixation capability or secretion of cellulase cause the improvement in plant development and yield. Unfortunately, the aforementioned biochemical properties as themselves, it appears, fail to make these bacteria as potent MGPB. Despite these listed potentialities assessment [31], the screening of MGPB for mushroom growing remains limited, and thus, a problem of feasible bacterial inoculants to be really applied for mushroom production process is still unresolved.

1.4   FUNGI–*AZOSPIRILLUM* INTERACTIONS

Investigations of the biologically active substances sourced from the xylotrophic macromycetes possess not only fundamental, but also practical significance. They promote the development of biotechnology of gaining valuable products from the mycelial biomass and culture liquid, as well as of using the fungal species in "white chemistry." Research on the environmentally safe biological methods of stimulating the growth of medicinal and edible mushrooms assists in the development of scientific foundations of culture technologies. To gain resistivity in the mushroom against a negative environmental impact seems to be feasible on account of the mushroom growing in combination with the growth-promoting bacteria. The soil bacteria from the *Azospirillum* genus attract attention as the microorganisms are capable of actively influencing the growth and development of agricultural crops. Thus, revealing and exploring the growth-promoting properties of different species of *Azospirillum* in respect of edible and/or medicinal mushrooms, as well as the properties promoting the suppression of contaminants in double culture, are believed to be actual and to possess undoubted novelty. Cultivation jointly with beneficial bacteria appears to be an effective biotechnological method for obtaining this valuable functional food faster, with greater and much better preserved mushroom yield. The future work consists in the problems of mushroom growth-promoting bacteria interaction with mushrooms, and in implementing these synthetic microbial associations in agriculture.

MHB include the representatives of *Azospirillum* genus, too [34–38]. For example, in the series of experiments on the crops evaluation, the inoculums of fungus *Glomus mosseae* and bacteria *Azospirillum brasilense* were introduced in soil to the plants under study. In the presence of one of the inoculums, an appreciable increase in plant biomass was observed; however, at the introduction of both inoculums, the plant biomass was much greater [39,40]. The literature data confirm that the nitrogen-fixing bacteria are associated with mycorrhizal fungi, spores and fruit bodies [41,42].

Inhabiting the root zone of soil, bacteria enter the interactions with mycorrhizal fungi, but the group of higher fungi which did not form mycorrhiza is also wide. Many medicinal and edible mushrooms are xylotrophs. Their associations with bacteria are studied poorly and incomparably worse than the mycorrhizal components being discussed in few works.

The feasibility of submerged joint fermentation of fungal–bacterial cultures, where bacteria are from *Azospirillum* genus (*A. baldaniorum* and *A. brasilense*) and fungi are xylotrophic macromycetes from *Flammulina*, *Ganoderma* and *Pleurotus* genera (*F. velutipes*, *G. lucidum* and *P. ostreatus*), was elucidated, and the appropriate parameters were screened and chosen [43–45]. Selection of mushroom objects is caused by the perfect properties of these cultivable basidiomycetes. Winter mushroom *Flammulina velutipes* serves as a source of valuable nutritional food ingredients [46,47] and provides a natural basis for developing high-performance pharmaceutical forms [48] and veterinary and medicinal preparations [49,50]. Numerous bioactive preparations and substances from lacquered polypore *Ganoderma lucidum* exhibit pharmacological efficacy. Comprehensive studies over a decade or more has proved the anticancer effect of low molecular weight compounds obtained from *Ganoderma*, mainly triterpenoids [51], and of biopolymers, mainly polysaccharides, of these mushrooms [52]. Formulations prepared with the implementation of mycelia, fruit bodies and fermentation media of *Ganoderma* species demonstrate a wide range of antiviral and bactericidal [53], immunomodulating [52], cytotoxic [54] and fungicidal [55] properties. Therewith not only lacquered polypore exhibits antibacterial activity in its extracts [56], but also other *Ganoderma* species are known to produce antimicrobials [52,54,57].

The xylotrophic basidiomycete *Pleurotus ostreatus* (oyster mushroom) has long been known for the excellent edible and medicinal properties [58]. Artificial growing of oyster mushroom is of great importance owing primarily to the high productivity of this culture exhibiting valuable nutritional properties. Mycelium of *P. ostreatus* contains a proteinaceous component at considerable proportion [59,60]. Therewith the fruit bodies could be obtained using the solid substrates unlikely applicable for any other purposes (non-food agricultural and industrial wastes) [61–63]. Contemporary studies on the basidiomycete *P. ostreatus* growth in liquid media have been conducted for the development of submerged culture technology to yield mycelial biomass for use as food supplements, fodders, various physiologically active preparations [64,65] and biotechnologically valuable products [66–68]. The submerged culture technique is commonly recognized as a rapid and efficient method of the seeding mycelium production for the wide-scale growing of mushroom [25,69].

When choosing bacteria applicable for joint cultivation with mushrooms, it seems reasonable to consider first those possessing the ability to favorably influence the plants. Such microorganisms include the bacteria from *Azospirillum* genus. Azospirilla as a representative of the rhizospheric bacteria group comprise the properties of promoting the growth and development of plants to different extents. Known is the bactericidal and fungicidal activity of azospirilla against several (phyto) pathogenic bacteria and microscopic fungi. Let us substantiate the choice of strains of bacteria. Lignin-containing substrates' decay by bacteria is catalyzed by the enzymes obviously analogous to those involved in lignin transformation by higher fungi [70]. For the dual-mode culture with xylotrophic basidiomycetes, we chose two strains, i.e., the model object for the PGPR–plant endophytic symbiosis studies, strain *A. baldaniorum* Sp245, and epiphytic, in respect of their symbiotic plants, strain *A. brasilense* SR80, for both of which the ability to degrade lignin was shown [71].

The parameters of growing mushrooms with the strains *A. brasilense* SR80 or *A. baldaniorum* Sp245 were selected. The temperature of cultivation was near 28°C. The most appropriate duration of fungus inoculum fermentation before joining it with azospirilla appeared to be 7 days. After this period of time, the bacterial suspension of *A. baldaniorum* or *A. brasilense* was added to the submerged liquid mushroom culture.

Different nutrient media, we tested for co-culture were diluted barley wort; monosaccharide-Asn media without salts (**I**) or with salts (**II**); modified malate–salt medium [72] (**III**); several formulations based on yeast extract: high-Glc medium (**IV**), low-Glc medium with peptone (**V**), and the medium (**V**) concentrated (**VI**). Both mushroom and bacterial cultures developed acceptable intensities of growth in the nutritive formulations (**I**) and (**II**), and to a lesser extent in (**IV**). Other media based on yeast extract, namely (**V**) and somewhat worse (**VI**), were featured by a normal growth of both bacterial strains. Barley wort appeared to be efficiently applicable for mycelium culture solely, and the liquid malate formulation (**III**) was implemented in obtaining the seeding bacterial suspensions.

Under the optimized conditions of liquid co-cultures, the azospirilla strains under study exhibited active motility and were in tight contact with fungal hyphae. *P. ostreatus* dried biomass exceeded this oyster mushroom's monoculture parameters by up to 54%. The cell concentration of bacterial strains, given other experimental conditions being equal, significantly influenced the binary culture growth. *F. velutipes* cultivation jointly with *A. baldaniorum* Sp245 at the optimal concentration of bacterial cells in inoculum allowed us to obtain a dry biomass value of 235% relative to control. The intensive growth of mushroom dual culture with azospirilla strains under study was favored by the formulation comprising D-glucose and D-fructose (mass ratio of 1:1) as a carbon source and the amino acid L-asparagine as a nitrogen source [45].

Edible/medicinal basidiomycetes cultivation under artificial conditions is related to definite problems, among which the most considerable is the low resistivity of the seeding mycelium vs the interfering microflora. Oyster mushroom cultivation in dual culture with the growth-promoting bacteria appeared to be an effective biotechnological method for obtaining the faster-growing seeding mycelium resistive to contaminants, which in turn allowed the mushroom yield to be greater and the process of fruit body formation to be less prolonged [44]. The characteristics of *P. ostreatus*

Pleurotus ostreatus

Pleurotus ostreatus

FIGURE 1.1   Scheme of the process of obtaining *Pleurotus ostreatus* fruit bodies. (Photos are by OT.)

fruit body development and contamination diminution caused by the implemented azospirilla were estimated. As an inoculating material, the binary submerged cultures of oyster mushroom with the azospirilla strain SR80 (Figure 1.1) appeared to possess the highest efficiency in parameters such as the accumulated biomass, the intensity of mycelium propagation when colonizing the specially treated wheat grain solid and the abundance of fruiting process development.

The implementation of PGPB in biocontrol of plant, but not mushroom, diseases was discussed in comprehensive reviews, e.g., [73]. The experiments with joint fungal–bacterial cultures, where the fungal component was *P. ostreatus* and the bacterial component was *A. brasilense* or *A. baldaniorum*, allowed us to approach important issues. Those are understanding the growth-promoting phenomenon and revealing the profoundly useful properties of the said dual cultures (compared to the oyster mushroom monoculture) manifested particularly in their capability of better withstanding to unwanted microorganisms, which inevitably accompany the non-aseptic process of mushroom fruiting body development [44]. The data gained are indicative of the great potentialities of binary mushroom–bacterial cultures application for efficiently obtaining mycelial biomass and fruiting of basidiomycetes.

1.5 CONCLUSIONS

Understanding the mechanisms and peculiarities of associative interactions at the bacterial–fungal communities could assist in the development of more effective and productive methods of growing the agricultural crops, including mushrooms. Edible mushroom cultivation jointly with the growth-promoting bacteria occurs to be an effective biotechnological method for obtaining the faster-growing seeding mycelium resistive to contaminants, which in turn allowed the mushroom yield to be greater and much better preserved. However, the investigations related to xylotrophic basidiomycetes are manifested in few isolated papers. The principal lines of future research into the mushroom–bacterial interactions should be aimed at the following items: (i) the development of fundamental basis and methods for enhancing the xylotrophic basidiomycete resistivity against environmental stresses by means of culture in conjunction with growth-promoting bacteria; (ii) determining the feasibility and biochemical prerequisites for lowering the biotic influence of interfering microflora on mycelium under the conditions of dual culture of the basidiomycetes with growth-promoting bacteria; and (iii) characterizing the antimicrobial potentiality of the extracellular compounds in the mushroom cultures as influenced by the growth-promoting bacteria. In general, the content of future work should be intended to substantiate the prerequisites of applying the mixed culture of valuable mushrooms with MGPB as a kind of seeding material with high protective properties owing to the enlarged rate of mycelium expansion in the solid nutrient substrate. The PGPB have extensively been investigated by now, in contrast to MGPB, and are useful to exploit the cultivable mushrooms resources.

ACKNOWLEDGMENTS

The research (fungal metabolites studies) was completed in the framework of the theme No. 121031100266-3 of the program of the fundamental research of the Russian Academy of Sciences. The authors thank the researchers who kindly provided the bacterial strains from the rhizosphere microorganism collection of IBPPM RAS (WFCC number 975 and WDCM number 1021).

REFERENCES

1. Kim, S., Ha, B.S., and Ro, H.S. 2015. Current technologies and related issues for mushroom transformation. *Mycobiology* 43(1):1–8.
2. Ashraf, J., Ali, M.A., Ahmad, W., Ayyub, C.M., and Shafi, J. 2013. Effect of different substrate supplements on oyster mushroom (*Pleurotus* spp.) production. *Food Science and Technology* 1(3):44–51.
3. Rathore, H., Prasad, S., Kapri, M., Tiwari, A., and Sharma, S. 2019. Medicinal importance of mushroom mycelium: Mechanisms and applications. *Journal of Functional Foods* 56:182–193.
4. Teplyakova, T.V. and Kosogova, T.A. 2016. Antiviral effect of Agaricomycetes mushrooms (review). *International Journal of Medicinal Mushrooms* 18(5):375–386.
5. Steenhoudt, O. and Vanderleyden, J. 2000. Azospirillum, a free-living nitrogen-fixing bacterium closely associated with grasses: Genetic, biochemical and ecological aspects. *FEMS Microbiology Reviews* 24(4):487–506.
6. Red'kina, T.V. 1990. Fungistatic activity of bacteria of the genus *Azospirillum*. *Agrokemia es Talajtan (Agrochemistry and Soil Science)* 39(3–4):465–468.

7. Bashan, Y. and De-Bashan, L.E. 2002. Protection of tomato seedlings against infection by *Pseudomonas syringae* pv. *tomato* by using the plant growth-promoting bacterium *Azospirillum brasilense*. *Applied and Environmental Microbiology* 68(6):2637–2643.

8. Nikitina, V.E., Tsivileva, O.M., and Loshchinina, E.A. 2006. Interrelations of xylotrophic basidiomycetes and soil nitrogen-fixing bacteria from the genus *Azospirillum*. *Advances in Medical Mycology* 7:293–294.

9. Tsivileva, O.M., Pankratov, A.N., and Nikitina, V.E. 2010. Extracellular protein production and morphogenesis of *Lentinula edodes* in submerged culture. *Mycological Progress* 9(2):157–167.

10. Loshchinina, E.A., Nikitina, V.E., Tsivileva, O.M., Stepanova, L.V., Ponomareva, E.G., and Shelud'ko, A.V. 2006. Morphological-cultural characteristics of the basidiomycete *Lentinus edodes* on co-cultivation with bacteria of the genus *Azospirillum*. *Vestnik SGAU* 6(2):24–26.

11. Loshchinina, E.A., Tsivileva, O.M., Makarov, O.E., and Nikitina, V.E. 2012. Changes in carbohydrate and fatty-acid content of *Lentinus edodes* mycelium in dual cultures with *Azospirillum brasilense*. *Izvestiya Vuzov. Prikladnaya Khimiya i Biotekhnologiya* (*Proceedings of Universities. Applied Chemistry and Biotechnology*) 2(3):64–67.

12. Bhat, M.A., Rasool, R., and Ramzan, S. 2019. Plant growth promoting rhizobacteria (PGPR) for Sustainable and Eco-Friendly Agriculture. *Acta Scientific Agriculture* 3(1):23–25.

13. Kumar, A., Patel, J.S., Meena, V.S., and Ramteke, P.W. 2019. Plant growth-promoting rhizobacteria: Strategies to improve abiotic stresses under sustainable agriculture. *Journal of Plant Nutrition* 42(11–12):1402–1415.

14. Siyar, S., Inayat, N., and Hussain, F. 2019. Plant growth promoting rhizobacteria and plants' improvement-a mini-review. *PSM Biological Research* 4(1):1–5.

15. Goswami, M. and Deka, S. 2020. Plant growth-promoting rhizobacteria – alleviators of abiotic stresses in soil: A review. *Pedosphere* 30(1): 40–61.

16. Fukami, J., Cerezini, P., and Hungria, M. 2018. *Azospirillum*: Benefits that go far beyond biological nitrogen fixation. *AMB Express* 8(1): 73/1–73/12.

17. Labutova, N.M. 2009. Interactions between endomycorrhizal fungi and rhizosphere microorganisms. *Mikologiya i Fitopatologiya* 43(1):3–19.

18. Wang, R., Liu, P.G., Wan, S.P., and Yu, F.Q. 2015. Study on mycorrhization helper bacteria (MHB) of Tuber indicum. *Microbiology China* 42(12):2366–2376.

19. Fu, Y., Li, X., Li, Q., Wu, H., Xiong, C., Geng, Q., Sun, H., and Sun, Q. 2016. Soil microbial communities of three major Chinese truffles in southwest China. *Canadian Journal of Microbiology* 62(11):970–979.

20. Deveau, A., Brulé, C., Palin, B., Champmartin, D., Rubini, P., Garbaye, J., Sarniguet, A., and Frey-Klett, P. 2010. Role of fungal trehalose and bacterial thiamine in the improved survival and growth of the ectomycorrhizal fungus *Laccaria bicolor* S238N and the helper bacterium *Pseudomonas fluorescens* BBc6R8. *Environmental Microbiology Reports* 2(4):560–568.

21. Sundram, S., Meon, S., Seman, I.A., and Othman, R. 2011. Symbiotic interaction of endophytic bacteria with arbuscular mycorrhizal fungi and its antagonistic effect on *Ganoderma boninense*. *The Journal of Microbiology* 49(4):551–557.

22. Frey-Klett, P., Garbaye, J., and Tarkka, M. 2007. The mycorrhiza helper bacteria revisited. *New Phytologist* 176(1):22–36.

23. Aspray, T.J., Frey-Klett, P., Jones, J.E., Whipps, J.M., Garbaye, J., and Bending, G.D. 2006. Mycorrhization helper bacteria: A case of specificity for altering ectomycorrhiza architecture but not ectomycorrhiza formation. *Mycorrhiza* 16(8):533–541.

24. Obase, K. 2019. Bacterial community on ectomycorrhizal roots of *Laccaria laccata* in a chestnut plantation. *Mycoscience* 60(1):40–44.

25. Zhang, W.R., Liu, S.R., Kuang, Y.B., and Zheng, S.Z. 2019. Development of a novel spawn (block spawn) of an edible mushroom, *Pleurotus ostreatus*, in liquid culture and its cultivation evaluation. *Mycobiology* 47(1):97–104.

26. Cho, Y.S., Kim, J.S., Crowley, D.E., and Cho, B.G. 2003. Growth promotion of the edible fungus *Pleurotus ostreatus* by fluorescent pseudomonads. *FEMS Microbiology Letters* 218(2):271–276.

27. Jayasinghearachchi, H.S. and Seneviratae, G. 2010. A mushroom-fungus helps improve endophytic colonization of tomato by *Pseudomonas fluorescens* through biofilms formation. *Research Journal of Microbiology* 5(7):689–695.

28. Febriansyah, E., Saskiawan, I., Mangunwardoyo, W., Sulistiyani, T.R., and Widhiya, E.W. 2018. Potency of growth promoting bacteria on mycelial growth of edible mushroom Pleurotus ostreatus and its identification based on 16S rDNA analysis. *AIP Conference Proceedings* 2002(1):020023.

29. Ebadi, A., Alikhani, H.A., and Rashtbari, M. 2012. Effect of plant growth promoting bacteria (PGPR) on the morphophysiological properties of button mushroom *Agaricus bisporus* in two different culturing beds. *International Research Journal of Basic and Applied Sciences* 3(1):203–212.

30. Zarenejad, F., Yakhchali, B., and Rasooli, I. 2012. Evaluation of indigenous potent mushroom growth promoting bacteria (MGPB) on *Agaricus bisporus* production. *World Journal of Microbiology and Biotechnology* 28(1):99–104.

31. Xiang, Q., Luo, L., Liang, Y., Chen, Q., Zhang, X., and Gu, Y. 2017. The diversity, growth promoting abilities and anti-microbial activities of bacteria isolated from the fruiting body of *Agaricus bisporus*. *Polish Journal of Microbiology* 66(2):201–207.

32. Young, L.S., Chu, J.N., Hameed, A., and Young, C.C. 2013. Cultivable mushroom growth-promoting bacteria and their impact on *Agaricus blazei* productivity. *Pesquisa Agropecuária Brasileira* 48(6):636–644.

33. Young, L.S., Chu, J.N., and Young, C.C. 2012. Beneficial bacterial strains on *Agaricus blazei* cultivation. *Pesquisa Agropecuária Brasileira* 47(6):815–821.

34. Bianciotto, V., Andreotti, S., Balestrini, R., Bonfante, P., Perotto, S. 2001. Extracellular polysaccharides are involved in the attachment of *Azospirillum brasilense* and *Rhizobium leguminosarum* to arbuscular mycorrhizal structures. *European Journal of Histochemistry* 45(1):39–49.

35. Alarcon, A., Davies-Jr., F.T., Egilla, J.N., Fox, T.C., Estrada-Luna, A.A., and Ferrera-Cerrato, R. 2002. Short term effects of *Glomus claroideum* and *Azospirillum brasilense* on growth and root acid phosphatase activity of *Carica papaya* L. under phosphorus stress. *Revista Latinoamericana de Microbiología* 44:31–37.

36. Russo, A., Vettori, L., Felici, C., Fiaschi, G., Morini, S., and Toffanin, A. 2008. Enhanced micropropagation response and biocontrol effect of *Azospirillum brasilense* Sp245 on *Prunus cerasifera* L. clone Mr. S 2/5 plants. *Journal of Biotechnology* 134(3–4):312–319.

37. Choudhary, D.K., Kasotia, A., Jain, S., Vaishnav, A., Kumari, S., Sharma, K.P., and Varma, A. 2016. Bacterial-mediated tolerance and resistance to plants under abiotic and biotic stresses. *Journal of Plant Growth Regulation* 35(1):276–300.

38. Choudhary, D.K., Varma, A., and Tuteja, N. 2017. Mycorrhizal helper bacteria: Sustainable approach. In: Varma, A., Prasad, R., and Tuteja, N. (Eds.), *Mycorrhiza - Function, Diversity, State of the Art.* Fourth edition. Springer, Cham, pp. 61–64.

39. Sridevi, S. and Ramakrishnan, K. 2010. Effects of combined inoculation of AM Fungi and *Azospirillum* on the Growth and Yield of Onion (*Allium cepa* L.). *Journal of Phytology* 2(1): 88–90.

40. Bama, M.E. and Ramakrishnan, K. 2010. Effects of combined inoculation of azospirillum and AM fungi on the growth and yield of finger millet (Eleusine coracana Gaertn) Var. Co 12. *Journal of Experimental Sciences* 1(8):10–11.

41. Levy, A., Chang, B.J., Abbott, L.K., Kuo, J., Harnett, G., and Inglis, T.J.J. 2003. Invasion of spores of the arbuscular mycorrhizal fungus *Gigaspora* decipiens by *Burkholderia* spp. *Applied and Environmental Microbiology* 69(10):6250–6256.

42. Duponnois, R. and Lesueur, D. 2004. Sporocarps of *Pisolithus albus* as an ecological niche for fluorescent pseudomonads involved in *Acacia mangium* Wild – *Pisolithus albus* ectomycorrhizal symbiosis. *Canadian Journal of Microbiology* 50(9):691–696.

43. Shaternikov, A.N., Tsivileva, O.M., and Nikitina, V.E. 2018. Bacteria from the *Azospirillum* genus in the optimization of artificial culture of *Pleurotus ostreatus* and *Ganoderma lucidum* mushrooms. *Aktual'naya biotekhnologiya (Actual Biotechnology)* 3 (26):41–45.

44. Tsivileva, O.M., Shaternikov, A.N., and Nikitina, V.E. 2020a. Bacteria of the *Azospirillum* genus for the optimization of the artificial culture of xylotrophic mushrooms. *Biotekhnologiya* 36(2):16–25.

45. Tsivileva, O.M., Shaternikov, A.N., and Nikitina, V.E. 2020b. Culture of xylotrophic macromycete *Flammulina velutipes* mycelium with azospirilla. *Biomics* 12(2):232–241.

46. Su, A., Ma, G., Xie, M., Ji, Y., Li, X., Zhao, L., and Hu, Q. 2019. Characteristic of polysaccharides from *Flammulina velutipes in vitro* digestion under salivary, simulated gastric and small intestinal conditions and fermentation by human gut microbiota. *International Journal of Food Science & Technology* 54(6):2277–2287.

47. Nie, Y., Jin, Y., Deng, C., Xu, L., Yu, M., Yang, W., Li, B., and Zhao, R. 2019. Rheological and microstructural properties of wheat dough supplemented with *Flammulina velutipes* (mushroom) powder and soluble polysaccharides. *CyTA-Journal of Food* 17(1):455–462.

48. Huang, L.H., Lin, H.Y., Lyu, Y.T., Gung, C.L., and Huang, C.T. 2019. Development of a transgenic *Flammulina velutipes* oral vaccine for hepatitis B. *Food Technology and Biotechnology* 57(1):105–112.

49. Mahfuz, S., Song, H., Miao, Y., and Liu, Z. 2019. Dietary inclusion of mushroom (*Flammulina velutipes*) stem waste on growth performance and immune responses in growing layer hens. *Journal of the Science of Food and Agriculture* 99(2):703–710.
50. Zhao, R., Hu, Q., Ma, G., Su, A., Xie, M., Li, X., Chen, G., and Zhao, L. 2019. Effects of *Flammulina velutipes polysaccharide* on immune response and intestinal microbiota in mice. *Journal of Functional Foods* 56:255–264.
51. Wu, G.S., Guo, J.J., Bao, J.L., Li, X.W., Chen, X.P., Lu, J.J., and Wang, Y.T. 2013. Anti-cancer properties of triterpenoids isolated from *Ganoderma lucidum*-a review. *Expert Opinion on Investigational Drugs* 22(8):981–992.
52. Osińska-Jaroszuk, M., Jaszek, M., Mizerska-Dudka, M., Błachowicz, A., Rejczak, T.P., Janusz, G., Wydrych, J., Polak, J., Jarosz-Wilkołazka, A., and Kandefer-Szerszeń, M. 2014. Exopolysaccharide from *Ganoderma applanatum* as a promising bioactive compound with cytostatic and antibacterial properties. *BioMed Research International* 2014: 743812/1–743812/10.
53. Gao, Y., Zhou, S., Huang, M., and Xu, A. 2003. Antibacterial and antiviral value of the genus *Ganoderma* P. Karst. species (Aphyllophoromycetideae): A review. *International Journal of Medicinal Mushrooms* 5(3):1–12.
54. Pereira-Jr., J.A.S., Rodrigues, D.P., Peixoto-Filho, R.C., Bastos, I.V.G.A., De Oliveira, G.G., Araújo, J.M., and Melo, S.J. 2013. contribution to pharmacognostic and morphoanatomical studies, antibacterial and cytotoxic activities of *Ganoderma parvulum* Murrill (Basidiomycota, Polyporales, Ganodermataceae). *Latin American Journal of Pharmacy* 32(7):996–1003.
55. Sridhar, S., Sivaprakasam, E., Balakumar, R., and Kavitha, D. 2011. Evaluation of antibacterial and antifungal activity of *Ganoderma lucidum* (Curtis) P. Karst fruit bodies extracts. *World Journal of Science and Technology* 1(6):8–11.
56. Prasad, Y. and Wesely, W.E.G. 2008. Antibacterial activity of the bio-multidrug (*Ganoderma lucidum*) on Multidrug resistant *Staphylococcus aureus* (MRSA). *Journal of Advanced Biotechnology* 10:9–16.
57. Ofodile, L.N., Uma, N., Grayer, R.J., Ogundipe, O.T., and Simmonds, M.S.J. 2012. Antibacterial compounds from the mushroom *Ganoderma colossum* from Nigeria. *Phytotherapy Research* 26(5):748–751.
58. Roy, A. and Prasad, P. 2013. Therapeutic potential of *Pleurotus ostreatus*: A review. *Research Journal of Pharmacy and Technology* 6(9):937–940.
59. Majesty, D., Ijeoma, E., Winner, K., and Prince, O. 2019. Nutritional, anti-nutritional and biochemical studies on the oyster mushroom, *Pleurotus ostreatus*. *EC Nutrition* 14(1):36–59.
60. Mutukwa, I.B., Hall-III, C.A., Cihacek, L., and Lee, C.W. 2019. Evaluation of drying method and pretreatment effects on the nutritional and antioxidant properties of Oyster mushroom (*Pleurotus ostreatus*). *Journal of Food Processing and Preservation* 43(4):e13910.
61. Nam, W.L., Phang, X.Y., Su, M.H., Liew, R.K., Ma, N.L., Rosli, M.H.N.B., and Lam, S.S. 2018. Production of bio-fertilizer from microwave vacuum pyrolysis of palm kernel shell for cultivation of Oyster mushroom (*Pleurotus ostreatus*). *Science of the Total Environment* 624:9–16.
62. Phan, C.W., Wang, J.K., Tan, E.Y.Y., Tan, Y.S., Sathiya Seelan, J.S., Cheah, S.C., and Vikineswary, S. 2019. Giant oyster mushroom, *Pleurotus giganteus* (Agaricomycetes): Current status of the cultivation methods, chemical composition, biological, and health-promoting properties. *Food Reviews International* 35(4):324–341.
63. Sangeetha, K., Senthilkumar, G., Panneerselvam, A., and Sathammaipriya, N. 2019. Cultivation of oyster mushroom (*Pleurotus* sp.) using different substrates and evaluate their potentials of antibacterial and phytochemicals. *International Journal of Research in Pharmaceutical Sciences* 10(2): 997–1001.
64. Vetvicka, V., Gover, O., Karpovsky, M., Hayby, H., Danay, O., Ezov, N., Hadar, Y., and Schwartz, B. 2019. Immune-modulating activities of glucans extracted from *Pleurotus ostreatus* and *Pleurotus eryngii*. *Journal of Functional Foods* 54:81–91.
65. Zhu, B., Li, Y., Hu, T., and Zhang, Y. 2019. The hepatoprotective effect of polysaccharides from *Pleurotus ostreatus* on carbon tetrachloride-induced acute liver injury rats. *International Journal of Biological Macromolecules* 131:1–9.
66. Pozdnyakova, N., Dubrovskaya, E., Chernyshova, M., Makarov, O., Golubev, S., Balandina, S., and Turkovskaya, O. 2018. The degradation of three-ringed polycyclic aromatic hydrocarbons by wood-inhabiting fungus *Pleurotus ostreatus* and soil-inhabiting fungus *Agaricus bisporus*. *Fungal biology* 122(5):363–372.
67. Karthikeyan, V., Ragunathan, R., Johney, J., and Kabesh, K. 2019. Production, Optimization and Purification of Laccase Produced by *Pleurotus ostreatus* MH591763. *Research & Reviews: A Journal of Microbiology and Virology* 9(1):56–64.

68. Noman, E., Al-Gheethi, A., Mohamed, R.M.S.R., and Talip, B.A. 2019. Myco-remediation of xenobiotic organic compounds for a sustainable environment: A critical review. *Topics in Current Chemistry* 377(3):17/1–17/41.

69. Bamigboye, C.O., Oloke, J.K., Burton, M., Dames, J.F., and Lateef, A. 2019. Optimization of the process for producing biomass and exopolysaccharide from the King Tuber Oyster Mushroom, *Pleurotus tuber-regium* (Agaricomycetes), for biotechnological applications. *International Journal of Medicinal Mushrooms* 21(4):311–322.

70. Janusz, G., Pawlik, A., Sulej, J., Świderska-Burek, U., Jarosz-Wilkołazka, A., and Paszczyński, A. 2017. Lignin degradation: Microorganisms, enzymes involved, genomes analysis and evolution. *FEMS Microbiology Reviews* 41(6): 941–962.

71. Kupryashina, M.A., Petrov, S.V., Ponomareva, E.G., and Nikitina, V.E. 2015. Ligninolytic activity of bacteria of the genera *Azospirillum* and *Niveispirillum*. *Microbiology* 84(6):791–795.

72. Day, J.M. and Döbereiner, J. 1976. Physiological aspects of N-fixation by a *Spirillum* from *Digitaria* roots. *Soil Biology and Biochemistry* 8(1):45–50.

73. Compant, S., Duffy, B., Nowak, J., Clément, C., and Barka, E.A. 2005. Use of plant growth-promoting bacteria for biocontrol of plant diseases: Principles, mechanisms of action, and future prospects. *Applied and Environmental Microbiology* 71(9):4951–4959.

2 Microbial Biostimulants for the Management of Insect Pests and Diseases

Muhammad Raheel, Amir Riaz, Muhammad Naveed Aslam, Sajjad Ali and Qaiser Shakeel
The Islamia University of Bahawalpur

Hafiz Muhammad Aatif and Muhammad Zeeshan Mansha
Bahauddin Zakariya University, Bahadur sub-campus, Layyah

Fahim Nawaz
MNS University of Agriculture, Multan

CONTENTS

2.1 INTRODUCTION

In their habitats, plants are subjected to a variety of biotic and abiotic environmental stresses that lead to various diseases and yield reduction and are a matter of great concern for farmers and scientists (Drobeck et al., 2019). Abiotic environmental stress is a major crop production problem in various crop production zones. The crop quality and quantity are affected due to biotic and abiotic factors. The quality of the crops is determined through agronomic (yield, fruit size, yield and resistance against pathogens) and organoleptic (e.g., color, form and firmness) parameters, in addition to vitamin and nutrient content. Abiotic factors include drought, humidity, rain, pollution, excessive

DOI: 10.1201/9781003188032-2

salinity, soil quality, acidity, high and low temperatures, wind and ultraviolet radiation (Di Vittori et al., 2018). A substantial reduction in yield can be caused by unfavorable stimuli because to combat stress, plants use energy reserves instead of food production. Bacteria, fungi and viruses are classified into biotic factors that are responsible for causing numerous plant diseases. Infections caused by fungi and bacteria can reduce yield and even result in the loss of the entire harvest. Plant protection items of different kinds are used to avoid these pathogens (Drobeck et al., 2019). Mineral and chemical plant protection materials are to be gradually phased out in favor of natural preparations as guided by European Union instructions. The explanation for this is that chemical and mineral plant defense agents have a destructive effect on the environment and the health benefits of crops (Drobeck et al., 2019).

The usage of chemicals to combat plant diseases has widely been documented, but the residual effects of pesticide overuse necessitate the use of biological control agents, which play a vital role in the plant diseases management. Furthermore, chemical fertilizers are to blame for the eutrophication of many water bodies. As a result, dead zones are created, which are devoid of living organisms. Induced systemic resistance (ISR) is one of the processes by which pathogen-suppressing bacteria and fungi enhance plant development and resistance; symbiotic connections between various microorganisms and plant roots have also been demonstrated to have a substantial impact on their development and survival. Majority of these microorganisms have stimulant properties that can be used to improve nutrient absorption, pest control and crop protection from environmental stresses. Biostimulants are increasingly being incorporated into production systems in addition to these conventional methods, with the aim of optimizing plant physiological processes. In the last two and a half decades, plant biostimulants from natural materials have received a lot of interest from the scientific world and manufacturers (Hayat et al., 2010). Biostimulants are a potentially unique technique for governing metabolic processes in plants to promote development, reduce stress-induced limits and boost yield.

2.2 HISTORY, CONCEPT AND DEFINITIONS OF BIOSTIMULANTS

The word "biostimulant" was coined by horticulturists to describe compounds that enhance plant development without being insecticides, soil improvers or nutrients (Du Jardin, 2015). Grounds Maintenance, an Internet journal devoted to the maintenance of turfs, can be found in the first description of the term biostimulants. Zhang and Schmidt from the Virginia Polytechnic Institute and State University's Department of Crop, Soil and Environmental Sciences described them as "materials that, in minute amounts, promote plant growth" in this Web journal in 1997 (Du Jardin, 2015). The authors tried to differentiate biostimulants from soil amendments and nutrients that likewise encourage the growth of plants; however, their application is carried out in greater amounts. The authors describe this term by emphasizing on the phrase "minute quantities" (Du Jardin, 2015). The term biostimulant was first described in the scientific literature by Kauffman (2007) in a peer-reviewed paper: "biostimulants are materials, other than fertilizers, that promote plant growth when applied in small quantities." By adding a description, they tried to summarize them as: "Biostimulants come in a number of formulations and contain a variety of ingredients, but these are commonly divided into three categories based on their source and material. These comprise amino acid-containing products (AA), hormone-containing products (HCP), and humic substances (HS)."

Biostimulants can be used as a fertilizer additive to aid nutrient absorption, promote plant development and improve abiotic stress tolerance (Du Jardin, 2015). The term "biostimulants" has a broad meaning that isn't specific enough (Drobeck et al., 2019). Biostimulants vary in two ways from conventional growth and plant protection chemicals. Any natural ingredient, mixture of natural chemicals, or microbe that improves crop quality without creating undesirable side effects is referred to as a biostimulant (Drobeck et al., 2019). The EBIC (European Biostimulants Industry) described plant biostimulants in Europe as follows: "Plant biostimulants are substances or microorganisms which, once added to the rhizosphere or plants, stimulate natural procedures for improving nutrient absorption and production, abiotic stress tolerance, and crop quality." The Biostimulant Coalition in North America described biostimulants as "substances, including microorganisms, applied to plants, seed, soil, or other growing media that may improve the plant's ability to assimilate applied nutrients or provide benefits to plant growth." Although it is necessary to follow a definition of biostimulants for regulatory purposes, any definition should be based on scientific principles (Biostimulant, 2013). Several definitions for plant biostimulants have been proposed. Jardin suggests that "any concept of biostimulants should focus on the agricultural roles of biostimulants, not on the nature of their constituents or modes of action." He emphasizes the importance of the final effect on plant productivity. Thus, biostimulants can be classified based on their established mode of action and origin, or solely on their demonstrated positive effect on plant productivity. The multi-component and generally undefined composition of several biostimulant products, and the possibility that a biostimulant's function is not clarified by the existence of any single constituent, but rather by the involvement of several constituents in the product, add to the difficulties in creating a concept (Biostimulant, 2013; Du Jardin, 2015).

2.3   CLASSIFICATION OF BIOSTIMULANTS

Amino acids, micronutrients, enzymes, proteins and other substances are examples of biostimulants. Protein hydrolases, phenols, salicylic acid, humic and fulvic acids and phenols are all natural stimulants. Fungi and bacteria that modify the composition of organisms in the soil or plants are a significant category of microbial biostimulants. They can hasten the rate of degradation or limit the number of specific fungal and bacterial groups (Drobeck et al., 2019). *Glomus intraradices*, *Trichoderma atroviride*, *T. reesei* and *Heteroconium chaetospira* are some of the most commonly used fungi as biostimulants. The useful bacteria include *Ochrobactrum* spp., *Bacillus* spp., *Arthrobacter* spp., *Enterobacter* spp., *Acinetobacter* spp., *Pseudomonas* spp. and *Rhodococcus* spp. (Mukherjee et al., 2013; Zhao et al., 2018; Colla et al., 2015a). A possible definition of a microbial biostimulant is as follows: "a mixture of microorganisms whose purpose, when applied to seeds, plants, or the rhizosphere, is to stimulate natural processes in order to improve/benefit nutrient uptake, nutrient performance, abiotic stress tolerance, crop quality, and/or yield." PGPR (plant growth-promoting rhizobacteria) was the "old name" for microbial biostimulants, coined by J.W. Kloepper in 1980 to describe bacteria named *Pseudomonas fluorescens* that aids in pathogen biological control and plant growth. Later, Kapulnik expanded this concept to include rhizobacteria that can directly promote plant growth. Fungi, especially mycorrhizal fungi, one of the most important microbes in agriculture, are now included in the term microbial biostimulant.

2.3.1   HUMIC AND FULVIC ACIDS

Humic chemicals have long been recognized as the most potent plant biostimulants. Humic substances (HS) are naturally occurring organic matter components that result from the biological and chemical degradation of plant, animal and microbial leftovers, as well as microbial processes (Du Jardin, 2015). HS are sets of heterogeneous compounds that were initially classified into humins, humic acids and fulvic acids based on their molecular weights and solubility. These too exhibit complicated dynamics of dissociation/association into supramolecular colloids, which is affected by the protons release and exudates by plant roots (Du Jardin, 2015). Humic acid compounds are widely acknowledged as an essential component of soil physicochemical properties. They encourage root growth, which increases soil nutrient availability by increasing the area of soil–root interaction (Rose et al., 2014; Du Jardin, 2015).

The HS possess the capability to increase soil cation exchange potential while also neutralizing the pH of soil. They are reported to make insoluble iron accessible to plants by forming complexes with them. This is a crucial property since it enables plants to be supplied with micronutrients that are not readily accessible. Fulvic acids, which are abundant in humus compounds, may be taken by plants as a complex of cations because of their modest molecular mass (Du Jardin, 2015). HS have also been demonstrated to impede plasma membrane H+-ATPase activity, resulting in an increase in H+ secretion and a decrease in soil and root surface pH. Nutrient supply and absorption are supported by a lower pH. They also have the ability to influence stress reduction and secondary metabolite formation (Du Jardin, 2015). Due to their tendency to bind heavy metals, HS and fulvic acid are less likely to be digested by plants during nutrient absorption. The HS are obtained from natural humified organic matter vermicompost or mineral deposits and composts. Furthermore, rather than decomposing in the soil or composting, the agricultural by-products are open for regulated disintegration and oxidation through chemical means, resulting in "humic-like" materials, which are suggested to be a replacement for natural HS (Du Jardin, 2015).

2.3.2 Protein Hydrolysates Together with a Few Nitrogen-Containing Compounds

Another important class of plant biostimulants is protein hydrolysates (PH). Partial hydrolysis is used to create a mixture of amino acids, oligopeptides and polypeptides, from proteins (Calvo et al., 2014). Chemical and enzymatic PH were used to extract amino acids and peptide mixtures from agroindustrial by-products, including animal wastes and crop residues (Chalamaiah et al., 2012). They can be sprayed on the leaves or injected into the rhizosphere. These mixtures improve soil condition and respiration and are used for the regulation of soilborne microbial growth since they can use the PH as a carbon and nitrogen supply for microbes. Additionally, PH may chelate as well as complex soil micronutrients, making them more available to plants. PH have been shown to boost activity, fertility, biomass and respiration of soilborne microbes. Complexing and chelating events of peptides and various amino acids are thought to help roots access and acquire nutrients. In conformity with bioassays utilizing plants and yeasts as test entities, the protection of animal origin hydrolyzed proteins has been recently evaluated, and no phytotoxicity, ecotoxicity or genotoxicity was identified (Corte et al., 2014). Nonetheless, PH usage originated from animal by-products in the food web is raising safety concerns.

2.3.3 Plant Disease Management through Induced Systemic Resistance – Use of Microbial Biostimulants

Pathogenic viruses, nematodes, bacteria and fungi that cause plant diseases are major obstacles to preserving the health and yield of crops (Narasimhan and Shivakumar, 2015). For enhancing crop growth and managing numerous plant diseases, plant beneficial microorganisms are an effective and environmentally acceptable substitute for chemical bactericides, fungicides and nematicides (Adam et al., 2014). In soil and or on tissues of plants, pathogenic microbial growth is inhibited by a variety of microbial biostimulants, as well as the pathogens' negative effects on plants. Bacteria, fungi and yeasts have the ability to encourage plant growth, thereby allowing their use as biocontrol agents of plant pathogens. These microorganisms generate hormone-like substances that affect biological processes in the agroecosystem, such as plant physiology, metabolism, morphology and interactions. Continuous research into soil-based organisms is allowing for a better understanding of the dynamic environment of bacteria, leading to the development of new microbiome ecosystem categories. "Crop probiotics" are on the verge of becoming important new plant biostimulants. Associations of useful bacteria or fungi are also entered in the group of microbes-based biostimulants. The following species are included among the fungi used in plant cultivation: *Heteroconium chaetospira*, *Glomus intraradices*, *T. atroviride* and *T. reesei*.

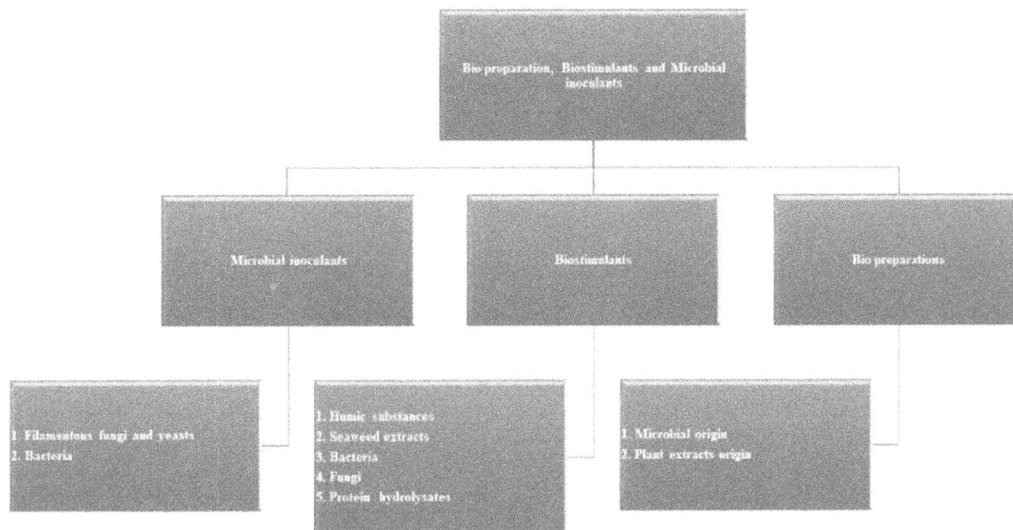

FIGURE 2.1 Classification of biopreparations, biostimulants and microbial inoculants. (Adopted from Pylak et al., 2019.)

Acinetobacter spp., *Arthrobacter* spp., *Enterobacter* spp., *Ochrobactrum* spp., *Pseudomonas* spp., *Bacillus* spp. and *Rhodococcus* spp. are among the bacterial species that increase plant development (Mukherjee et al., 2013; Colla et al., 2015a). PGPR *and Rhizobium* spp. make up the largest community of beneficial bacteria.

2.3.3.1 Fungi Strains as Biostimulants in the Management of Plant Diseases

2.3.3.1.1 Genus Trichoderma

The fungi belonging to the *Trichoderma* genus are one of the most important antagonists of plant pathogens. Being a part of few biopreparations/stimulants, *T. harzianum* was recognized as a widely used plant pathogen antagonist. It can not only function as a mycoparasite, but can also produce antibiotics and is an outstanding nutrient competitor due to its fast growth rate. Furthermore, it has the ability to activate plant defense mechanisms (Benítez et al., 2004). *Trichoderma* spp. produces antimicrobial agents, and the fungi can produce gliotoxins, sesquiterpenes, viriden, peptaibols and isonitriles. Those toxic substances are used by fungi to prevent other rivals in the ecological niche from growing. *Trichoderma* fungi have the ability to function as mycoparasites by degrading pathogen cell walls due to their ability to manufacture chitinase. *Trichoderma* spp. is infective to numerous plant pathogens, including *Armillaria, Phytophthora, Pythium, Rhizoctonia, Botrytis, Fusarium, Helminthosporium, Colletotrichum, Chondrostereum, Dematophora, Monilia, Endothia, Fulvia, Fusicladium, Macrophomina, Nectria, Phoma, Plasmopara, Pseudoperonospora, Venturia, Sclerotinia, Sclerotium, Rhizopus* and *Verticillium* (Pylak et al., 2019). While some *Trichoderma* fungi produce cellulases, such as strain G79/11 of *T. atroviride*, they may also manufacture additional enzymes, making them suitable candidates for antifungal biopreparations. *Trichoderma* spp. illustrates the more frequent mechanisms of antagonism against pathogenic fungus. The five primary strategies engaged in combating plant pathogenic fungi plus encouraging growth of plants are manufacturing of inhibitory materials, contest for space and food, fungal parasitism, induced resistance and pathogen enzymes inactivation (Oszust et al., 2017a,b; Yedidia et al., 1999). *Trichoderma* spp. demonstrates the most important procedure of antagonism against fungi (Figure 2.2).

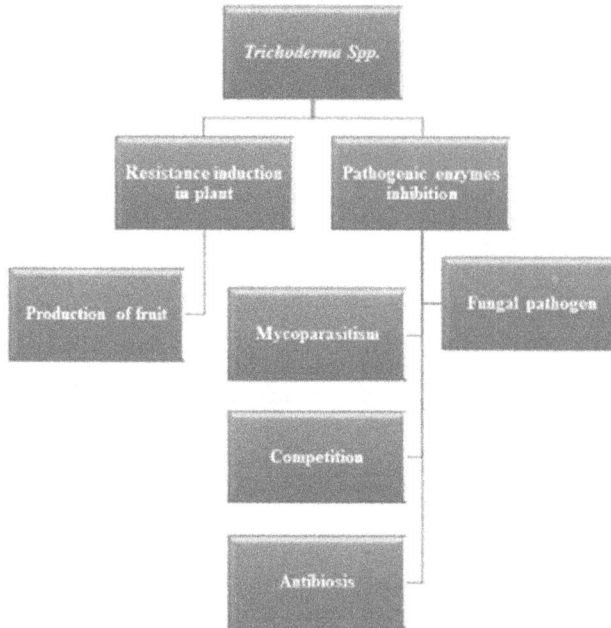

FIGURE 2.2 Processes involved in attacking other fungi and promoting plant growth. (Adopted from Pylak et al., 2019.)

2.3.3.1.2   Aureobasidium pullulans

Aureobasidium pullulans is another microorganism widely used in strawberry farming around the world due to its high effectiveness in protecting strawberries from *Botrytis cinerea* and *Rhizopus stolonifer* (Mounir et al., 2007). It's a fungus that colonizes plants in their natural environment, similar to yeast. A range of reasons affect the growth of *A. pullulans*, including nutrient availability, temperature and pH. Many scientists have confirmed its effectiveness in strawberry defense (Wagner and Hetman, 2016). These produced contest for nutrients and space with other fungi, act as mycoparasites and develop antibiotics and antimicrobial enzymes. The antipathogenic effect and decreased disease severity are not well known and are thought to be a result of a variety of reactions. The contest for space and resources weakens the cells of pathogen; resultantly, they become more susceptible to antibiotics and enzymes of host or opponent (Adikaram et al., 2002). Both *B. cinerea* and *R. stolonifer* were found to be resistant to A. pullulans L47 (Pylak et al., 2019).

2.3.3.2   Use of Bacterial Biostimulants for the Management of Plant Diseases

2.3.3.2.1   Bacillus spp.

When there is a lack of water, salinity and heavy metals build up in the soil, and *Bacillus* spp. siderophores and exopolysaccharides are produced, which inhibit toxic ions from flowing through plant tissues and manage water transport and ionic balance while regulating pathogenic bacteria populations (Radhakrishnan, 2017). *Bacillus* spp. produce 1-aminocyclopropane-1-carboxylate deaminase, gibberellic acid and indole-3-acetic acid, which regulate phytohormone metabolism within the cells and promote tolerance in plants against stress. *Bacillus* spp. release several compounds such as cellulase, lipopeptides, chitosanase, protease, glucanase and hydrogen cyanide that degrade cell wall of plant pathogens, including fungi, bacteria, viruses, nematodes and insect pests. *Bacillus lentimorbus*, *B. megaterium*, *B. pumilus* and *B. subtilis* are bacteria that can stop *B. cinerea* from growing in vitro conditions. These not only hinder their growth, but can also reduce germination of conidiospores up to 80% on strawberry fruits (Radhakrishnan, 2017).

The inoculation of *Bacillus* spp. guards the plants from various bacterial diseases by suppressing their development. *Pseudomonas savastanoi*, *Ralstonia solanacearum* and *Xanthomonas axonopodis* are some examples of bacterial pathogens that are suppressed by *Bacillus* spp. (Yi et al., 2013; Krid et al., 2012). *Bacillus spp.* produce biofilms on the root surface and emit several toxins that kill harmful bacteria and reduce disease incidence in plants. These toxins include bacillomycin, fengycin, iturin, macrolactin and surfactin. The secretions of *Bacillus* spp. destroy the plant pathogen by destroying the cell walls of pathogenic bacteria and alter their cell morphology (Elshakh et al., 2016). In plant protection systems, pathogenic bacteria such as *X. oryzae* and *R. solanacearum* reduce the activities of defense enzymes such as lipoxygenase and phenylalanine ammonia-lyase, but after *Bacillus* spp. are administered, these enzymatic actions are enhanced in infected plants (Radhakrishnan, 2017).

Antagonistic activity of *Bacillus* spp. inhibits mycelial development and, therefore, prevent plant fungal disease (Radhakrishnan, 2017; Akram et al., 2016). *Bacillus* spp. binds to cell walls of fungi and release several enzymes such as cellulase, chitosanase, glucanase and protease as well as HCN and siderophores that break and distort the hyphae, causing vacuolation and protoplast leakage, resulting in altered cell structure and functions. Furthermore, some plant pathogenic fungi are eliminated in rhizospheres due to various antifungal peptides. These peptides are produced by bacteria and include mixirin, pumilacidin, iturin, fengycin and surfactin (Han et al., 2015).

Plant viruses are the second most common cause of plant disease, after fungi. Chemical therapies have proven to be the most successful form of virus containment. Long-term chemical use results in soil deposits and increases plant pathogen drug resistance (Zhao et al., 2017). Antiviral compounds are produced by certain *Bacillus* spp. ISR is observed in plants as a result of contact with *Bacillus* spp., resulting in increased growth of plants during infection due to cucumber mosaic virus (Radhakrishnan, 2017). *Bacillus amyloliquefaciens*, plantarum biofilm formation and surfactin production protect plants from viral diseases by activating ISR machinery (Radhakrishnan, 2017; Chowdhury et al., 2015). By suppressing the synthesis of capsids and enhancing the expression of signaling genes for resistance, *Bacillus* spp. establish systemic resistance to tobacco mosaic virus disease (Radhakrishnan, 2017). Phytoparasitic nematodes are serious plant parasites that pose a great threat to plant growth by damaging crop yield and quality. Root-knot nematodes (RKN) on a global scale have been classified as the very harmful parasite as compared to other nematodes. The RKN has a broad host range as they can affect up to 5,500 species of plants (Trudgill and Blok, 2001). These nematodes can be managed by inoculating with bacteria that keep nematode populations under control. *Bacillus* spp., for example, reduce knot formation and the number of egg masses in RKN-infected roots, thereby improving plant resistance. Bacteriocins and antimicrobial peptides produced by *Bacillus* spp. exhibit nematode growth inhibition (Adam et al., 2014). In *B. amyloliquefaciens*, the PZN gene cluster having nematicidal performance was found. Likewise, *Bacillus* spp. release Cry5B and Cry6A (crystal proteins) that regulate the development of RKN (*Meloidogyne hapla*) as well as free-living nematode (*Caenorhabditis elegans*) (Chowdhury et al., 2015; Liu et al., 2013).

2.3.3.2.2 Pseudomonas Fluorescens

PGPR, such as fluorescent *Pseudomonas*, have a significant part in promoting the growth of plants, inducing systemic resistance and biocontrol of pathogen, among other things (Ganeshan and Kumar, 2007). Many strains of this species have been observed for promoting plant growth and minimizing severity of plant diseases. Several species of *P. fluorescens* are potential biocontrol options against fungi causing root and seeds infection, hence suppressing plant diseases (Ganeshan and Kumar, 2007). They have repeatedly been reported for decreasing the disease severity due to fungal pathogens (Hoffland et al., 1996). Several species of *P. fluorescens* were recognized as antagonistic to *Stemphylium vesicarium* causing brown spot of pear in a variety of environments (Bonaterra et al., 1998). Before sowing, the seed treatment of paddy seeds with *P. fluorescens* Pf1 formulation induced resistance in paddy crop against bacterial blight of rice. This bacterial treatment was responsible for lowering the disease incidence to 1.2% from 6.8% (Vidhyasekaran et al., 2001). Various laboratory and field studies have revealed that *P. fluorescens* RPB14 demonstrated possible biocontrol against *Phaseolus vulgaris* (Ganeshan and Kumar, 2007). The gram crop is infected by a disease named as dry root rot caused by *Macrophomina phaseolina,* which can also be managed by using Pf1 strain of *P. fluorescens* because the bacterial strain was effectively inhibiting the growth of *M. phaseolina* (Jayashree et al., 2000). Seed treatment with *P. fluorescens* was also effective in managing the disease complex due to *Heterodera cajani* (cyst nematode) and *M. phaseolina.* There was a reduction in incidence of root rot and population of nematode, as well as enhanced pod production. A study conducted in Australia on broccoli found that *Alternaria tenuis, Helminthosporium tetramera, M phaseolina, Sclerotinea rolscii* and *Fusarium solani* were antagonistic to *P. fluorescens* (Atef, 2000).

2.3.3.2.3   Streptomyces spp.

Actinomycetes dominate the microbial population in the soil. The genus *Streptomyces*, which contains the most antibiotics, extracellular enzymes and bioactive compounds, is the most common and potentially the most important actinomycetes (van Dissel et al., 2014). This genus has shown over time that it possesses a lot of potential for improving agriculture's future. Its primary biocontrol capacity stems from its high synthesis of metabolites and antibiotics that aid in its pathogen inhibitory activity, such as chitinase from *S. coelicolor* and siderophores from *S. aureus* (Som et al., 2017; Gherbawy et al., 2012). In YH27A strain of *S. violaceusniger*, the antibiotic geldanamycin and antifungal nigericin were identified (Shekh and Naim, 2017; Errakhi et al., 2016). It seems that tomato plants are protected from the bacteria *Pectobacterium carotovorum* subsp. *brasiliensis* by using *Streptomyces* spp. Furthermore, six Streptomyces spp. isolates produce volatiles that enhance tomato root development.

2.3.4   MICROBIAL BIOSTIMULANTS FOR THE MANAGEMENT OF INSECT PESTS

Insects have been reported for a long time to become infected with various entomopathogenic microorganisms, which are a major cause of their natural death. In other circumstances, diseases are accountable for natural breakouts of certain epidemics within populations of several species of insects (El-Husseini, 2006). As a result, under such circumstances, they may have a significant hand in the extinction of an insect species. A scientist, "Bassi" was motivated by this occurrence to introduce the concept of using insect pathogenic microorganisms to combat agricultural insect pests (Bassi, 1838). In Russia, the first effective and extensive microbial application by means of conidiospores of *Metarhizium anisopliae* fungus occurred in order to combat the beet weevil, *Bothynoderes punctiventris.* For this purpose, large quantities of pure conidiospores were developed in the laboratory. During the last five decades, the management of insect pests and plant diseases using microbes has progressed rapidly, with impressive results in controlled laboratory settings often accompanied by poor results in field applications. As a result, we must comprehend the significant concepts needed to create entomopathogens that are stable, effective and safe for microbial control (El-Husseini, 2006).

2.3.4.1   Entomopathogenic Fungi

More than 400 species of entomopathogenic fungi (EPF) have been identified, with only around 20 species having the capacity to be used in the microbial control of insect pests. *Neozygites, Erynia, Lagnidium, Metarhizium, Beauveria, Entomophthora, Ashersonia, Verticillium, Nomuraea, Hirsutella* and *Paecilomyces* are the 11 genera that include the majority of them (Roberts and Wraight, 1986). The first successful application of an EPF, *Metarhizium anisopliae,* in sugar beet fields dates back to 1878 when Russian scientist Metchnikoff used it to combat wheat leaf beetle, *Anisoplia austriaca,* and the beet weevil, *Cleonus punctiventris* (Metchnikoff, 1879; Müller-Kögler, 1965). Efficient open-field applications were effective only in places with elevated temperatures and humidity, such as tropical and subtropical countries (El-Hady, 2004). *Beauveria bassiana* and *M. anisopliae* were found to be very successful in managing cacao coleopteran and lepidopteran insect pests in Brazil (El-Husseini, 2006; El-Kady et al., 1983). In crops whose vegetation leads to excessive relative humidity in the micro-climate among plants, such as clover fields in Mediterranean region and sugar beet fields in Northern Asia and Europe, fungi were also effective biocontrol agents (El-Husseini, 1981). The combined application of Rhabdionvirus oryctes viruses *and M. anisopliae* has extensively been used in the Pacific to kill populations of rhinoceros beetles on the coconut palm (Flexner et al., 1986; El-Husseini, 2006). Limited field trials were conducted in Arab countries, most of which are located in arid or semi-arid areas, to test the effectiveness of EPF against various pests in various crops (Marie, 2004; Seufi and Osman, 2005). Low levels of RH may be to blame for entomopathogenic fungi's poor efficiency in open fields in dry and semi-arid locations, as an elevated RH is required by attached conidiospores for their germination and penetration into the insect body (El-Husseini, 2006).

Concerns about the security of entomopathogenic microorganisms employed in microbial control have grown in response to the rising demand for biological control agents in the usage of integrated management systems for insect pests (El-Husseini, 2006). Registration of bioproducts dependent on microbes pays special attention to the following aspects: (i) genetic recombination and natural strain displacement, (ii) toxic metabolite threats, (iii) allergic properties and (iv) impact on non-target species (biodiversity). These topics have been thoroughly discussed. Fungi, on the other hand, exhibit less detail, posing a threat to beneficial organisms such as parasitoids, predatory insects, pollinators and bees. The degree of precision needed to protect is determined by the characteristics of the targeted region as well as the taxonomic classes of the species involved (El-Husseini, 2006). The majority of commercially produced EPF for field application in microbial pest control demonstrated little infectivity in humans or other vertebrates. *Hirsutella thompsonii, Lagenidium giganteum, Nomuraea rileyi* and *V. lecanii* safety studies revealed negative results in a variety of mammalian and bird species (Ignoffo, 1973; El-Husseini, 2006).

2.3.4.2 Entomopathogenic Bacteria

The Lactobacillaceae, Micrococcaceae, Pseudomonadaceae and Bacillaceae families of bacteria contain bacteria that cause diseases in insects. Based on non-target organism and human protection, Bacillaceae members of Eubacteriales order are the maximum investigated, marketed and most beneficial for the management of insects belonging to Diptera, Coleopteran and Lepidoptera orders (El-Husseini, 2006). Japanese beetle (*Popillia japonica*) larvae, an insect that came from East Asia to North America, became the first microbial control therapy using an entomopathogenic bacterium (EPB). By injecting *Bacillus popilliae* (responsible for causing milky disease in host) into grubs collected from the fields, the bacterium *B. popilliae* was found to be responsible for causing mortality of grubs (El-Husseini, 2006). Formulations that had been processed were converted into dirt, or their field application was carried out. Only a single application of the pathogen controlled the population of insects for more than a decade. The deceased larvae became a cause of infection for other fit population in the soil. In soil, the existence of *B. popilliae* spores far from the sun's disruptive impact may be credited for this singularly effective example. The *B. popilliae var. melolonthae* was also active in controlling the coleopteran *Melolontha melolontha* in Western Europe (El-Husseini, 2006). Dr. Edward Steinhaus in California, USA, was drawn to *B. thuringiensis* (Bt.) after its detection in 1915 by Dr. Ernest Berliner. During the late 1950s in California, *B. thuringiensis* (Bt.) was successfully adopted on a vast area to monitor the larval stages of the Luzerne moth (*Goliath philodice*). Many forms of Bt. have been isolated since then and have shown varying effectiveness against a variety of dipteran, lepidopteran and coleopteran members, based on exotoxins as well as endotoxins released from the Bt. isolate (Brown et al., 1958). Since the toxin produced by Bt. is not host specific, it poses threats to humans as well as the environment; resultantly, the manufacture of exotoxins reduced its use in the early 1960s. The discovery of *B. thuringiensis kurstaki* HD-1 (a new subspecies of Bt.), which lacks exotoxins, paved the way for the strain's global commercialization. As a result, several writers from all over the world have recorded the effectiveness of Bt. *kurstaki* in managing various agricultural lepidopteran pests. Arab countries, for example Egypt, began using Bt. in cotton crop for managing cotton leafworm (*Spodoptera littoralis*) young larval stages in 1960. Similarly, in clover fields, *Trifolium alexandrinum* L, this pest was also successfully handled (El-Husseini, 1981; Mohamed et al., 2005).

The potency of Bt. was greatly boosted when a molasses- or sugar-based nourishing stimulant was added to the sprayed solution, and chitinase significantly improved the potency of Bt. In Egypt, the local Bt. agent "Protecto" is commonly used in date palms to manage *Ephestia* spp. in addition to lepidopteran insects in grapes, tomatoes and agronomic crops.

As a result of the widespread use of *B. sphaericus*, which caused resistance in mosquito larvae, the discovery of the Bt. *israelensis* subspecies, specifically used against larval stages of Diptera, provided a latest and healthy substitute to chemical pesticides application in water sources to combat immature stages of mosquito (Baumann et al., 1991; Goldberg and Margalit, 1977). The discovery of the Bt. *tenebrionis* subspecies, an important and specialized bacterium against Coleopteran larvae, broadened Bt. host's range to include pests of the Coleoptera order (Hermstadt et al., 1986, 1987)

2.3.4.3   Entomopathogenic Viruses

These are obligate parasites of insects with RNA or RNA in capsids to create virions or nucleocapsids (El-Husseini, 2006). Their efficacy as biocontrol agents is influenced by their natural survival and methods of dispersion in the surroundings. The insect *Oryctes rhinoceros* (the coconut palm beetle) was successfully treated with the Rhabdionvirus oryctes due to the virus's vertical transmission and teratological effects (Monty, 1974; Padidam, 1991; El-Husseini, 2006). Infected adults release the virus in their feces at feeding locations near breeding sites in damaged palm logs as well as at the top of coconut palms where larvae are deposited, infecting the subsequent larval progenies (Young, 1974). Induced horizontal dispersal via application methods, especially when used by planes, and in certain instances, the action of diseased larvae are principally linked to the effective management of numerous larvae of Lepidoptera order using occluded viruses belonging to Baculoviridae family. Infected larvae of certain species migrate to higher elevations, allowing the virus to spread horizontally through food contamination and light. *Diprion hercyniae*, a European origin sawfly, swiftly established itself as a major forest pest in the USA and Canada (van Driesche and Bellows, 1996). The pest was not controlled any chemical-based conventional insecticides. *Borrelinavirus diprion*, a microbial control virus, was the only successful measure that minimized the population of pest.

2.4   CONCLUSIONS

Plant pathogens and insect pests are responsible for both quantitative and qualitative crop losses. To counter these pests, pesticides are used to manage insects, bacteria and fungi and mitigate infestations, all of which have harmful impacts on the environment and mammalian toxicity. Moreover, the widespread use of chemical pesticides has resulted in the development of resistance in insects, fungi and oomycete pathogens against these pesticides. Therefore, a lot of work is being put into the development of new and eco-friendly pest control solutions such as microbial biostimulants. The use of microbial biostimulants in agriculture has a huge amount of potential for dealing with the issues of insect pests and diseases.

REFERENCES

Adam, M., H. Heuer, and J. Hallmann. 2014. Bacterial antagonists of fungal pathogens also control root-knot nematodes by induced systemic resistance of tomato plants. *PLoS ONE*. 9: 2. doi:10.1371/journal.pone.0090402.

Adikaram, N.K.B., D.C. Joice, and L.A. Terry. 2002. Biocontrol activity and induced resistance as a possible mode of action for Aureobasidium pullulans against Grey Mould of Strawberry fruit. *Australas. Plant Pathol.* 31: 223–229. doi:10.1071/ap02017.

Akram, W., T. Anjum, and B. Ali. 2016. Phenylacetic acid is ISR determinant produced by Bacillus Fortis Iags162, which involves extensive re-modulation in metabolomics of tomato to protect against Fusarium wilt. *Front. Plant Sci.* 7: 498. doi:10.3389/fpls.2016.00498.

Atef, N.M. 2000. In Vitro Antagonistic action of eggplant and sweet potato Phylloplane Bacteria to some parasitic fungi. *Phytopathol Mediter.* 39: 366–375.

Bassi, A. 1838. Controlling the wheat beetle with metarhizium anisopliae. In: Mtarhizium anisopliae – an Entomopathogenic Fungus. *Pfanzenschutz-Nachrichten Bayer* 45: 113–128 (1992)

Baumann, P., M.A. Clark, L.B. Bumann, and A.H. Broodwell. 1991. Bacillus sphaericus as a mosquito pathogen: Properties of the organism and its toxins. *Microbiol. Rev.* 55: 425–436.

Benítez, T., M. Rincón, and A.C. Codón. 2004. Biocontrol mechanisms of *Trichoderma* Strains. *Int. Microbiol.* 7: 249–260.

Biostimulant Coalition, 2013. What are biostimulants? http://www.biostimulantcoalition.org/about/.

Bonaterra, A., J. Alemany, and E. Montesinos. 1998. Biochemical and MRFLP genomic characterization of antagonistic *Pseudomonas fluorescens* involved in biological control of brown spot of pear. In *Molecular Approaches in Biological Control*, Delemont, Switzerland, 15–18 September 1997. Bulletin OILB SROP vol. 21, pp. 273–278.

Brown, E.R., M.D. Mady, E.L. Treece, and C.W. Smith. 1958. Differential diagnosis of *Bacillus cereus*, *Bacillus anthrax* and *Bacillus cereus* var. Mycoides. *J. Bacteriol.* 75: 499–509.

Calvo P., L. Nelson, and J.W. Kloepper. 2014. Agricultural uses of plant biostimulants. *Plant Soil.* 383–341. doi:10.1007/s11104-014-2131-8.

Chalamaiah, M., B.D. Kumar, R. Hemalatha, and T. Jyothirmayi. 2012. Fish protein Hydrolysates: Proximate composition, amino acid composition, antioxidant activities and applications: A review. *Food Chem.* 135: 3020–3038. doi:10.1016/j.foodchem.2012.06.100.

Chowdhury, S.P., A. Hartmann, X. Gao, and R. Borriss. 2015. Biocontrol mechanism by root-associated *Bacillus amyloliquefaciens* FZB42-a review. *Front. Microbiol.* 6: 780.

Colla, G., S. Nardi, M. Cardarelli, A. Ertani, L. Lucini, R. Canaguier, and Y. Rouphael. 2015a. Protein hydrolysates as biostimulants in horticulture. *Sci Hortic.* 196: 28–38.

Colla, G., Y. Rouphael, and M. Cardarelli. 2015b. Co-Inoculation of *Glomus intraradices* and *Trichoderma atroviride* acts as a biostimulant to promote growth, yield and nutrient uptake of vegetable crops. *J. Sci. Food Agric.* 95: 1706–1715.

Corte, L., M.T. Dellàbate, A. Magini, M. Migliore, B. Felici, L. Roscini, R. Sardella, B. Tancini, C. Emiliani, G. Cardinali, and A. Benedetti. 2014. Assessment of safety and efficiency of Nitrogen organic fertilizers from animal-based protein hydrolysates-a laboratory multidisciplinary approach. *J. Sci. Food Agric.* 94: 235–245.

Di Vittori, L., L. Mazzoni, M. Battino, and B. Mezzetti. 2018. Pre-harvest factors influencing the quality of berries. *Sci. Hortic.* 233: 310–322.

Drobeck, M., M. Frac, and J. Sybulaska. 2019. Plant biostimulants: Importance of the quality and yield of horticultural crops and the improvement of plant tolerance to abiotic stress a review. *Agronomy* 9(6): 335. doi:10.3390/agronomy9060335.

Du Jardin, P. 2015. Plant biostimulants: Definition, concept, main categories and regulation. *Sci. Hortic.* 196: 3–14. doi:10.1016/j.scienta.2015.09.021.

El-Hady, M.M. 2004. Susceptibility of the Citrus brown mite, *Eutetranychus orientalis* (Klein) to the Entomopathogenic fungi, *Verticillium lecanii* and *Metarhizium anisopliae*. Egypt. *J. Biol. Pest Control* 14(2): 409–410.

El-Husseini, M.M. 1981. New approach to control the cotton leafworm, *Spodoptera littoralis* (boid.) by *Bacillus thuringiensis* in clover fields. *Bull. Entomol. Soc. Egypt Economic Seri.* 12: 1–6.

El-Husseini, M.M. 2006. Microbial control of insect pests: Is it an effective and environmentally safe alternative? *Arab. J. Pl. Prot.* 24: 162–169.

El-Kady, M.K., L.S. Xara, P.F. De Matos, J.V.N. Da Rocha, and D.P. De Oliveira. 1983. Effect of the entomopathogen metarhizium anisopliae in guinea pigs and mice. *Environ. Entomol.* 12: 37–42.

Elshakh, A.S.A., S.I. Anjum, W. Qiu, A.A. Almoneafy, W. Li, and Z. Yang. 2016. Controlling and defence-related mechanisms of Bacillus strains against bacterial leaf blight of rice. *J. Phytopathol.* 164: 534–546. doi:10.1111/jph.12479.

Errakhi, R., F. Bouteau, and F. Mathieu. 2016. Isolation and characterization of antibiotics produced by Streptomyces J-2 and their role in biocontrol of plant diseases, especially Grey Mould. In *Biocontrol of Major Grapevine Diseases: Leading Research.* ISBN: 9781780647128. doi:10.1079/9781780647128.0076.

Flexner, J.L., B. Lighthart, and B.A. Croft. 1986. The effects of microbial pesticides on non-target beneficial Arthropods. *Agric. Ecosyst. Environ.* 16: 203–254.

Gherbawy, Y, H. Elhariry, and G. Khiralla. 2012. Molecular screening of streptomyces isolates for antifungal activity and family 19 Chitinase Enzymes. *J. Microbiol.* 50(3): 459–468.

Goldberg, L.J., and V.F. Margalit. 1977. A bacterial spore demonstrating rapid larvicidal activity against Anopheles sergentii, Uranotaenia unguiculata, Culex univitatus, Aedes aegyptii and Culex pipiens. *Mosquito News.* 37: 355–358.

Ganeshan, G., and A.M. Kumar. 2007. Pseudomonas fluorescens, a potential bacterial antagonist to control plant disease. *J. Plant Interact.* 1(03): 123–134.

Han, Y., B. Zhang, and P. Li. 2015. Purification and identification of two antifungal cyclic peptides produced by *Bacillus amyloliquefaciens* l-h15. *Appl. Biochem. Biotechnol.* 176: 2202–2212. doi:10.1007/s12010-015-1708-x.

Hayat, R., S. Ali, and I. Ahmed. 2010. Soil beneficial bacteria and their role in plant growth promotion: A Review. *Ann. Microbiol.* 60: 579–598.

Hermstadt, C., G.G. Soares, E.R. Wilcox, and D.L. Edwards. 1986. A new strain of *Bacillus thuringiensis* with activity against coleopteran insects. *Biotechnology* 4: 305–308.

Hermstadt, C., F. Gaertner, W. Gelernter, and D.L. Edwards. 1987. *Bacillus thuringiensis* isolate with activity against Coleoptera. In: *Biotechnology in Invertebrate Pathology and Cell Culture.* K. Maramorosch (ed.). Academic Press, New York, pp. 101–113.

Hoffland, E., J. Halilinen, and J.A. Van Pelt. 1996. Comparison of systemic resistance induced by avirulent and nonpathogenic *Pseudomonas* Species. *Phytopathology* 86: 757–762.

Ignoffo, C.M. 1973. Effects of entomopathogens on invertebrates. *Ann. N. Y. Acad. Sci.* 217: 141–164.

Jayashree, K., V. Shanmugam, T. Raguchander, A. Ramanathan, and R. Samiyappan. 2000. Evaluation of *Pseudomonas fluorescens* (pf1) against blackgram and sesame root-rot disease. *J Biol. Contr.* 14: 55–61.

Kauffman, G.L. 2007. Effects of a biostimulant on the heat tolerance associated with photosynthetic capacity, membrane thermostability, and polyphenol production of perennial ryegrass. *Crop Sci.* 47: 261–267.

Krid, S., M.A. Triki, and A. Rhouma. 2012. Biocontrol of olive knot disease by *Bacillus subtilis* isolated from olive leaves. *Ann. Microbiol.* 62: 149–154. doi:10.1007/s13213-011-0239-0.

Liu, Z., A. Budiharjo, and R. Borriss. 2013. The highly modified microcin peptide plantazolicin is associated with nematicidal activity of *Bacillus amyloliquefaciens* FZB42. *Appl. Microbiol. Biotechnol.* 97: 10081–10090.

Marie, S.S. 2004. Laboratory bioassay and field application of *Beauveria bassiana* (bals.) Vuillemin against the Tortoise Beetle, *Cassida vittata* vill. In Sugar beet. *Egypt. J. Biol. Pest Control* 14(2): 375–378.

Metchnikoff, E. 1879. Effect of a microbial insecticide, *Bacillus thuringiensis kurstaki* on non-target Lepidoptera in a spruce budworm infected forest. *J. Res. Lepid.* 29: 267–276.

Mohamed, E.H., S.A. Abd El-Halim, and M.M. El-Husseini. 2005. Efficacy and Residual Effect of *Bacillus thuringiensis* against larvae of the cotton leafworm, *Spodoptera littoralis* (Boisd.) in Egyptian clover fields. *Egypt. J. Biol. Pest Control* 15(2): 81–84.

Monty, J. 1974. Teratological effects of the Virus Rhabdionvirus oryctes on *Oructes rhinoceros* (L.) (Coleoptera: Scarabaeidae). *Bull. Entomol. Res.* 64: 633–636.

Mounir, R., A. Durieux, and A.H. Jijakli. 2007. Production, formulation and antagonistic activity of the biocontrol like-yeast *Aureobasidium pullulans* against *Penicillium Expansum. Biotechnol. Lett.* 29: 553–559. doi:10.1007/s10529-006-9269-2.

Mukherjee, P.K., B.A. Horwitz, and C.M. Kenerley. 2013. *Trichoderma* research in the genome era. *Annu. Rev. Phytopathol.* 51: 105–129.

Müller-Kögler, E. 1965. *Pilzkrankheiten bei insekten. Anwendungen zur biologischen schädlingsbekämpfung und grundlagen der insektenmycologie.* Paul PerrymVerlag, Berlin & Hamburg. 444 pp.

Narasimhan, A., and S. Shivakumar. 2015. Evaluation of *Bacillus Subtilis* (Jn032305) Biofungicide to control Chili anthracnose in pot controlled conditions. *Biocontrol Sci. Technol.* 25: 543–559.

Oszust, K., A. Pawlik, G. Janusz, and M. Frac. 2017a. Characterization and influence of a multi-enzymatic biopreparation for biogas yield enhancement. *BioResources* 12: 6187–6206. doi:10.15376/biores.12.3.6187-6206.

Oszust, K., A. Pawlik, A. Siczek, and M. Frac. 2017b. Efficient cellulases production by *Trichoderma atroviride* G79/11 in submerged culture based on Soy flour-cellulose-lactose. *BioResources* 12: 8468–8489. doi:10.15376/biores.12.4.8468-8489.

Padidam, M. 1991. Rational deployment of *Bacillus thuringiensis* strains for control of insect pests In India. *Curr. Sci.* 60: 464–465.

Pylak, M., K. Oszust, and M. Frac. 2019. Review report on the role of bioproducts, biopreparations, biostimulants and microbial inoculants in organic production of fruit. *Rev. Environ. Sci. Biotechnol.* 18: 597–616.

Radhakrishnan, R. 2017. *Bacillus*: A biological tool for crop improvement through Bio-molecular changes in adverse environments. *Front Physiol.* 8: 667. doi:10.3389/fphys.2017.00667.

Roberts, D.W., and S.P. Wraight. 1986. Current status on the use of Insect pathogens as Biocontrol agents in Agriculture: Fungi. In: R.A. Samson, J.M. Vlak, and D. Peters (eds.) *Fundamental and applied aspects of Invertebrate Pathology.* International Colloquium of Invertebrate Pathology, Veldhoven, pp. 510–513.

Rose, M.T., A.F. Patti, K.R. Little, A.L. Brown, W.R. Jackson, and T.R. Cavagnaro. 2014. A Meta-analysis and review of plant-growth response to humic substances: Practical implications for agriculture. *Adv. Agron.* 124: 37–89.

Seufi, A.M., and G.E. Osman. 2005. Comparative susceptibility of the Egyptian cotton leafworm, *spodoptera littoralis* (boisd.) to some baculovirus isolates. *Egypt. J. Biol. Pest Control* 15(1&2): 21–26.

Shekh, M., and A. Naim. 2017. Isolation and Characterization of antibacterial metabolites from *Streptomyces* species. Daffodil International University. Project report. DSpace Repository.

Som, N.F., D. Heine, N. Holmes, and M.I. Hutchings. 2017. The MtrAB two-component system controls antibiotic production in *Streptomyces Coelicolor* A3 (2). *Microbiology* 163(10): 1415–1419.

Trudgill, D. L., and V.C. Blok. 2001. Apomictic, polyphagous root-knot nematodes: Exceptionally successful and damaging biotrophic root pathogens. *Annu. Rev. Phytopathol.* 39: 53–77.

van Dissel, D., D. Claessen, and G.P. van Wezel. 2014. Morphogenesis of streptomyces in submerged cultures. In: S. Sima, and G.M. Gadd (eds.) *Advances in Applied Microbiology.* Vol. 89, Academic Press, Burlington, pp. 1–45.

Van Driesche, R.G., and T.S. Bellows. 1996. *Biological Control.* Chapman & Hall Publisher, New York, 539 pp.

Vidhyasekaran, P., N. Kamala, A. Ramanathan, K. Rajappan, V. Paranidharan, and R. Velazhahan. 2001. Induction of systemic resistance by *pseudomonas fluorescens* pf1 against *Xanthomonas oryzae* pv. *Oryzae* in rice leaves. *Phytoparasitica* 29: 155–166.

Wagner, A., and B. Hetman. 2016. Effect of some biopreparations on health status of Strawberry (Fragaria Ananassa Duch.). *J. Agric. Sci. Technol. B* 6: 295–302. doi:10.17265/2161-6264/2016.05.002.

Yedidia, I., N. Benhamou, and I. Chet. 1999. Induction of defense responses in cucumber plants (cucumis sativus l.) by the biocontrol agent *Trichoderma harzianum*. *Appl. Environ. Microbiol.* 65: 1061–1070. doi:10.1007/s00410-001-0323-8.

Yi, H.S., J.W. Yang, and C.M. Ryu. 2013. ISR meets SAR outside: Additive action of the endophyte *Bacillus pumilus* INR7 and the chemical inducer, benzothiadiazole, on induced resistance against bacterial spot in field-grown pepper. *Front. Plant Sci.* 4: 122. doi:10.3389/fpls.2013.00122.

Young, E.C. 1974. The epizootiology of two pathogens of the coconut palm rhinoceros beetle. *J. Invertebr. Pathol.* 24: 82–92.

Zhao, L., C. Feng., K. Wu, and X. Hao. 2017. Advances and prospects in biogenic substances against plant virus: A review. *Pest. Biochem. Physiol.* 135: 15–26.

Zhao, D., H. Zhao, and L. Chen. 2018. Isolation and identification of bacteria from rhizosphere soil and their on plant growth promotion and root-knot nematode disease. *Biol. Control* 119: 12–19.

3 Managing Insect Pests Using Microbial Biostimulants as Insecticides

Sajjad Ali, M. Anjum Aqueel, Asif Sajjad, Saqib Ajmal,
Maqshoof Ahmad, Qaiser Shakeel, and Muhammad Raheel
The Islamia University of Bahawalpur

Kaleem Tariq
Abdul Wali Khan University Mardan

CONTENTS

3.1 BACKGROUND

Nowadays, modern agricultural production system is confronting with new challenges to deal with crop yield-reducing agents by utilizing new ecological and molecular tactics to achieve the higher yields with minimized negative environmental impacts. Recent estimates indicate that 30% and 50% losses are expected in global agriculture production owing to biotic and abiotic stresses, respectively (Kumar and Verma, 2018). In this situation, improving plant growth and plant resistance and biocontrol of insect pests are key approaches to maximizing the crop produce (Kessler and Halitschke, 2009).

The plant–insect interactions are very complex and involve multiple factors. Herbivorous insects constitute 50% of the insects. Around 18%–26% global crop production losses are caused by pest insects (Wielkopolan and Obrepalska-Steplowska, 2016). Plants have established not only direct defenses against insect herbivore pests, but also indirect defenses to boost the efficiency of natural enemies of these herbivore insect pests. Henceforth, plants are the mediator of these multitrophic interactions among insect assailants and beneficial microbial organisms (Pieterse and Dicke, 2007). Numerous soil-dwelling microbial entities such as plant growth-promoting rhizobacteria (PGPR) and mycorrhizal fungi assist plants to cope with abiotic and biotic stresses by promoting plant growth and inducing pest resistance in the host plants. Plants serve as a linking corridor between such beneficial below-ground microbial community interacting in a bidirectional way with the above-ground herbivore, their natural enemies and pollinator insects (Pineda et al., 2010).

DOI: 10.1201/9781003188032-3

3.1.1 PLANT–ENDOPHYTE INTERACTIONS

Plants have developed strong symbiotic relationships with beneficial microbes, especially mycor-rhizal fungi, PGPR and other endophytes. Extensive bioecological research on different functions and survival of endophytes reveals that around 0.3 million plant species show intimate endophytic associations (Tabbene et al., 2009). Endophytes are those microbes which are present asymptomati-cally in host plants causing no disease in them. Most of these beneficial microbes are living in the rhizosphere zone. These symbiotic (endophytic) microbial bodies may transport soil nutrients to plants, promote plant growth and development, intensify plant tolerance to stress, counter pathogen virulence and increase plant resistance to insect and diseases, and affect the development of contes-tant plant species in the neighborhood. Thus, these microbes may change the biology, metabolism and behavior of the insects and may have a substantial influence on the plant–insect interactions (Bezemer and van Dam, 2005). Mostly, they are nutritional and defense mutualistic partners of their host plants. The plant growth mediated by microorganism stimulation is explicated by improved nutrition leading toward tolerance to abiotic and biotic stresses (van Loon, 2007). The promotion of plant growth has traditionally been thought of as the chief mechanism behind the microorganism–plant–insect interactions. However, the worth of induced plant defenses has recently been highlighted, but much is to be explored about the induced systemic resistance (ISR) developed in such tritrophic interactions. So, plant resistance mechanisms can be improved with these special beneficial micro-organisms by effective utilization of plant–microbe–insect interactions, in addition to plant growth promotion benefits (Bennett et al., 2006; Lata et al., 2018).

3.1.2 BIOSTIMULANTS AND MICROBIAL BIOSTIMULANTS

These endophytic microbes serve as biostimulants for the host plants to produce and fill them with certain compounds reducing insect pest herbivory and thus strengthening the host plant defense against the insect herbivores (Panaccione et al., 2014). These microbial endophytes directly assist the plant hosts by producing certain immune compounds with their intimate interactions during the plant tissues colonization (Bamisile et al., 2018). So, these endophytes, associated with plants, act as bioinsecticides.

The biostimulants are heterogeneous, grouped with amino acids, chitosan, humic substances, protein hydrolysates, seaweeds extracts, biopolymers and some inorganic molecules. Also, the sub-group, the microbial biostimulants, is composed of beneficial microbial organisms (i.e., eubacteria, fungi and yeast) with the capability to enhance the plant growth and production through the nutrient uptake promotion and improved abiotic and biotic stress tolerance (Du Jardin, 2015). In comparison with xenobiotic agrochemicals, microbial biostimulants are not accumulating for longer, possess lower toxicity and are less prone to apt resistance development in pest strains. Therefore, they are safer for the environment, ecosystem and humans. Henceforth, the biostimulants-based product market is progressively increasing in the recent past. Europe is the global leader in this sector, with over 580 million Euros sale of biostimulants during 2015 (Sangiorgio et al., 2020). Now, microbial biostimulants have got higher interest because plants carry a wide array of microbial entities in their endosphere, rhizosphere and phyllosphere zones. The rhizosphere zone serves as the key area for plant–microbe interactions in root region because microbe attraction, during stress time, is medi-ated by varied root exudations having 11%–40% of photosynthesized carbon of the plants (Hassan and Mathesius, 2012).

3.1.3 PLANT-PROTECTING MECHANISM OF MICROBIAL BIOSTIMULANTS

Microbial biostimulants are involved in the triggering of ISR in plant hosts to deter the herbivory injuries in them. ISR is prompted by defense genes priming with higher expressions, systemically in plant leaves due to insect attack which frequently includes responsiveness of plant hormones such as ethylene (ET) and jasmonic acid (JA) (Figure 3.1). ISR is linked with signaling pathways depending

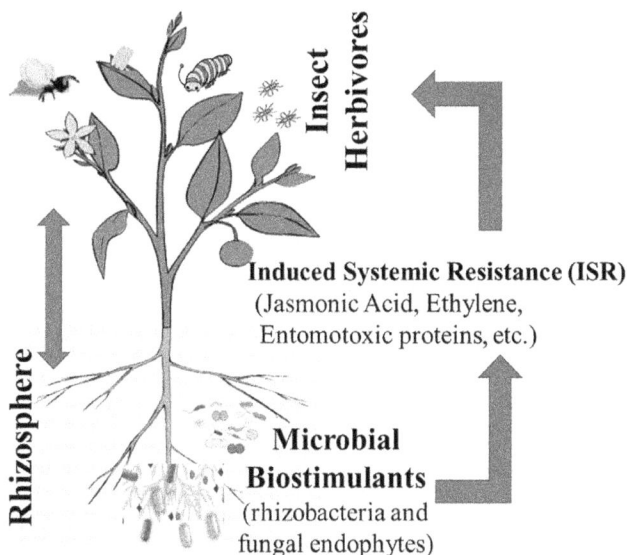

FIGURE 3.1 Multitrophic interactions between microbial biostimulants and insect herbivores.

on ET, JA and salicylic acid (SA) (Howe and Jander, 2008). JA, ET and SA are the vital mediators of plant resistance responses against plant attackers. The signaling pathways controlling the ISR and plant defenses against herbivore insects partially overlap (Pieterse et al., 2012).

After the initiation of plat defense signaling pathways, host plant is able to manufacture a group of antinutritional proteins reducing the insects' ability to digest the dietary plant materials. These antinutritional proteins belong to alpha-amylase inhibitors (a-AIs), chitinases, lectins, polyphenol oxidases (PPO) and protease inhibitors (PIs). These proteins are upregulated frequently during insect–plant interactions (Wielkopolan and Obrepalska-Steplowska, 2016). So, the endophytic microbial biostimulants join in the production of numerous significant secondary metabolites with diverse modes of action and structures such as alkaloids, benzopyranones, flavonoids, phenolic acids, quinones, steroids, terpenoids, tetralones and xanthones showing excellent defense performance against the destructive plant attackers (Aravind et al., 2009). Some endophytic microbial biostimulants can directly affect the insect fitness by manufacturing certain toxins. For example, *Bacillus thuringiensis* (Bt.) produce crystalline proteins serving as insecticides by making pores in midgut epithelial cells (Vachon et al., 2012). Also, bacteria deploy additional toxins and numerous effectors interfering with the insect immunity system to promote the infection process (Nielsen-LeRoux et al., 2012). Furthermore, endophytic microbial stimulants may also facilitate herbivore-induced release of plant volatiles that attract natural enemies of the herbivore insect pests (Takabayashi and Dicke, 1996). Comprehensively, endophytic microbial biostimulants produce bioactive metabolites (insecticidal molecules), for which the plant serves as a delivery system for the translocation of these molecules to the target insect pest, so microbial biostimulants act as biopesticides (Jaber and Ownley, 2018). Moreover, endophytic microbes may have substantially longer efficacy period than non-endophytic microorganisms because of their ability to survive at least for the entire growing season of the treated crops (Harman et al., 2008).

3.2 MICROBIAL BIOSTIMULANTS PROTECTING PLANTS AGAINST INSECT PESTS

In the early 1980s, Webber (1981) reported that microbes can protect plants from herbivorous insects. A plant protection example was given by the endophyte fungus *Phomopsis oblonga*, which confers protection to elm tree against the beetle *Physocnemum brevilineum*. *P. oblonga* reduced the causal agent (*Ceratocystis ulmi*) of Dutch elm disease by regulating its vector (*P. brevilineum*) by producing certain secondary metabolites (Claydon et al., 1985). Afterward, enormous studies have depicted that endophytic fungi and bacteria are playing a vital role in the plant defenses systems against insect herbivory, which are discussed below.

Mostly, entomopathogenic fungi (EPF) are not able to develop and set themselves in the living plant tissues. However, endophytic insect pathogenic fungi (EIPF) have evolved with this character in the fungal order Hypocreales (Ascomycota), with the generalist insect pathogen *Beauveria* (Cordycipitaceae) and *Metarhizium* (Clavicipitaceae) as the most well-described EIPF genera and considered as a typical pathogenicity model for EIPF infections (Spatafora et al., 2007; Moonjely et al., 2016; Zhang et al., 2018). So, EIPF are both plant mutualists and insect pathogens at the same time, making a three-party interaction to allow nutrient transfers across the fungus, insect and plant (Branine et al., 2019). Various EIPF establish themselves in specific plant parts. *Metarhizium* is chiefly found in plant roots, while *Beauveria* is found within many tissues of host plants (Behie et al., 2015). Particularly, the biocontrol potential of endophytic fungi, *Beauveria bassiana*, *Sarocladium strictum*, *Metarhizium anisopliae* and *M. robertsii*, against various insect pests was demonstrated for the adverse effects on *Aphis gossypii*, *Bemisia tabaci*, *Chortoicetes terminifera*, *Helicoverpa armigera*, *H. zea*, *Nesidiocoris tenuis*, *Spodoptera exigua*, *S. littoralis*, *Tuta absoluta* and *Trialeurodes vaporariorum* in tomato and other crops such as maize, cotton, banana, faba bean, common bean, white jute and poppy (Branine et al., 2019; Sinno et al., 2020). The suppression of the pests by these fungal endophytes can be credited to their manufacturing of secondary mycotoxigenic metabolites, in the host plants, toxic to the insect pests (Gurulingappa et al., 2010). Mycorrhizal fungal biostimulants usually have negative effects on generalist insect chewers and impartial or positive effect against the specialist insect chewers and phloem feeder insects (Pozo and Azcon-Aguilar, 2007). Details of the fungal biostimulants used against different insect herbivores are given in Table 3.1.

Bacterial biostimulants are well known among the plant growth-promoting bacteria (PGPB) or PGPR. The latter (PGPR) is, explicitly, colonizing the rhizosphere zone. In this regard, the most studied bacterial genera are *Bacillus*, *Burkholderia*, *Pseudomonas*, *Streptomyces* and *Serratia* (Bhattacharyya and Jha, 2012; Farrar et al., 2014). These genera are renowned for their wide array of secondary metabolites, including anticancer, antibiotics, antifungal, antiviral and insecticidal agents. Particularly, various species of the genus *Bacillus* delivered a significant contribution (up to 50%) toward the commercially formulated biocontrol products. Among the numerous bacterial endophytes, *B. amyloliquefaciens* is ranked as one of the most capable entomopathogenic bacterial endophytes (Chen et al., 2009, 2013), which naturally produce the lipopeptides (metabolic compounds) responsible for the development of induced plant resistance against the fall armyworm (*Spodoptera frugiperda*) (Boning and Bultman, 1996; Ongena and Jacques, 2008; Choudhary and Johri, 2009). *Enterobacter cloacae* is another endophytic bacterium reported as a virtuous biocontrol agent against the whitebacked planthopper (*Sogatella furcifera*), a damaging pest of paddy rice, by producing Pinellia ternata agglutinin (PTA) protein (Zhang, 2011). Another endophyte bacterium *Clavibacter xyli* was genetically transformed by incorporating cry1A (crystal endotoxin) gene isolated from *B. thuringiensis* (Torres and Ruberson, 2008), and these transgenic bacteria exhibited substantial insecticidal actions against many lepidopteran insect pests, especially the European corn borer, *Ostrinia nubilalis*. Diamondback moth (DBM) (*Plutella xylostella*) is a major yield-reducing lepidopterous insect pest in broccoli, cabbage and cauliflower vegetables around the globe (Dickson et al., 1990). Several studies showed the effective management of this insect pest by using the endophytic bacteria. *Enterobacter cloacae* (ENF14), *Alcaligenes piechaudii* (EN5) and *K. ascorbate* (EN4) isolates were observed for effective control actions against the diamondback moth (Thuler et al., 2006; Thuler and Bortoli, 2007). Some of the bacterial biostimulants used against different insect herbivores are given in Table 3.2.

A more in-depth investigation of bacterial and fungal plant endophytes may add supplementary endophytes to be used in various crops to lessen the insect pests by improving plant tolerance to herbivore feeding. The identification of diverse action mechanisms can bring about the combinations of different isolates to control a broader range of various insect pests (Lutz et al., 2004).

TABLE 3.1

Endophytic Fungal Biostimulants Mediating Negative Effects on Insect Herbivores

Fungal Endophytes	Target Herbivore Insects	References
Beauveria bassiana	*Ostrinia nubilalis*	Bing and Lewis (1992)
Acremonium alternatum	*Plutella xylostella*	Raps and Vidal (1998)
B. bassiana	*Sesamia calamistis*	Cherry et al. (2004)
Metarhizium anisopliae	*Agriotes obscurus*	Kabaluk and Ericsson (2007)
B. bassiana	*Cosmopolites sordidus*	Akello et al. (2008)
B. bassiana	*Iraella luteipes*	Quesada-Moraga et al. (2006)
B. bassiana	*Chilo partellus*	Reddy et al. (2009)
Purpureocillium lilacinum and *B. bassiana*	*Helicoverpa zea*	Powell et al. (2009); Castillo Lopez and Sword (2015)
Acremonium strictum	*Helicoverpa armigera, Trialeurodes vaporariorum* and *Aphis fabae*	Jaber and Vidal (2009,2010)
B. bassiana, P. lilacinum and *Lecanicillium lecanii*	*Aphis gossypii* and *Chortoicetes terminifera*	Gurulingappa et al. (2010); Castillo Lopez et al. (2014)
B. bassiana and *M. anisopliae*	*Acyrthosiphon pisum* and *Aphis fabae*	Akello and Sikora (2012)
B. bassiana	*Rhynchophorus ferrugineus*	Arab and El-Deeb (2012)
M. brunneum and *B. bassiana*	*Bemisia tabaci*	El-Deeb et al. (2012)
M. anisopliae and *B. bassiana*	*Liriomyza huidobrensis*	Akutse et al. (2013)
M. anisopliae	*Plutella xylostella*	Batta (2013)
B. bassiana	*Apion corchori*	Biswas et al. (2013)
B. bassiana	*Phaedrotoma scabriventris* and *Diglyphus isaea*	Akutse et al. (2014)
Clonostachys rosea	*Thrips tabaci*	Muvea et al. (2015)
M. robertsii, M. anisopliae and *B. bassiana*	*Sesamia nonagrioides*	Mantzoukas et al. (2015)
B. bassiana	*Spodoptera exigua*	Shrivastava et al. (2015)
M. pingshaense	*Amala cincta*	Peña-Peña et al. (2015)
B. bassiana	*Helicoverpa armigera*	Qayyum et al. (2015); Vidal and Jaber (2015)
B. bassiana	*Costelytra zealandica*	Lefort et al. (2016)
M. anisopliae and *B. bassiana*	*Ophiomyia phaseoli*	Mutune et al. (2016)
B. bassiana	*Tuta absoluta*	Klieber and Reineke (2016)
M. brunneum and *B. bassiana*	*Spodoptera littoralis*	Resquín-Romero et al. (2016); Sánchez-Rodríguez et al. (2018)
B. bassiana and *M. brunneum*	*Myzus persicae*	Jaber and Araj (2017)
B. bassiana	*Planococcus ficus*	Rondot and Reineke (2018)

3.3   SCOPE OF TRANSGENIC MICROBIAL BIOSTIMULANTS FOR INSECT PEST MANAGEMENT

Transgenic endophytic microbial biostimulants would be a valuable approach and an effective alternative to the genomic manipulations of plant hosts (Li et al., 2017). Virulent genes introduced into the endophytic microbial bodies can confer new characteristics, suitable for the biocontrol of the insect pests and promotion of growth in host plants. For instance, the endophyte bacterium *Clavibacter xyli*, colonizing xylem tissues of several plants, was genetically modified to prompt the *B. thuringiensis* gene, encoding entomotoxin, to control various insect pests (Glandorf et al., 2001). In another example, the endophyte *Burkholderia pyrrocinia* was transmuted with the Bt. toxin genes to produce entomotoxic proteins against the instar larvae of *Bombyx mori* (silkworms) (Li et al., 2017). Moreover, *Pseudomonas putida* was also genetically modified by adding an antifungal gene and incorporated into wheat plants, which resulted in the reduced fungal population in soil

TABLE 3.2
Endophytic Bacterial Biostimulants Mediating Negative Effects on Insect Herbivores

Bacterial spp.	Target Herbivore Insects	References
Serratia marcescens	*Acalymma vittatum* and *Diabrotica undecimpunctata*	Zehnder et al. (1997a)
Acremonium alternatum	*Plutella xylostella*	Raps and Vidal (1998)
Bacillus pumilus, *Flavimonas oryzihabitans* and *Pseudomonas putida*	*Acalymma vittatum, Bemisia argentifolii* and *Diabrotica undecimpunctata*	Zehnder et al. (1997a,b); Murphy et al. (2000)
A. strictum	*A. fabae, H. armigera* and *Trialeurodes vaporariorum*	Jaber and Vidal (2009, 2010)
Rhizobium leguminosarum	*S. littoralis* and *Myzus persicae*	Kempel et al. (2009)
B. amyloliquefaciens	*B. argentifolii* and *M. persicae*	Murphy et al. (2000); Herman et al. (2008)
B. subtilis	*B. argentifolii, B. tabaci* and *M. persicae*	Murphy et al. (2000); Herman et al. (2008); Valenzuela-Soto et al. (2010)
P. fluorescens	*Aphis gossypii, Amrasca biguttula biguttula, Cnaphalocrocis medinalis, Spodoptera exigua, M. persicae* and *Pieris rapae*	Commare et al. (2002); De Vos et al. (2007); Van Oosten et al. (2008)

(Li et al., 2017). It looks likely that the future efforts, to achieve crop protection success, would involve the exploitation of transgenic modifications of the microbial endophytes. However, because of the agility of endophytic microbial biostimulants, it is difficult to restraint these microbes to specific plant organisms.

3.4   LIMITATIONS AND FUTURE VIEWPOINTS FOR MICROBIAL BIOSTIMULANTS APPLICATION

Developing new microbial biostimulants could be under some specific complications.

First of all, registration procedure for commercial use is generally complex due to the lack of international legislations (Backer et al., 2018). Secondly, the product efficiency largely depends on crop plant treated because of phenological state of host plants. The development of phytostimulants requires assessing the associated microorganisms' establishment with host plants. An enduring plant colonization, by the endophyte, is a crucial requirement for biostimulants to be effective. Lastly, the best formulation for assured product effectiveness and conservation is to be well defined, for minimum influence due to cultural and environmental conditions (Bashan et al., 2014). Gram-positive bacteria-based microbial biostimulants permit powdered formulations with long-lasting stability and drought tolerance because of spores production by bacterial bodies (Tabassum et al., 2017). At large, the higher production expenses related to commercial biostimulants and the inconsistent efficacy are observed under field conditions (Nadeem et al., 2014). These are the major deterrents for developing biostimulant products. Resultantly, less commercialized microbial biostimulant products with limited diffusion in agricultural practices are available.

The increased demand for biocontrol agents in the utilization of IPM systems for livestock, veterinary and medicinal insect pests has raised concerns about the safety of entomopathogenic microorganisms used in microbial control. Registration of bioproducts dependent on microbes pays special attention to the following aspects: (i) allergic properties, (ii) toxic metabolite threats, (iii) genetic recombination and natural strain displacement and (iv) impact on biodiversity (El-Husseini, 2006). The degree of precision required for protection is determined by the characteristics of the targeted region as well as the taxonomic classes of the species involved. Due to the above issues, a lot of work is being put into identifying new pathogen control solutions that are safe, are cost-effective, and have a low environmental impact. To conclude, developing synthetic microbial communities, incorporating several microorganisms, with diverse plant growth-promoting (PGP) functions, is an inimitable opportunity to enhance the effectiveness and reliability of microbial biostimulant products, even though manufacturing such microbial community would face significant challenges.

3.5 CONCLUSIONS

Microbial biostimulants can be a potential tool for sustainable and reliable approach to reducing the biotic stresses such as insect herbivory damage. Microbial biostimulants can be deployed directly to enhance the plant growth, health and productivity in commercial crops. Microbial biostimulants can benefit their host plants directly or indirectly against herbivorous injuries. Genetically modified endophytic microbial biostimulants would be a valuable approach and effective alternative to genomic manipulations of their host plants to defend them against herbivore insect pests (Li et al., 2017). The current research exertions to find microbial biostimulants would bring a substantial decrease in synthetic chemicals application in practical agricultural production. Therefore, endophytic microbial biostimulants would be helpful in cultivating crops with lesser fertilizer, fungicide, herbicide and insecticide application. In future, we foresee transformed agricultural practices with optimized relationship of plants with soil microbes and endophytes to have maximized agricultural production. It's critical to develop specific balanced methodologies so that these approaches can be used without endangering the environment or humans. It is essential to look into the proper doses, the stage of phenological growth of the plant to allow the application, and the specific types of microbial biostimulants to bring greater benefits. Further research investigations about the insecticidal potential of various EIPF ancestries for agricultural biocontrol applications as well as the detection of innovative fungal secondary metabolic compounds with effective insecticidal potential will surely lead to modernizations of biological control of insect pests.

REFERENCES

Akello, J., T. Dubois, D. Coyne, and S. Kyamanywa. 2008. Effect of endophytic *Beauveria bassiana* on populations of the banana weevil, *Cosmopolites sordidus*, and their damage in tissue-cultured banana plants. *Entomol. Exp. Appl.* 129: 157–165.

Akello, J., and R. Sikora. 2012. Systemic acropedal influence of endophyte seed treatment on *Acyrthosiphon pisum* and *Aphis fabae* offspring development and reproductive fitness. *Biol. Control* 61: 215–221.

Akutse, K.S., N.K. Maniania, K.K.M. Fiaboe, J. Van den Berg, and S. Ekesi. 2013. Endophytic colonization of *Vicia faba* and *Phaseolus vulgaris* (Fabaceae) by fungal pathogens and their effects on the life-history parameters of *Liriomyza huidobrensis* (Diptera: Agromyzidae). *Fungal Ecol.* 6: 293–301.

Akutse, K.S., K.K.M. Fiaboe, J. Van den Berg, S. Ekesi, and N.K. Maniania. 2014. Effect of endophyte colonization of *Vicia faba* (Fabaceae) plants on the life-history of leafminer parasitoids *Phaedrotoma scabriventris* (Hymenoptera: Braconidae) and *Diglyphus isaea* (Hymenoptera: Eulophidae). *PLoS ONE* 9: e109965.

Arab, Y.A., and H.M. El-Deeb. 2012. The use of endophyte *Beauveria bassiana* for bio-protection of date palm seedlings against red palm weevil and *Rhizoctonia* root-rot disease. *Sci. J. King Faisal Univ.* 13: 91–100.

Aravind, R., A. Kumar, S.J. Eapen, and K.V. Ramana. 2009. Endophytic bacterial flora in root and stem tissues of black pepper (*Piper nigrum* L.) genotype: Isolation, identification and evaluation against *Phytophthora capsici*. *Appl. Microbiol.* 48(1): 58–64.

Batta, Y.A. 2013. Efficacy of endophyic and applied *Metarhizium anisopliae* (Metch.) Sorokin (Ascomycota: Hypocreales) against larvae of *Plutella xyllostella* L. (Yponomeutidae: Lepidoptera) infesting *Brassica napus* plants. *Crop Prot.* 44: 128–134.

Backer, R., J.S. Rokem, G. Ilangumaran, J. Lamont, D. Praslickova, E. Ricci, S. Subramanian, and D.L. Smith. 2018. Plant growth-promoting rhizobacteria: Context, mechanisms of action, and roadmap to commercialization of biostimulants for sustainable agriculture. *Front. Plant Sci.* 9: 1473.

Bamisile, B.S., C.K. Dash, K.S. Akutse, R. Keppanan, O.G. Afolabi, M. Hussain, M. Qasim, and L. Wang. 2018. Prospects of endophytic fungal entomopathogens as biocontrol and plant growth promoting agents: An insight on how artificial inoculation methods affect endophytic colonization of host plants. *Microbiol. Res.* 217: 34–50.

Bashan, Y., L.E. de-Bashan, S.R. Prabhu, and J.P. Hernandez. 2014. Advances in plant growth promoting bacterial inoculant technology: Formulations and practical perspectives (1998–2013). *Plant Soil*. 37: 1–33.

Behie, S.W., S.J. Jones, and M.J. Bidochka. 2015. Plant tissue localization of the endophytic insect pathogenic fungi *Metarhizium* and *Beauveria*. *Fungal Ecol*. 13: 112–119.

Bennett, A.E., J. Alers-Garcia, and J.D. Bever. 2006. Three-way interactions among mutualistic mycorrhizal fungi, plants, and plant enemies: Hypotheses and synthesis. *Am. Nat*. 167: 141–152.

Bezemer, T.M., and N.M. van Dam. 2005. Linking aboveground and belowground interactions via induced plant defenses. *Trends Ecol. Evol*. 20: 617–624.

Bhattacharyya, P.N., and D.K. Jha. 2012. Plant growth-promoting rhizobacteria (PGPR): Emergence in agriculture. *World J. Microb. Biot*. 28: 1327–1350.

Bing, L.A., and L.C. Lewis. 1992. Temporal relationships between *Zea mays*, *Ostrinia nubilalis* (Lep.: Pyralidae) and endophytic *Beauveria bassiana*. *Entomophaga* 37: 525–536.

Biswas, C., P. Dey, S. Satpathy, P. Satya, and B. Mahapatra. 2013. Endophytic colonization of white jute (*Corchorus capsularis*) plants by different *Beauveria bassiana* strains for managing stem weevil (*Apion corchori*). *Phytoparasitica* 41: 17–21.

Branine, M., A. Bazzicalupo, and S. Branco. 2019. Biology and applications of endophytic insect-pathogenic fungi. *PLoS Pathog*. 15(7): e1007831.

Boning, R.A., and T.L. Bultman. 1996. A test for constitutive and induced resistance by tall Fescue (*Festua arundinacea*) to an insect herbivore: Impact of fungal endophyte *Acremonium coenophialum*. *Am. Midl. Nat. J*. 136: 328–335.

Castillo Lopez, D., K. Zhu-Salzman, M.J. Ek-Ramos, and G.A. Sword. 2014. The entomopathogenic fungal endophytes *Purpureocillium lilacinum* (formerly *Paecilomyces lilacinus*) and *Beauveria bassiana* negatively affect cotton aphid reproduction under both greenhouse and field conditions. *PLoS ONE* 9: e103891.

Castillo Lopez, D., and G.A. Sword. 2015. The endophytic fungal entomopathogens *Beauveria bassiana* and *Purpureocillium lilacinum* enhance the growth of cultivated cotton (*Gossypium hirsutum*) and negatively affect survival of the cotton bollworm (*Helicoverpa zea*). *Biol. Control* 89: 53–60.

Chen, X.H., R. Scholz, M. Borriss, H. Junge, G. Mögel, and S. Kunz. 2009. Difficidin and bacilysin produced by plant associated *Bacillus amyloliquefaciens* are efficient in controlling fire blight disease. *J. Biotechnol*. 140: 38–44.

Chen, Y.T., Q. Yuan, L.T. Shan, M.A. Lin, D.Q. Cheng, and C.Y. Li. 2013. Antitumor activity of bacterial exopolysaccharides from the endophyte *Bacillus amyloliquefaciens* isolated from *Ophiopogon japonicus*. *Oncol. Lett*. 5: 1787–92.

Choudhary, D.K., and B.N. Johri. 2009. Interactions of *Bacillus* spp. and plants with special reference to induced systemic resistance (ISR). *Microbiol. Res*. 164: 493–513.

Cherry, A.J., A. Banito, D. Djegui, and C. Lomer. 2004. Suppression of the stem-borer *Sesamia calamistis* (Lepidoptera; Noctuidae) in maize following seed dressing, topical application and stem injection with African isolates of *Beauveria bassiana*. *Int. J. Pest Manage*. 50: 67–73.

Claydon, N., J.F. Grove, and M. Pople. 1985. Elm bark beetle boring and feeding deterrents from *Phomopsis oblonga*. *Phytochem* 24: 937–943.

Commare, R.R., R. Nandakumar, A. Kandan, S. Suresh, M. Bharathi, T. Raguchander, and R. Samiyappan. 2002. *Pseudomonas fluorescens* based bioformulation for the management of sheath blight disease and leaffolder insect in rice. *Crop Prot*. 21: 671–677.

De Vos, M., V.R. Van Oosten, G. Jander, M. Dicke, and C.M. Pieterse. 2007. Plants under attack: multiple interactions with insects and microbes. *Plant Signal. Behav*. 2: 527–529.

Dickson, M.H., A.M. Shelton, S.D. Elgenbrode, M.L. Vamosy, and M. Mora. 1990. Selection for resistance to diamondback moth (*Plutella xylostella*) in cabbage. *Hortic. Sci*. 25(12): 1643–1646.

Du Jardin, P. 2015. Plant biostimulants: Definition, concept, main categories and regulation. *Sci. Hortic*. 19: 3–14.

El-Deeb, H.M., S.M. Lashin, and Y.A. Arab. 2012. Reaction of some tomato cultivars to tomato leaf curl virus and evaluation of the endophytic colonization with *Beauveria bassiana* on the disease incidence and its vector, *Bemisia tabaci*. *Arch. Phytopathol. Plant Prot*. 45: 1538–1545.

El-Husseini, M.M. 2006. Microbial control of insect pests: Is it an effective and environmentally safe alternative? *Arab. J. Pl. Prot*. 24:162–169.

Farrar, K., D. Bryant, and N. Cope-Selby. 2014. Understanding and engineering beneficial plant–microbe interactions: Plant growth promotion in energy crops. *Plant Biotechnol. J*. 12: 1193–1206.

Glandorf, D.C.M., P. Verheggen, T. Jansen, J.W. Jorritsma, E. Smit, and P. Leeflang. 2001. Effect of genetically modified *Pseudomonas putida* WCS358r on the fungal rhizosphere microflora of field-grown wheat. *Appl. Environ. Microbiol*. 67: 3371–3378.

Gurulingappa, P., G.A. Sword, G. Murdoch, and P.A. McGee. 2010. Colonization of crop plants by fungal entomopathogens and their effects on two insect pests when in planta. *Biol. Control* 55: 34–41.

Harman, G.E., T. Bjorkman, K.L. Ondik, and M. Shoresh. 2008. Changing paradigms on the mode of action and uses of *Trichoderma* spp. for biocontrol. *Outlooks Pest. Manag.* 19: 24–29.

Herman, M.A.B., B.A. Nault, and C.D. Smart. 2008. Effects of plant growth-promoting rhizobacteria on bell pepper production and green peach aphid infestations in New York. *Crop Prot.* 27: 996–1002.

Hassan, S., and U. Mathesius. 2012. The role of flavonoids in root–rhizosphere signalling: opportunities and challenges for improving plant–microbe interactions. *J. Exp. Bot.* 63: 3429–44.

Howe, G.A., and G. Jander. 2008. Plant immunity to insect herbivore: A dynamic interaction. *New Phytol.* 156: 145–169.

Jaber, L.R., and S. Vidal. 2009. Interactions between an endophytic fungus, aphids and extrafloral nectaries: Do endophytes induce extrafloral-mediated defences in *Vicia faba*? *Funct. Ecol.* 23: 707–714.

Jaber, L.R., and S. Vidal. 2010. Fungal endophyte negative effects on herbivory are enhanced on intact plants and maintained in a subsequent generation. *Ecol. Entomol.* 35: 25–36.

Jaber, L.R., and S.E. Araj. 2018. Interactions among endophytic fungal entomopathogens (Ascomycota: Hypocreales), the green peach aphid *Myzus persicae* Sulzer (Homoptera: Aphididae), and the aphid endoparasitoid *Aphidius colemani* Viereck (Hymenoptera: Braconidae). *Biol. Control* 116: 53–61.

Jaber, L.R., and B.H. Ownley. 2018. Can we use entomopathogenic fungi as endophytes for dual biological control of insect pests and plant pathogens? *Biol. Control* 116: 36–45.

Kabaluk, J.T., and J.D. Ericsson. 2007. *Metarhizium anisopliae* seed treatment increases yield of field corn when applied for wireworm control. *Agron. J.* 99: 1377–1381.

Kempel, A., R. Brandl, and M. Schädler. 2009. Symbiotic soil microorganisms as players in aboveground plant-herbivore interactions-the role of rhizobia. *Oikos* 118: 634–640.

Kessler, A., and R. Halitschke. 2009. Testing the potential for conflicting selection on floral chemical traits by pollinators and herbivores: Predictions and case study. *Funct. Ecol.* 23: 901–912.

Klieber, J., and A. Reineke. 2016. The entomopathogenic *Beauveria bassiana* has epiphytic and endophytic activity against the tomato leafminer *Tuta absoluta*. *J. Appl. Entomol.* 140: 580–589.

Kumar, A., and J.P. Verma. 2018. Does plant–Microbe interaction confer stress tolerance in plants: A review? *Microbiol. Res.* 207: 41–52.

Lata, R., S. Chowdhury, S.K. Gond, and J.F. White. 2018. Induction of abiotic stress tolerance in plants by endophytic microbes. *Let Appl. Microbiol.* 66: 268–276.

Lefort, M.C., A.C. McKinnon, T.L. Nelson, and T.R. Glare. 2016. Natural occurrence of the entomopathogenic fungi *Beauveria bassiana* as a vertically transmitted endophyte of *Pinus radiata* and its effect on above- and below-ground insect pests. *N. Z. Plant Prot.* 69: 68–77.

Li, Y., C. Wu, Z. Xing, B. Gao, and L. Zhang. 2017. Engineering the bacterial endophyte *Burkholderia pyrrocinia* JK-SH007 for the control of Lepidoptera larvae by introducing the cry218 genes of *Bacillus thuringiensis*. *Biotech Equip.* 31: 1167–1172.

Lutz, M.P., S. Wenger, M. Maurhofer, G. Défago, and B. Duffy. 2004. Signaling between bacterial and fungal biocontrol agent in a strain mixture. *FEMS Microbiol. Ecol.* 48(3): 447–455.

Mantzoukas, S., C. Chondrogiannis, and G. Grammatikopoulos. 2015. Effect of three endophytic entomopathogens on sweet sorghum and on the larvae of the stalk borer *Sesamia nonagrioides*. *Entomol. Exp. Appl.* 154: 78–87.

Moonjely, S., L. Barelli, and M.J. Bidochka. 2016. Insect pathogenic fungi as endophytes. In: St Leger R.J. (ed.). *Advances in Genetics*. Academic Press, Cambridge, MA, pp. 107–135.

Murphy, J.F., G.W. Zehnde, D.J. Schuster, E.J. Sikora, J.E. Polston, and J.W. Kloepper. 2000. Plant growth-promoting rhizobacterial mediated protection in tomato against Tomato mottle virus. *Plant Dis.* 84: 779–784.

Mutune, B., S. Ekesi, S. Niassy, V. Matiru, C. Bii, and N.K. Maniania. 2016. Fungal endophytes as promising tools for the management of bean stem maggot *Ophiomyia phaseoli* on beans *Phaseolus vulgaris*. *J. Pest Sci.* 89: 993–1001.

Muvea, A.M., R. Meyhöfer, N.K. Maniania, H.M. Poehling, S. Ekesi, and S. Subramanian. 2015. Behavioral responses of *Thrips tabaci* Lindeman to endophyte-inoculated onion plants. *J. Pest Sci.* 88: 555–562.

Nadeem, S.M., M. Ahmad, Z.A. Zahir, A. Javaid, and M. Ashraf. 2014. The role of mycorrhizae and plant growth promoting rhizobacteria (PGPR) in improving crop productivity under stressful environments. *Biotechnol. Adv.* 32: 429–448.

Nielsen-LeRoux, C., S. Gaudriault, N. Ramarao, D. Lereclus, and A. Givaudan. 2012. How the insect pathogen bacteria *Bacillus thuringiensis* and *Xenorhabdus/Photorhabdus* occupy hosts. *Curr. Opin. Microbiol.* 15: 220–231.

Ongena, M., and P. Jacques. 2008. *Bacillus* lipopeptides: Versatile weapons for plant disease biocontrol. *Trends Microbiol.* 16: 115–125.

Panaccione, D.G., W.T. Beaulieu, and D. Cook. 2014. Bioactive alkaloids in vertically transmitted fungal endophytes. *Funct. Ecol.* 28: 299–314.

Peña-Peña, A.J., M.T. Santillán-Galicia, J. Hernández-López, and A.W. Guzmán-Franco. 2015. *Metarhizium pingshaense* applied as a seed treatment induces fungal infection in larvae of the white grub *Anomala cincta. J. Invertebr. Pathol.* 130: 9–12.

Pieterse, C.M.J., and M. Dicke. 2007. Plant interactions with microbes and insects: From molecular mechanisms to ecology. *Trends Plant Sci.* 12: 564–569.

Pieterse, C.M.J., D. Van der Does, C. Zamioudis, A. Leon-Reyes, and S.C.M. van Wees. 2012. Hormonal modulation of plant immunity. *Annu. Rev. Cell Dev. Bi.* 28: 489–521.

Pineda, A., S.J. Zheng, J.A. van Loon, C.M.J. Pieterse, and M. Dicke. 2010. Helping plants to deal with insects: the role of beneficial soil-borne microbes. *Trends Plant Sci.* 15: 507–514.

Powell, W.A., W.E. Klingeman, B.H. Ownley, and K.D. Gwinn. 2009. Evidence of endophytic *Beauveria bassiana* in seed-treated tomato plants acting as a systemic entomopathogen to larval *Helicoverpa zea* (Lepidoptera: Noctuidae). *J. Entomol. Sci.* 44: 391–396.

Pozo, M.J., and C. Azcon-Aguilar. 2007. Unraveling mycorrhiza-induced resistance. *Curr. Opin. Plant Biol.* 10: 393–398.

Qayyum, M.A., W. Wakil, M.J. Arif, S.T. Sahi, and C.A. Dunlap. 2015. Infection of *Helicoverpa armigera* by endophytic *Beauveria bassiana* colonizing tomato plants. *Biol. Control* 90: 200–207.

Quesada-Moraga, E., B.B. Landa, J. Muñoz-Ledesma, R.M. JiménezDíaz, and C. Santiago-Álvarez. 2006. Endophytic colonization of opium poppy, *Papaver somniferum*, by an entomopathogenic *Beauveria bassiana* strain. *Mycopathologia* 161: 323–329.

Raps, A., and S. Vidal. 1998. Indirect effects of an unspecialized endophytic fungus on specialized plant-herbivorous insect interactions. *Oecologia* 114: 541–54.

Reddy, N.P., A.P. Khan, U.K. Devi, H.C. Sharma, and A. Reineke. 2009. Treatment of millet crop plant (*Sorghum bicolor*) with the entomopathogenic fungus (*Beauveria bassiana*) to combat infestation by the stem borer, *Chilo partellus* Swinhoe (Lepidoptera: Pyralidae). *J. Asia Pac. Entomol.* 12: 221–226.

Resquín-Romero, G., I. Garrido-Jurado, C. Delso, A. Ríos-Moreno, and E. Quesada- Moraga. 2016. Transient endophytic colonizations of plants improve the outcome of foliar applications of mycoinsecticides against chewing insects. *J. Invertebr. Pathol.* 136: 23–31.

Rondot, Y., and A. Reineke. 2018. Endophytic *Beauveria bassiana* in grapevine *Vitis vinifera* (L.) reduces infestation with piercing-sucking insects. *Biol. Control.* 116: 82–89.

Sangiorgio, D., A. Cellini, I. Donat, C. Pastore, C. Onofrietti, and F. Spinelli. 2020. Facing climate change: Application of microbial biostimulants to mitigate stress in horticultural crops. *Agronomy* 10: 794.

Sánchez-Rodríguez, A.R., S. Raya-Díaz, A.M. Zamarreño, J.M. García-Mina, M.C. Del Campillo, and E. Quesada-Moraga. 2018. An endophytic *Beauveria bassiana* strain increases spike production in bread and durum wheat plants and effectively controls cotton leafworm (*Spodoptera littoralis*) larvae. *Biol. Control* 116: 90–102.

Shrivastava, G., B.H. Ownley, R.M. Augé, H. Toler, M. Dee, A. Vu, T.G. Köllner, and F. Chen. 2015. Colonization by arbuscular mycorrhizal and endophytic fungi enhanced terpene production in tomato plants and their defense against a herbivorous insect. *Symbiosis* 65: 65–74.

Sinno, M., M. Ranesi, L. Gioia, and G. d'Errico, and S.L. Woo. 2020. Endophytic fungi of tomato and their potential applications for crop improvement. *Agriculture* 10: 587.

Spatafora, J.W., G.H. Sung, J.M. Sung, N.L. Hywel-Jones, and J.F. White. 2007. Phylogenetic evidence for an animal pathogen origin of ergot and the grass endophytes. *Mol. Ecol.* 16: 1701–1711.

Tabassum, B., A. Khan, M. Tariq, M. Ramzan, M.S.I. Khan, N. Shahid, and K. Aaliya, 2017. Bottlenecks in commercialization and future prospects of PGPR. *Appl. Soil Ecol.* 121: 102–117.

Tabbene, O., I.B. Slimene, F. Bouabdallah, M.L. Mangoni, M.C. Urdaci, and F. Limam. 2009. Production of antimethicillin-resistant staphylococcus activity from *Bacillus subtilis* sp. strain B38 newly isolated from soil. *App. Biochem. Biotechnol.* 157: 407–419.

Takabayashi, J., and M. Dicke. 1996. Plant-carnivore mutualism through herbivore-induced carnivore attractants. *Trends Plant Sci.* 1: 109–113.

Thuler, R.T., R. Barros, R.L.R. Mariano, and J.D. Vendramim. 2006. Efeito de bactérias promotoras do crescimento de plantas (BPCP) no desenvolvimento de *Plutella xylostella* (L.) (Lepidoptera: Plutellidae) em couve. *Científica* 34(2): 217–222.

Thuler, R.T., and S.A. Bortoli. 2007. Hoffmann-Campo CB. Classificação de cultivares de brássicas com relação resistência à traçadas-crucíferas e à presença de glucosinolatos. *Pesqui. Agropecu. Bras.* 42(4): 467–474.

Torres, J.B., and J.R. Ruberson. 2008. Interactions of *Bacillus thuringiensis* Cry1Ac toxin in genetically engineered cotton with predatory heteropterans. *Transgenic Res.* 17(3) 345–354.

Vachon, V., R. Laprade, and J.L. Schwartz. 2012. Current models of the mode of action of *Bacillus thuringiensis* insecticidal crystal proteins: A critical review. *J. Invertebr. Pathol.* 111: 1–12.

Valenzuela-Soto, J.H., M.G. Estrada-Hernández, and E. Ibarra-Laclette. 2010. Inoculation of tomato plants (*Solanum lycopersicum*) with growth-promoting *Bacillus subtilis* retards whitefly *Bemisia tabaci* development. *Planta* 231: 397–410.

van Loon, L.C. 2007. Plant responses to plant growth-promoting rhizobacteria. *Eur. J. Plant Pathol.* 119: 243–254.

van Oosten, V.R., N. Bodenhausen, P. Reymond, J.A. Van Pelt, L.C. Van Loon, M. Dicke, and C.M. Pieterse. 2008. Differential effectiveness of microbially induced resistance against herbivorous insects in *Arabidopsis*. *Mol. Plant-Microbe Interact.* 21: 919–930.

Vidal, S., and L.S. Jaber. 2015. Entomopathogenic fungi as endophytes: plant-endophyte- herbivore interactions and prospects for use in biological control. *Curr. Sci.* 109: 46–54.

Webber, J. 1981. A natural control of Dutch elm disease. *Nature* 292: 449–451.

Wielkopolan, B., and A. Obrepalska-Steplowska. 2016. Three-way interaction among plants, bacteria, and coleopteran insects. *Planta* 244: 313–332.

Zehnder, G., J. Kloepper, C. Yao, and G. Wei. 1997a. Induction of systemic resistance in cucumber against cucumber beetles (Coleoptera: Chrysomelidae) by plant growth-promoting rhizobacteria. *J. Econ. Entomol.* 90: 391–396.

Zehnder, G., J. Kloepper, S. Tuzun, C. Yao, G. Wei, O. Chambliss, and R. Shelby. 1997b. Insect feeding on cucumber mediated by rhizobacteria-induced plant resistance. *Entomol. Exp. Appl.* 83: 81–85.

Zhang, X.F. 2011. Insecticidal effect of recombinant bacterium containing Pinellia ternate agglutinin against *Sogatella furcifera*. *Crop Prot.* 30: 1478–1484.

Zhang, W., X. Zhang, K. Li, C. Wang, L. Cai, and W. Zhuang. 2018. Introgression and gene family contraction drive the evolution of lifestyle and host shifts of hypocrealean fungi. *Mycology* 9: 176–188.

4 Microbial Biostimulants as Fungicides against Root-Borne Pathogens

Qaiser Shakeel, Rabia Tahir Bajwa, Muhammad Raheel, and Sajjad Ali
The Islamia University of Bahawalpur

Yasir Iftikhar and Mustansar Mubeen
University of Sargodha

Ifrah Rashid
The Islamia University of Bahawalpur

CONTENTS

DOI: 10.1201/9781003188032-4

4.1   MICROBIAL BIOSTIMULANTS

Dr. Patrick du Jardin defined a biostimulant as any microorganism or substance that is proficient in enhancing the efficacy of plants in the absorption and assimilation of essential nutrients, tolerance to pathogenic or non-pathogenic stress, vigor and productivity of plants. Microbial biostimulants include PGPR (plant growth-promoting rhizobacteria) with the aim of promoting plant growth and improving plant health, and biofertilizers and biocontrol agents with the aim of increasing crop productivity, protecting plants from diseases or minimizing the disease severity and disease incidence along with enhancement of plant growth. At present, the application of microbial biostimulants or biofertilizers is gaining a lot of consideration involving the fact that they modify the microbial biota of the soil or the plants (du Jardin, 2015).

Due to lack of virtuous sources of resistance in host plants and complications in controlling the pathogen inoculum, the control of root-borne pathogens with extensive persistent survival structures becomes problematic (Azcón-Aguilar and Barea, 1997). So, alternate schemes based on either adding or influencing microbes are used to protect plants or to control pathogen infections (Grosch et al., 2005). Furthermore, the repeated use of chemicals leads to a serious risk to plant production system, humans, animals and the environment. Application of microbes having a protective effect against root-borne diseases as an alternative method to the use of chemicals to reduce the negative impacts of chemicals is being promoted by many researchers (Owen et al., 2015). Interestingly, the application of microbial biostimulants is an alternative method to replace the damages due to traditional agricultural practices. For example, microbial biostimulants can replace the flooding in the field for the accumulation of essential plant nutrients (i.e., nitrogen, phosphorus, potassium, calcium, sulfur and magnesium). Atmospheric nitrogen can be fixed by the species of Rhizobium, Sinorhizobium or Azotobacter. Some species of the genus Pseudomonas or Trichoderma are very efficient in solubilizing phosphorous, and the species of *Paenibacillus, Bacillus* and *Acidithiobacillus* are efficient in the solubilization of potassium (Verma et al., 2019). Moreover, in addition to compensating for the damage due to flooding or over-irrigation, microbial biostimulants also serve as an alternative of fertilizing techniques. Likewise, the application of some microbes including *Bacillus thuringiensis* is an alternative to various pesticides used for killing particular plant pathogens of various species of nematodes, Lepidoptera, Coleoptera and Diptera, which is a fully eco-friendly approach (Vilchez and Manzanera, 2011). This chapter focuses on the efficacy of microbial biostimulants against soil-borne diseases associated with roots or root-borne diseases.

4.2   HUMAN AND ENVIRONMENTAL SAFETY INDEX

Some researchers revealed the fact that some of these microbes can influence the biodiversity of the ecosystem or may produce infections in plants, animals and humans, posing a negative impact on the environment. The most suitable criteria to release a microbe in an eco-friendly manner should be based on the examination of the isolated microbes and their evaluation on other model organisms for the assessment of their impact on animals, plants, human health and all over the environment. Therefore, the most appropriate method to test the nature of microbes is by using *in vitro* developed bioassays (Vílchez et al., 2016). This safety index reveals the impact of microbes in query on plants and animals and acts as a pathogenicity model that permits forecasting the microbe's impact on the health of human beings and the environment. Although this scheme is not perfect, it allows for testing the bacterial impact on model organisms of soil network at various trophic levels. Microbes are applied for the assessment of their effect on survival and viability in a model organism *Escherichia coli*, for studying the effect on metabolism in *Aliivibrio fischeri* and for investigating the fertility, mortality and growth rate of nematodes (*Caenorhabditis elegans*), Neuropteran arthropods (*Adalia bipunctata*), Coleoptera (*Chrysoperla carnea*), Annelids (*Eisenia fetida*), plants (*Capsicum annuum*) and organisms of other environments including aquatic lives, i.e., Daphnia magna. A probit model is used to calculate the score in the range from 0 to 100, and the application of the microbes or biostimulants is considered appropriate for the environment when the score is a minimum 50 (Barros-Rodríguez et al., 2020).

4.3 CATEGORIES OF MICROBIAL BIOSTIMULANTS

Regardless of the current efforts to elucidate the regulatory status of biostimulants, there is no legal or official definition of biostimulants in the whole world, including the well-developed countries such as European Union and the USA. Due to this situation, it becomes difficult to list and categorize the biostimulants and this objective is obtained by the concept. In spite of this, many scientists, researchers and regulators documented some major categories of biostimulants regarding both microbes and substances (du Jardin, 2012; Calvo et al., 2014; Halpern et al., 2015). The major categories of microbial biostimulants involve beneficial fungi, beneficial bacteria and biocontrol agents. The microbial biostimulants can be endosymbiotic, rhizospheric and free living. A detailed discussion of these categories is documented below.

4.4 BENEFICIAL FUNGI

The roots of the plants interact with the fungi in various ways, including mutual symbiosis (i.e., living in direct contact with the host plant and exhibiting mutual beneficial relationship) and parasitism (Behie and Bidochka, 2014). Since the origin of terrestrial plants, fungi and plants coexist and the concept of mutual symbiosis and parasitism is suitable to deliberate the extensive range of relationships established over the evolutionary period (Bonfante and Genre, 2010; Johnson and Graham, 2013). Mycorrhizal fungi belonging to a heterogeneous group of taxa are a main constituent of natural resource of agriculture that develop symbiotic association with about 90% of terrestrial plant species. 'Mycorrhiza' is a Greek term, which means 'fungus roots,' and this term was introduced by Frank in 1885 (Frank, 1885). There are seven well-known types of mycorrhizae, including arbuscular mycorrhizae, orchidaceous mycorrhizae, ericoid mycorrhizae, arbutoid mycorrhizae, ectomycorrhizae, ectendomycorrhizae and monotropoid mycorrhizae (Singh et al., 2019).

4.5 ARBUSCULAR MYCORRHIZAL FUNGI (AMF)

The association of mycorrhizal fungi modifies the structures and functions; however, the most common association is arbuscular mycorrhizal fungi (AMF). AMF belong to the Glomeromycota phylum (Schussler et al., 2001), and they require a host for the completion of their life cycle. AMF develop a mutual symbiotic association with the host plant, in which the fungus facilitates the plant in nutrients uptake (particularly phosphorus), protects the plant from biotic or abiotic stresses, increases the yield and crop productivity and in return receives carbon from the plant. A very important role is played by the AMF in the suppression of various plant diseases (Whipps, 2004). The antagonistic effect of AMF against many soil- or root-borne pathogens (including *Phytophthora* species, *Pythium ultimum* (Trotta et al., 1996; Cordier et al., 1996), Macrophomina species, *Aphanomyces* species, *Fusarium* species, *Verticillium* species (Singh et al., 2019) and *Rhizoctonia solani* (Yao et al., 2002)) has been documented by many researchers, and because of their antagonistic effect, these biostimulants can be applied instead of fungicides or as a biocontrol agent or bioprotectant (Gianinazzi et al., 1995). Additionally, AMF suppress some bacterial diseases (for example, *Ralstonia solanacearum* is suppressed by *Glomus mosseae* and bacterial wilt on tomato is also reduced by AMF (Tahat, 2009)). Products based on AMF are applied to many crops as a biostimulant to obtain many benefits from crop productivity to protection from diseases. It becomes problematic to use these fungi on a large scale because due to the obligatory biotrophic character of AMF, their propagation in a large scale is a big challenge, additionally, the limited information about host plant specificities determinants and population subtleties of mycorrhizal microbes in agroecological system (Dalpé and Monreal, 2004). Due to their negative antagonistic effect against root-borne plant pathogens, they can be utilized as a biocontrol agent and as a plant growth promoter because of their capability to facilitate or increase the nutrients uptake by the plant (Veerabhadraswamy and Rajkumar, 2011). The discussion about the different mechanisms of actions by AMF is as follows.

4.6   MECHANISM OF DISEASE SUPPRESSION BY AMF

AMF reduce the root-borne diseases by colonizing the plant roots through various mechanisms, viz. improving the uptake of plant nutrition, producing antibiotics, compensating for the root damages, competing with the pathogens (for photosynthates and nutrients, colonizing the sites of infection), inducing antioxidative or antimicrobial substances (i.e., phytoalexins), inducing hydrolase enzymes, variation in the population of microbes colonizing the mycorrhizosphere, triggering defense-related (DR) enzymes, increasing pathogenesis-related (PR) proteins and stimulating the defense mechanism in the host plant. The plant becomes more vigorous due to the improved nutrients uptake facilitated by the symbiotic relationship with the AMF; as a result, the plant becomes more tolerant or resistant to biotic diseases (Azcón-Aguilar and Barea, 1996; Singh et al., 2019). Since a long time, the injurious or hazardous effects of the root-borne pathogens (including various fungi, straminopiles, nematodes and bacteria) after colonizing the plant roots with AMF were recounted to be effectively reduced (Gerdemann, 1974; Whipps, 2004).

4.7   INCREASED UPTAKE OF NUTRITION

AMF colonizing the plant roots increase the nutrients availability to the plants, hence promoting plant growth and productivity. Some researchers reported that the modifications in plant roots exudation induced by phosphorus have the ability to suppress the pathogenic spore germination (Graham, 1982; Sharma et al., 2007). In non-mycorrhizal plants, tolerance or resistance to pathogen infection in host plant can be enhanced by AMF through increasing uptake of deficient essential nutrients (in addition to phosphorus) (Gosling et al., 2006). The spores of AMF germinate and penetrate the thick-walled hyphae into the roots of the host plant to cause internal infection. After penetration, inter- or intracellular spread of hyphae occurs in the cortex of the roots without disturbing the consistency of the plant cells (Strack et al., 2003). Many studies suggest that the mechanism of enhanced nutrients uptake accounts for the highly vigorous plants, which are more resistant or tolerant to various pathogens (including fungi or nematodes) of mycorrhizal plants (Linderman, 1994), but some reports contradictory to this have also been found (Hooker et al., 1994). The biocontrol mediated by AMF is not much dependent on increased nutrient availability to the host plant, and other mechanisms are involved for this function (Toussaint et al., 2008).

4.8   INTERACTION OF SOIL MICROBIAL POPULATION

The interaction of AMF and the pathogen was first studied by Safir in 1968. The role played by AMF in increasing plant nutrients uptake and their interaction with soil microbes was studied regarding the growth of the host plant. There are limited data about in what ways the structure of the soil is affected by the interaction of AMF and pathogen (Schreiner and Bethlenfalvy, 1995). In the roots of non-mycorrhizal plants, the quality of respiration rate and the quantity of root exudates vary in comparison with the mycorrhizal host plant due to the alteration in soil microbial population (Marschner et al., 2001). In the presence of bacteria, hyphae germinating from the spores are more elongated and branched and form small vesicles compared to those in the absence of bacteria. Modification in the soil microbial community affects the health and growth of a mycorrhizal plant (Azcón-Aguilar et al., 2002). This influence has not been particularly considered as a biocontrol mechanism by AMF, but it is indicated that such a mechanism is also practicable for operating biocontrol (Linderman, 1994). In mycorrhizosphere, the alteration in the composition of functional groups of rhizospheric microbes (including the population and antagonistic activity of the pathogen) by AMF is reported in some studies (Secilia and Bagyaraj, 1987). In the rhizosphere of mycorrhizal tomato plant, the population of *Fusarium oxysporum* was reduced compared with non-mycorrhizal tomato plant (Johansson et al., 2004). The AMF-colonized root system of the host plant influences the composition of bacterial population in the soil (Burke et al., 2002). Some biotic and abiotic factors (including the genotype of the host, soil content, soil moisture, inoculum level of MF (mycorrhizal fungi), virulence of MF species, inoculation time of AMF and the potential of pathogen inoculum and microflora of the rhizosphere) play a key role in the assessment of efficacy of AMF as a biocontrol agent (Singh et al., 2000). AMF provide systemic bioprotection to barley plant against all root-borne diseases by colonizing the host roots at high degree (Khaosaad et al., 2007). The resistance in plants against wilt diseases, particularly *Fusarium oxysporum*, is increased by using AMF (Dugassa et al., 1996). Two zones of interactions of soil microbial community can be noticed, i.e., the rhizosphere and mycosphere (Bansal and Mukerj, 1994). The efficacy of MF to colonize the roots of host plant is certainly enhanced by mycorrhization helper bacteria (Fitter and Grabaye, 1994).

4.9   COMPENSATION OF ROOT DAMAGE

The loss of biomass and functional roots caused by root-borne pathogens (including fungi and nematodes) is prevented by AMF via compensation, and tolerance in the host plant is increased against the pathogen infection (Linderman, 1994; Cordier et al., 1996). By the mechanism of compensation, an indirect contribution is elucidated for biocontrol via conserving the root system. The hyphae of AMF rise out in the soil performing the function of enhancing the surface area of the roots for absorption and maintaining the activity of root cells by forming branched structures called arbuscules (sites for the exchange of nutrients) (Gianinazzi et al., 1995).

4.10   COMPETITION WITH THE PATHOGEN FOR SPACE

The mechanism to understand the association between mycorrhizal fungi and rhizosphere microbes is the physical competition between AMF and soil microbes to occupy more space for the colonization of the host plant roots (Bansal and Mukerji, 1994). In tomato plants, the association between mycorrhizal fungi (AMF) and the pathogen (Phytophthora) has revealed that the pathogen is unable to penetrate into the cells containing arbuscular fungi (Cordier et al., 1998). AMF and root-borne pathogens compete to colonize in the same cortical cells of the host and develop in separate tissues elucidating competition for space (Dehne, 1982). Due to the space competition, the multiplication and survival of the pathogen becomes difficult. The rapid development of pathogenic symptoms on host plant is modified by the colonization of AMF, suggesting the host plant to form induced systemic resistance (ISR) (Azcón-Aguilar, 2002).

4.11   COMPETITION WITH THE PATHOGEN FOR HOST PHOTOSYNTHATES

AMF obtain carbon from the host plant; thus, the competition for photosynthates can be a reason for the pathogenic reduction particularly in the mycorrhizal host plant. Once AMF primarily colonize the tissues, the higher demand for carbon can suppress the pathogenic growth (Azcón-Aguilar and Barea, 1996). AMF obtain 4%–20% of the total photosynthate from the host plant, although limited data are available to support this mechanism (Smith and Read, 2008).

4.12   AMF AS A BIOPROTECTANT

The efficacy of AMF as a biocontrol agent against various phytopathogens has been demonstrated by many researchers. The AMF have also been applied and considered as an important method for plant disease control (Singh et al., 2000; Kasiamdari et al., 2002; Garmendia et al., 2005;

Garcia-Garrido, 2009; Xavier et al., 2014). The mode of action of AMF against the pathogens is divided into two subdivisions as follows.

I. Interaction between AMF and Phyto-mycopathogens

The antagonistic efficacy of AMF has been observed against a broad range of phytopathogens, mainly the root-borne mycopathogens responsible for causing many root rots and wilt diseases in plants. Moreover, the effective biocontrol has also been noticed in some above-ground pathogens, including *Alternaria solani* causal organism of early leaf spot in tomato (Jung et al., 2012; Singh et al., 2019). Both types of pathogens, either necrotrophs or biotrophs, can be controlled by AMF via direct or indirect competition for nutrition and root colonization with root-borne pathogens in the rhizosphere and root system of the plant (Larsen and Bodker, 2001). Combination of various species of mycorrhizal fungi including *Gigaspora gigantea*, *G. margarita*, *Glomus intraradices*, *G. mosseae* and *G. clarum* restricted the growth of *Sclerotium cepivorum* causing white rot disease in onion. Additionally, the same combination has effectively been used against Fusarium root rot disease, resulting in a successful reduction in disease severity and incidence. The clover plant in symbiotic association with *Glomus mosseae* was observed to be completely prevented by the *Pythium ultimum* infection (El-Haddad et al., 2004; Carlsen et al., 2008). The activation of tolerance by *Glomus intraradices* and *G. claroideum* in pea plants against pea root rot disease caused by *Aphanomyces euteiches* has been reported by Thygesen et al. (2004).

II. Interaction between AMF and Phytopathogenic Nematodes

Under *in vitro* experimentation, the extensive network of *Glomus intraradices* has been found to directly suppress the nematodes growth, particularly *Pratylenchus coffeae* and *Radopholus similis*, as well as cease the conidial growth of *Fusarium oxysporum* f.sp. *chrysanthemi* in the roots of infected plants. It is reported that the population of phyto-burrowing nematode *Radopholus similis* is successfully suppressed by AMF in the roots of different genotypes of banana (Elsen et al., 2003a–c). The mycorrhizae-induced resistance was described first by Elsen et al. (2008) against phytoparasitic nematodes, which is induced systemically in the roots of banana. In mycorrhizal roots, the root growth increases, ultimately enhancing the host tolerance against phytoparasitic nematodes (Pinochet et al., 1996).

Currently, the AMF (endomycorrhizal fungi) are considered as a biostimulant on the basis of all the above factors and can be used as a biocontrol agent or as a biopesticide (biofertilizers or biofungicides) against many root-borne infections instead of chemical control.

4.13 BENEFICIAL BACTERIA

All the possible ways of bacterial interaction with the plants include the spread of bacterial niches into the plant cell from the soil within the rhizosphere; the phyto-bacterial linkage can be temporary or permanent; additionally, some bacteria can be transported vertically by the seed. The life of the host plant is influenced by the bacteria due to bacterial contribution in biogeochemical reactions, nutrients supply, inducing disease tolerance or resistance, enhancing nutrients uptake ability, increasing tolerance of host plant against abiotic stresses and inflection of morphogenesis by PGR (plant growth regulators) (Ahmad et al., 2008). Regarding the utilization of biostimulants in agriculture, two major types of bacteria are considered in this diversity, i.e., mutualistic symbiotic Rhizobium bacteria and mutualistic rhizospheric PGPR. Rhizobium is commercially available as biofertilizers to facilitate nutrients uptake by the plants. The life of the host plant is altered by PGPR in all aspects, including growth, morphogenesis and development, nutrients uptake, reaction against biotic and abiotic stresses and interaction with other microbial communities in the rhizosphere

(Ahmad et al., 2008; Babalola, 2010; Berendsen et al., 2012; Berg et al., 2014; Bhattacharyya and Jha, 2012; Gaiero et al., 2013; Philippot et al., 2013; Vacheron et al., 2013).

4.14 RHIZOBIA

Rhizobia attract the keen interest of researchers because of their efficacy to fix the atmospheric nitrogen in the leguminous crops (Vasileva and Ilieva, 2012). Furthermore, in addition to fixing the atmospheric nitrogen, the antagonistic efficacy against many root-borne pathogens has been observed in the Rhizobia (Noreen, 2016). Some strains of the Rhizobium are known to show antagonistic effect by limiting the disease severity or incidence caused by various root-borne pathogens including *Phytophthora clandestine*, *Pythium* species, *Fusarium solani*, *Fusarium oxysporum* and *Rhizoctonia bataticola* (Nautiyal, 1997; Simpfendorfer et al., 1999; Ozkoc and Deliveli, 2001; Al-Ani et al., 2012). Under *in vivo* conditions, the application of *Bradyrhizobium japonicum*, *Rhizobium leguminosarum* and *Sinorhizobium meliloti* as soil drenches or seed dressing resulted in disease suppression of *Fusarium* spp., *Rhizoctonia solani* and *Macrophomina phaseolina* in both the host plants, i.e., leguminous and non-leguminous (Ehteshamul-Haque and Ghaffar, 1993).

4.15 RHIZOBIA AS A BIOCONTROL AGENT

Phytophthora megasperma causing root rot in soybean has been observed to be effectively suppressed by Rhizobium. The antagonistic effect of 49 different strains of *Sinorhizobium meliloti* was studied against *Fusarium oxysporum* by Antoun et al. (1978), and the results showed a reduction in pathogen irrespective of their symbiotic efficacy. The inhibition percentage of pathogen growth varied from 5% to 50% with different strains of *S. meliloti*. Some strains of *Bradyrhizobium japonicum* producing rhizobitoxine were found to protect soybean plants from being infected by *Macrophomina phaseolina* causing charcoal rot in leguminous plants (Chakraborty and Purkayastha, 1984). The isolation of Rhizobia from the root nodules of prickly moses (*Acacia pulchella*) indicated that the survival of zoospores of *Phytophthora cinnamon* was restricted by these bacteria (Malajczuk et al., 1984). The antagonistic efficacy of *Rhizobium leguminosarum* bv. *phaseoli* against *Fusarium solani* f.sp. *phaseoli* was assessed by inoculating bean seeds with the respective strain. The host plants grown in sterilized soil infested with *Fusarium solani* showed significant suppression of the pathogen (Buonassisi et al., 1986). The Rhizobia having antagonistic efficacy exhibited higher proficiency in root hair system of plant compared to the Rhizobia not having the antagonistic effect (Deshwal et al., 2003). Rhizobia can also interact and colonize the non-leguminous plants, so they can also be utilized as a biocontrol agent against the diseases of non-leguminous plants.

4.16 BIOCONTROL MECHANISMS INVOLVED BY RHIZOBIA

Rhizobia involve some sort of mechanisms for biocontrol, including the production of HCN, antibiotics and siderophores. The defense mechanism of the host plant can be influenced by Rhizobia by triggering the phytoalexins production through the plants. Microscopic studies revealed that the interaction of Rhizobia with the pathogens including *Sclerotinia sclerotiorum*, *M. phaseolina*, *R. solani* and *F. oxysporum* resulted in the deformation of the hyphal tip, abnormal swelling, cytoplasmic disintegration and lysis of the fungal hyphae.

4.17 PRODUCTION OF ANTIBIOTICS

Rhizobia produced antibiotics playing a key role in the reduction in plant disease. R. *leguminosarum* bv. *trifolii* has been reported to secrete trifolitoxin, a peptide antibiotic. By the direct mechanism of antibiotic rhizobitoxine action, *B. japonicum* inhibits the *M. phaseolina* to protect the soybean crop from being infected (Chakraborty and Purkayastha, 1984).

4.18   PRODUCTION OF HCN

Many microbes produce secondary metabolites, including HCN, which have a poisonous effect on the growth and development of some microorganisms. Some rhizospheric microbes including Rhizobia are known to produce HCN for the reduction in the pathogenic growth, resulting in the protection of their host plant from infection. However, Rhizobia are not much efficient in producing HCN; only 3% and 12.5% rhizobial strains were observed to have the ability to produce HCN by Antoun et al. (1998) and Beauchamp et al. (1991), respectively.

4.19   PRODUCTION OF SIDEROPHORES

Siderophores are produced extracellularly by almost each and every facultative anaerobic microbe to conquer the iron restriction for their better growth. Iron is solubilized and chelated by the siderophores, and the complexes of ferri-siderophore are occupied by the cells (Neilands, 1982). A large amount of siderophores have been found to be produced by the Rhizobia, including citrate (*B. japonicum*), rhizobactin (*S. meliloti*), Bradyrhizobium (*Arachis hypogaea*), anthranilate (*R. leguminosarum* bv. *viciae*), catechol (*R. leguminosarum*), *R. leguminosarum* bv. *trifolii*), Bradyrhizobium (*Vigna unguiculata*) and vicibactin (*R. leguminosarum* bv. *viciae*) (Rioux et al., 1986; Patel et al., 1988; Guerinot et al., 1990). Under the stress condition of iron, the production of siderophores provides an extra benefit to Rhizobia, i.e., elimination of the pathogen because of iron starvation.

4.20   PRODUCTION OF PHYTOALEXINS

The resistant pea plants against *F. solani* f.sp. *pisi* showed the contribution of a phytoalexin, i.e., 4-hydroxy-2,3,9-tri-methoxypterocarpan (Pueppke and VanEtten, 1974, 1975). It has been reported that *R. leguminosarum* bv. *viciae* prevents the root colonization of *F. solani* f.sp. *pisi* in pea plants (Chakraborty and Chakraborty, 1988). This protection is stimulated due to the enhanced production of phytoalexin 4-hydroxy-2,3,9-tri-methoxypterocarpan by host pea.

4.21   BIOCONTROL AGENTS

Regarding microbial biostimulants, biocontrol agents are those microbes that protect the host plant from biotic stresses to an acceptable level. The mechanisms involved in the inhibition of the pathogen by biocontrol agents include antibiosis, competition, parasitism and ISR facilitated by the host plant. Due to various modes of mechanisms, the results become practically unpredictable (Arora et al., 2011; Brahmaprakash and Sahu, 2012). Regardless of this, the use of bacterial biostimulants is rising in the world market and the inoculation of PGPR is considered as probiotics for plants; i.e., the effective involvement has been found in plant nutrition and immunity (Berendsen et al., 2012). Among fungi, Trichoderma and, among bacteria, Pseudomonas and Bacillus attain most of the attention as a biocontrol agent and are being widely used against many root-borne pathogens.

4.22   TRICHODERMA

Unlike mycorrhizal fungi, Trichoderma lives a part of its life cycle in the host plant and spends the other part of its life without the host plant. Trichoderma is used as a model organism for the dissection of the nutrient transfer mechanism between the host plant and myco-endosymbionts (Behie and Bidochka, 2014). *Trichoderma* species have extensively been investigated and are commercially utilized as biocontrol (induces resistance in the host plants against pests), biopesticide (mycoparasitism) and biofertilizer (promoting the plant growth) (Mukherjee et al., 2012; Nicolás et al., 2014). Trichoderma is very effective against several diseases, including wilt, damping off, foot rot, collar rot, stem rot, root rot and blight leaf spot of many crops such as pulses oil seeds, Cucurbitaceae family (ridge gourd, bottle gourd and cucumber) and solanaceous crops (such as chili, tomato, eggplant and capsicum).

4.23 MECHANISM OF BIOCONTROL BY TRICHODERMA

There are three major mechanisms of biocontrol mediated by Trichoderma: competition for nutrition or space, antibiosis and mycoparasitism.

4.24 COMPETITION

Biocontrol agents compete with the pathogen for nutrients and space in the host plant, so the effect of one microbe that colonizes the host roots first and utilizes many resources is harmful on the other microbe as it becomes unable to acquire proper nutrition and space. Hence, the pathogenic growth ceases or is limited due to the high demand for and less availability of nutrients and space. As a seed treatment, biocontrol agents are very effective because due to their root colonization of the host plant, enhanced plant health and mass of the roots increase the plant yield or productivity (Mukhopadhyay and Pan, 2012b).

4.25 ANTIBIOSIS

Antibiosis is one of the main features in determining the saprophytic capability of the fungus. Trichoderma produces a number of antibiotics against various plant pathogens, specifically root-borne ones (Weindling, 1934). The production of a range of antibiotics by *T. harzianum*, viz. suzukacillin, alamethicin and trichodermin, alters the physiological and morphological structures to facilitate its penetration. A wide range of root-borne fungi are suppressed by the *Trichoderma* spp., including *Sclerotinia, Macrophomina, Sclerotium, Rhizoctonia, Pythium, Phytophthora, Verticillium* and *Fusarium* (Zaher et al., 2013; Ragab et al., 2015; Chen et al., 2016). The production of volatile and non-volatile antibiotics has been found in Trichoderma spp. having antagonistic effect against a wide range of phyto-mycopathogens (Mukhopadhyay and Kaur, 1990). Studies revealed that *T. harzianum* and *T. viride* have a great potential for inhibiting the hyphal growth of *S. rolfsii* due to the production of volatile and non-volatile antibiotics (Rao and Kulkarni, 2003). The non-volatile antibiotics were found to be more efficient as compared to volatile antibiotics (Bunker and Mathur, 2001).

4.26 MYCOPARASITISM

The antagonistic activity of mycoparasitism by Trichoderma is considered as one of the most important mechanisms involved against several phyto-mycopathogens (Haran et al., 1996). Kumar et al. studied the mycoparasitism of *Trichoderma harzianum* in biocontrol of wood decay fungi and revealed that within 95 hours of contact with the pathogen, Trichoderma has been seen to be slightly coiling around *F. solani* and form its appressorium over the hyphae of the pathogen. The complete inhibition of the pathogen has been recorded within 6 days, while Trichoderma continued multiplying involving the process of conidiogenesis. Microscopic examination of the antagonist-pathogen hyphal interaction has revealed that Trichoderma parallel has different mode of interaction against pathogenic hyphea i.e. penetration into the pathogenic hyphae, coiling of the biocontrol agent around the pathogenic hyphae through the production of appressorium (knob- or hook-like structures) (Mukhopadhyay and Pan, 2012a).

REFERENCES

Ahmad, I., Pichtel, J., and Hayat, S. 2008. Plant-Bacteria Interactions. Strategies and Techniques.

Al-Ani, R.A., Adhab, M.A., Mahdi, M.H., and Abood, H.M. 2012. Rhizobium japonicum as a biocontrol agent of soybean root rot disease caused by *Fusarium solani* and *Macrophomina phaseolina*. *Plant Protec. Sci.* 48(4): 149–155.

Antoun, H., Beauchamp, C.J., Goussard, N., Chabot, R., and Lalande, R. 1998. Potential of Rhizobium and *Bradyrhizobium* species as plant growth promoting rhizobacteria on non- legumes: Effect on radishes (*Raphanus sativus* L.) *Plant Soil* 204: 57.

Antoun, H., Bordeleau, L.M., and Gagnon, C. 1978. Antagonisme entre Rhizobium meliloti et Fusarium oxysporum en relation avee l efficacité symbiotique. *Can. J. Plant Sci.* 58: 75.

Arora, N.K., Khare, E., and Maheshwari, D.K. 2011. Plant growth promoting rhizobacteria: Constraints in bioformulation, commercialization, and future strategies. In: Maheshwari, D.K. (Ed.), *Plant Growth and Health Promoting Bacteria.* Springer, Heidelberg, pp. 97–116.

Azcón-Aguilar, C., and Barea, J.M. 1996. Arbuscular mycorrhizas and biological control of soil-borne plant pathogens–an overview of the mechanisms involved. *Mycorrhiza* 6: 457–464.

Azcón-Aguilar, C., and Barea, J.M. 1997. Applying mycorrhiza biotechnology to horticulture: Significance and potentials. *Sci. Hortic.* 68: 1–24.

Azcon-Aguilar, C., Jaizme-Vega, M.C. and Calvet, C. 2002. The contribution of arbuscular mycorrhizal fungi for bioremediation. In: Gianinazzi, S., H. Schuepp, J.M. Barea and K. Haselwandter (Eds.), *Mycorrhizal Technology in Agriculture: From Genes to Bioproducts.* Birkhauser Verlag, Berlin, pp. 187–197. ISBN-10: 0-89054-245-71.

Babalola, O.O. 2010. Beneficial bacteria of agricultural importance. *Biotechnol. Lett.* 32: 1559–1570.

Bansal, M., and Mukerj, K.G. 1994. Positive correlation between AM-induced changes in root exudation and mycorrhizosphere mycoflora. *Mycorrhiza* 5: 39–44.

Barros-Rodríguez, A., Rangseekaew, P., Lasudee, K., Pathom-Aree, W., and Manzanera, M. 2020. Regulatory risks associated with bacteria as biostimulants and biofertilizers in the frame of the European Regulation (EU) 2019/1009. *Sci. Total Environ.* 740: 140239.

Beauchamp, C.J., Dion, P., Kloepper, J.W., and Antoun, H. 1991. Physiological characterization of opine-utilizing rhizobacteria for traits related to plant growth- promoting activity, *Plant Soil* 132: 273.

Behie, S.W., and Bidochka, M.J. 2014. Nutrient transfer in plant-fungal symbioses. *Trends Plant Sci.* 19: 734–740.

Berendsen, R.L., Pieterse, C.M., and Bakker, P.A. 2012. The rhizosphere microbiome and plant health. *Trends Plant Sci.* 17: 1360–1385.

Berg, G., Grube, M., Schloter, M., and Smalla, K. 2014. Unraveling the plant microbiome: looking back and future perspectives. *Front. Microbiol.* 5: 1–7, Article 148.

Bhattacharyya, P.N., and Jha, D.K. 2012. Plant growth-promoting rhizobacteria (PGPR): Emergence in agriculture. *World J. Microbiol. Biotechnol.* 28: 1327–1350.

Bonfante, P., and Genre, A. 2010. Interactions in mycorrhizal symbiosis. *Nat. Commun.* 1: 1–11.

Brahmaprakash, G.P., and Sahu, P.K. 2012. Biofertilizers for sustainability. *J. Indian Inst. Sci.* 92: 37–62.

Bunker, R.N, and Mathur, K. 2001. Antagonism of local biocontrol agents to *Rhizoctonia solani* inciting dry root rot of chilli. *J. Mycol. Plant Pathol.* 31: 337–353.

Buonassisi, A.J, Copeman, R.J., Pepin, H.S., and Eaton, G.W. 1986. Effect of Rhizobium spp on Fusarium solani f.sp. phaseoli. *Can. J. Plant Pathol.* 8: 140.

Burke, D.J., Hamerlynck, E.P., and Hahn, D. 2002. Interactions among plant species and microorganisms in salt march sediments. *Appl. Environ. Microbiol.* 68: 1157–1164.

Calvo, P., Nelson, L., and Kloepper, J.W. 2014. Agricultural uses of plant biostimulants. *Plant Soil* 383(1): 3–41.

Carlsen, S.C.K., Understrup, A., Fomsgaard, I.S., Mortensen, A.G., and Ravnskov, S. 2008. Flavonoids in roots of white clover: Interaction of arbuscular mycorrhizal fungi and a pathogenic fungus. *Plant Soil* 302: 33–43.

Chakraborty, U., and Chakraborty, B.N. 1988. Interaction of Rhizobium leguminosarum and Fusarium solani f.sp. pisi on pea affecting disease development and phytoalexin production, *Can. J. Botany* 67: 1698.

Chakraborty, U., and Purkayastha, R.P. 1984. Role of rhizobitoxine in protecting soybean roots from Macrophomina phaseolina, *Can. J. Microbiol.* 30: 285.

Chen, J., Sun, S., Miao, C., Wu, K., Chen, Y., Xu, L., Guan, H., and Zao, L. 2016. Endophytic *Trichoderma gamsii* YIM PH30019: A promising biocontrol agent with hyperosmolar, mycoparasitism and antagonistic activities of induced volatile organic compounds on root rot pathogenic fungi of *Panax notoginseng. J. Ginseng. Res.* 40: 315–324.

Cordier, C., Pozo, M.J., Barea, J.M., Gianinazzi, S., and Gianinazzi-Pearson, V. 1998. Cell defense responses associated with localized and systemic resistance to Phytophthora parasitica induced in tomato by an arbuscular mycorrhizal fungus. *Mol. Plant Microbe Interact.* 11: 1017–1028.

Cordier, C., Gianinazzi, S., and Gianinazzi-Pearson, V. 1996. Colonisation patterns of root tissues by *Phytophthora nicotianae* var. parasitica related to reduced disease in mycorrhizal tomato. *Plant Soil* 185: 223–232.

Dalpé, Y., and Monreal, M. 2004. Arbuscular mycorrhiza inoculum to support sustainable cropping systems. Online. Symposium Proceeding. Crop Management network.

Dehne, H.W. 1982. Interaction between vesicular-arbuscular mycorrhizal fungi and plant pathogens. *Phyto-pathology* 72: 1115–1119.

Deshwal, V.K., Dubey, R.C., and Maheshwari, D.K. 2003. Isolation of plant growth-promoting strains of Bradyrhizobium (Arachis) sp. with biocontrol potential against Macrophomina phaseolina causing charcoal rot of peanut. *Curr. Sci.* 84: 443.

du Jardin, P. 2012. The Science of Plant Biostimulants—A bibliographic analysis. Ad hoc Study Report to the European Commission.

du Jardin, P. 2015. Plant biostimulants: Definition, concept, main categories and regulation. *Sci. Hortic.* 196: 3–14.

Dugassa, G.D., von Allen, H., and Schonbeck, F. 1996. Effect of Arbuscular Mycorrhiza (AM) on health of *Linum usitatissimum* L. infected by fungal pathogen. *Plant Soil* 185: 173–182.

Ehteshamul-Haque, S., and Ghaffar, A. 1993. Use of rhizobia in the control of root rot diseases of sunflower, okra, soybean and mungbean. *J. Phytopathol.* 138: 157–63.

El-Haddad, S.A., Abd El-Megid, M.S., and Shalaby, O.Y. 2004. Controlling onion white rot by using Egyptian formulated endo-mycorrhiza (Multi-VAM). *Ann. Agric. Sci.* 49: 733–745.

Elsen, A., Baimey, H., Swennen, R., and De Waele, D. 2003a. Relative mycorrhizal dependency and mycorrhiza-nematode interaction in banana cultivars (Musa spp.) differing in nematode susceptibility. *Plant Soil* 256: 303–313.

Elsen, A., Beeterens, R., and Swennen, R. 2003b. Effects of an arbuscular mycorrhizal fungus and two plant-parasitic nematodes on Musa genotypes differing in root morphology. *Biol. Fertil. Soils* 38: 367–376.

Elsen, A., Declerck, S., and Waele, D.D. 2003c. Use of root organ cultures to investigate the interaction between Glomus intraradices and Pratylenchus coffeae. *Appl. Environ. Microbiol.* 69: 4308–4311.

Elsen, A., Gervacio, D., Swennen, R., and De Waele, D. 2008. AMF induced biocontrol against plant parasitic nematodes in Musa sp.: A systemic effect. *Mycorrhiza* 18: 251–256.

Fitter, A.H., and J. Grabaye 1994. Interactions between mycorrhizal fungi and other soil organisms. *Plant Soil* 159: 123–132.

Frank, A.B. 1885. Über die auf Wurzelsymbiose beruhende Ernährung gewisser Bäume durch unterirdische Pilze (On the nourishing, via root symbiosis, of certain trees by underground fungi). *Berichte der Deutschen Botanischen Gesellschaft (in German)* 3: 128–145.

Gaiero, J.R., McCall, C.A., Thompson, K.A., Dayu, N.J., Best, A.S., and Dunfield, K.E. 2013. Inside the root microbiome: Bacterial root endophytes and plant growth promotion. *Am. J. Bot.* 100: 1738–1750.

García-Garrido, J.M., Lendzemo, V., Castellanos-Morales, V. et al. Strigolactones, signals for parasitic plants and arbuscular mycorrhizal fungi. Mycorrhiza 19, 449–459 (2009). https://doi.org/10.1007/s00572-009-0265-y.

Garmendia, J., Frankel, G., & Crepin, V. F. (2005). Enteropathogenic and enterohemorrhagic Escherichia coli infections: translocation, translocation, translocation. Infection and immunity, 73(5), 2573–2585. https://doi.org/10.1128/IAI.73.5.2573-2585.2005.

Gerdemann, J.W. 1974. *Vesicula-arbuscular Mycorrhiza.* Academic Press, New York.

Gianinazzi, S., Trouvelot, A., Lovato, P., van Tuinen, D., Franken, P., and Gianinazzi-Pearson, V. 1995. Arbuscular mycorrhizal fungi in plant production of temperate agroecosystems. *Crit. Rev. Biotechnol.* 15: 305–311.

Gosling, P., Hodge, A., Goodlass, G., and Bending, G.D. 2006. Arbuscular mycorrhizal fungi and organic farming. *Agric. Ecosyst. Environ.* 113: 17–35.

Graham, J.H. 1982. Effect of citrus root exudates on germination of chlamydospores of vesicular-arbuscular mycorrhizal fungus *Glomus epigaeum. Mycologia* 74: 831–835.

Grosch, R., Lottmann, J., Faltin, F., and Berg, G. 2005. Use of bacterial antagonists to control diseases caused by *Rhizoctonia solani. Gesunde Pflanzen,* 57: 199–205.

Guerinot, M.L., Meidl, E.J., and Plessner, O. 1990. Citrate as a siderophore in Bradyrhizobium japonicum, *J. Bacteriol.* 172: 3298.

Halpern, M., Bar-Tal, A., Ofek, M., Minz, D., Muller, T., and Yermiyahu, U. 2015. The use of biostimulants for enhancing nutrient uptake. In: Sparks, D.L. (Ed.), *Advances in Agronomy,* Vol. 129, pp. 141–174. https://www.cabdirect.org/cabdirect/abstract/20153092279.

Haran, S., Schickler, H., and Chet, I. 1996. Molecular mechanisms of lytic enzymes involved in the biocontrol activity of *Trichoderma harzianum. Microbiology* 142(9): 2321–2331.

Hooker, J.E., Jaizme-Vega, M., and Atkinson, D. 1994. Biocontrol of plant pathogens using arbuscular mycorrhizal fungi. In: Gianinazzi, S., and Schüepp, H. (Eds.), *Impact of Arbuscular Mycorrhizas on Sustainable Agriculture and Natural Ecosystems.* Birkhäuser, Basel, pp. 191–200.

Johansson, J.F., Paul, L.R., and Finlay, R.D. 2004. Microbial interactions in the mycorrhizosphere and their significance for sustainable agriculture. *FEMS Microbiol. Ecol.,* 48: 1–13.

Johnson, N.C., and Graham, J.H. 2013. The continuum concept remains a useful framework for studying mycorrhizal functioning. *Plant Soil* 363: 411–419.

Jung, S.C., Martinez-Medina, A., Lopez-Raez, J.A., and Pozo, M.J. 2012. Mycorrhiza-induced resistance and priming of plant defenses. *J. Chem. Ecol.* 38: 651–664.

Kasiamdari, R.S., Smith, E.S., Scott, E.S., Smith, F.A. (2002). Identification of binucleate Rhizoctonia as a contaminant in pot cultures of arbuscular mycorrhizal fungi and development of a PCR-based method of detection. Mycol. Res. 106 (12), 1417–1426.

Khaosaad, T., Garcia-Garrido, J.M., Steinkellner, S., and Vierheilig, H. 2007. Take-all disease is systemically reduced in roots of mycorrhizal barley plants. *Soil Biol. Biochem.* 39: 727–734.

Larsen, J., and Bodker, L. 2001. Interactions between pea root-inhabiting fungi examined using signature fatty acids. *New Phytol.* 149: 487–493.

Linderman, R.G. 1994. Role of VAM fungi in biocontrol. In: Pfleger, F.L. and R.G. Linderman (Eds.), *Mycorrhizae and Plant Health*. The American Phytopathological Society, St. Paul, MN, pp. 1–27. ISBN: 978-0-89054-158-2.

Malajczuk, N., Pearse, M., and Litchfield, R.T. 1984. Interactions between Phytophthora cinnamoni and Rhizobium isolates, *Trans. Br. Mycol. Soc.* 82: 491.

Marschner, P., Crowley, D., and Lieberei, R. 2001. Arbuscular mycorrhizal infection changes bacterial 16s DNA community composition in the rhizosphere of maize. *Mycorrhiza* 11: 297–302.

Mukherjee, P.K., Horwitz, B.A., Herrera-estrella, A., Schmoll, M., and Kenerley, C.M. 2012. Trichoderma research in the genome era. *Annu. Rev. Phytopathol.* 51: 105–129.

Mukhopadhyay, A.N., and Kaur, N.P. 1990. Biological control of chickpea wilt complex by *Trichoderma harzianum*. In: *Proceedings of Third International Conference on Plant Protection in the Tropics*, Malaysia, pp. 20–23.

Mukhopadhyay, R., and Pan, S.K. 2012a. Isolation and selection of some antagonistic *Trichoderma*species from different new alluvial zones of Nadia district, West Bengal. *J. Bot. Soc. Bengal* 66(2): 149–152.

Mukhopadhyay, R., and Pan, S.K. 2012b. Effect of biopriming of radish (*Raphanus sativus*) seeds with some antagonistic isolates of *Trichoderma*. *J. Bot. Soc. Bengal* 66(2): 157–160.

Nautiyal, C.S. 1997. Rhizosphere competence of *Pseudomonas* sp., NBR 19926 and *Rhizobium* sp. NBRI9513 involved in the suppression of chickpea (*Cicer arietinum* L.) pathogenic fungi. *FEMS Microbiol. Ecol.* 23: 145–58.

Neilands, J.B. 1982. Microbial envelope proteins related to iron. *Ann. Rev. Microbiol.* 36: 285.

Nicolás, C., Hermosa, R., Rubio, B., Mukherjee, P.K., and Monte, E. 2014. Trichoderma genes in plants for stress tolerance-status and prospects. *Plant Sci.* 228: 71–78.

Noreen, R. 2016. Role of mungbean root nodule associated fluorescent *Pseudomonas* and rhizobia in suppressing the root rotting fungi and root knot nematode affecting chickpea (*Cicer arietinum* L.). *Pak. J. Bot.* 48: 2139–2145.

Owen, D., Williams, A., Griffith, G.W., and Withers, P.J.A. 2015. Use of commercial bio inoculants to increase agricultural production through improved phosphorus acquisition. *Appl. Soil Ecol.* 86: 41–54.

Ozkoc, I., and Deliveli, M.H. 2001. In vitro inhibition of mycelial growth of some root rot fungi by *Rhizobium leguminosarum* biovar *phaseoli*. *Turk J. Biol.* 25: 435–45.

Patel, H.N., Chakraborty, R.N., and Desai, S.B. 1988. Isolation and partial characterisation of phenolate siderophore from Rhizobium leguminosarum IARI102, *FEMS Microbiol. Lett.* 56: 131.

Philippot, L., Raaijmakers, J.M., Lemanceau, P., and Putten, W.H.V.D. 2013. Going back to the roots: The microbial ecology of the rhizosphere. *Nat. Rev. Microbiol.* 11: 789–799.

Pinochet, J., Calvet C., Camprubi, A., and Fernandez, C. 1996. Interactions between migratory endoparasitic nematodes and arbuscular mycorrhizal fungi in perennial crops: A review. *Plant Soil* 185: 183–190.

Pueppke, S.G., and VanEtten, H.D. 1975. Identification of three new pterocarpans (6a, 11a-dihydro-6H-benzofuro [3,2-c] [1] benxopyrans) from Pisum sativum infected with Fusarium solani f.sp. pisi. *J. Chem. Soc. Perkin Trans.* 1: 946.

Pueppke, S.G., and VanEtten, H.D. 1974. Pisatin accumulation and lesion development in peas infected with Aphanomyces euteiches, Fusarium solani f.sp. pisi or Rhizoctonia solani. *Phytopathology* 64(1974): 1433.

Ragab, M.M.M., Abada, K.A., Abd-El-Moneim, M.L., and Abo-Shosha, Y.Z. 2015. Effect of different mixtures of some bioagents and *Rhizobium phaseoli* on bean damping-off under field condition. *Inter. J. Sci. Eng. Res.* 6(7): 1009–1106.

Rao, S.N., and Kulkarni, S. 2003. Effect of Trichoderma spp. on the growth of Sclerotium rolfsii Sacc. *J. Biol. Control* 17(2): 181–184.

Rioux, C.R., Jordan, D.C., and Rattray, J.B.M. 1986. Iron requirement of Rhizobium leguminosarum and secretion of anthranilic acid during growth on an iron-deficient medium. *Arch. Biochem.* 248: 175.

Safir, G. 1968. The influence of vesicular-arbuscular mycorrhiza on the resistance of onion to *Pyrenochaeta terrestris*. M.Sc. Thesis, University of Iuinois, Urban.

Schreiner, R.P., and Bethlenfalvy, G.J. 1995. Mycorrhizal interactions in sustainable agriculture. *Crit. Rev. Biotechnol.* 15: 271–287.

Schussler, A., Schwarzott, D., and Walker, C. 2001. A new fungal phylum the Glomeromycota, phylogeny and evolution. *Mycol. Res.* 105: 1413–1421.

Secilia, J., and Bagyaraj, D.J. 1987. Bacteria and actinomycetes associated with pot cultures of vesicular-arbuscular mycorrhizas. *Can. J. Microbiol.* 33: 1069–1073.

Sharma, M.P., Gaur, A., and Mukerji, K.G. 2007. Arbuscular mycorrhiza mediated plant pathogen interactions and the mechanisms involved. In: Sharma, M.P., Gaur, A., and Mukerji, K.G. (Eds.), *Biological Control of Plant Diseases.* Haworth Press, Binghamton, NY, pp. 47–63.

Simpfendorfer, S., Harden, T.J., and Murray, G.M. 1999. The in vitro inhibition of *Phytophthora clandestine* by some rhizobia and the possible role of *Rhizobium trifolii* in biological control of *Phytophthora* root rot of subterranean clover. *Aust. J. Agric. Res.* 50: 1469–1473.

Singh, R., Adholega, A., and Mukerji, K.G. 2000. Mycorrhiza in control of soil borne pathogens. In: Mukerji, K.G., Chamola, B.P., and Singh, J. (Eds.), *Mycorrhizal Biology.* Kluwer Academic/Plenum Publishers, New York, pp. 173–196.

Singh, V., Naveenkumar, R., and Muthukumar, A. 2019. Arbuscular mycorrhizal fungi and their effectiveness against soil borne diseases. *Management* 183: 199.

Smith, S.E., and Read, D.J. 2008. Mineral nutrition, toxic element accumulation and water relations of arbuscular mycorrhizal plants. In: *Mycorrhizal Symbiosis*, 3rd Edn., Academic Press, London, pp. 145–148. ISBN-10: 0123705266.

Strack, D., Fester, T., Hause, B., Schliemann, W., and Walter, M.H. 2003. Arbuscular mycorrhiza: Biological, chemical and molecular aspects. *J. Chem. Ecol.* 9: 1955–1979.

Tahat, M.M. 2009. Mechanisms involved in the biological control of tomato bacterial wilt caused by *Ralstonia solanacearum* using arbuscular mycorrhizal fungi. Ph.D. Thesis, Universiti Putra Malaysia.

Thygesen, K., Larsen, J., and Bodker, L. 2004. Arbuscular mycorrhizal fungi reduce development of Pea root-rot caused by Aphanomyces euteiches using oospores as pathogen inoculum. *Eur. J. Plant Pathol.* 110: 411–419.

Toussaint J.P., Kraml, M., Nell, M., Smith, S.E., Smith, F.A., Steinkellner, S., Schmiderer, C., Vierheilig, H., and Novak, J. 2008. Effect of Glomus mosseae on concentrations of rosmarinic and caffeic acids and essential oil compounds in basil inoculated with Fusarium oxysporum f.sp. basilici. *Plant Pathol.* 57: 1109–1116.

Trotta, A., Vanese, G.C., Gnavi, E., Fascon, A., Sampo, S., and Berta, G. 1996. Interaction between the soil-borne root pathogen *Phytophthora nicotianae* Var *parasitica* and the arbuscular mycorrhizal fungus *Glomus mosseae* in tomato plant. *Plant Soil* 185: 199–209.

Vacheron, J., Desbrosse, G., Bouffaud, M.-L., Touraine, B., Moënne-Loccoz, Y., Muller, D., Legendre, L., Wisniewski-Dye, P., and Rigent-Combaret, C. 2013. Plant growth-promoting rhizobacteria and root system functioning. *Front. Plant Sci.* 4: 1–19. Article 356.

Vasileva, V., and Ilieva, A. 2012. *Nodulation and Nitrogen Assimilation in Legumes under Elements of Technology.* Lap Lambert Academic Publishing, Saarbrucken, 120 p.

Veerabhadraswamy, A.L., and Rajkumar, H.G. 2011. Effect of arbuscular mycorrhizal fungi in the management of black bundle disease of Maize caused by Cephalosporium acremonium. *Sci. Res. Rep.* 1: 96–100.

Verma, M., Mishra, J., and Arora, N.K. 2019. Plant growth-promoting rhizobacteria: Diversity and applications. In: Sobti, R.C., Arora, N.K., and Kothari, R. (Eds.), *Environmental Biotechnology: For Sustainable Future.* Springer, Singapore, pp. 129–173.

Vílchez, J.I., Navas, A., González-López, J., Arcos, S.C., and Manzanera, M. 2016. Biosafety test for plant growth-promoting bacteria: Proposed environmental and human safety index (EHSI) protocol. *Front. Microbiol.* 6: 1514.

Vilchez, S., and Manzanera, M. 2011. Biotechnological uses of desiccation-tolerant microorganisms for the rhizoremediation of soils subjected to seasonal drought. *Appl. Microbiol. Biotechnol.* 91: 1297–1304.

Weindling, R. 1934. Studies on lethal principle effective in the parasitic action of *Trichoderma lingorum* on *Rhizoctonia solani* and other soil fungi. *Phytopathology* 24: 1153–1179.

Whipps, J.M. 2004. Prospects and limitations for mycorrhizas in biocontrol of root pathogens. *Can. J. Bot.* 82: 1198–1227.

Xavier, J. C., Patil, K. R., and Rocha, I. (2014). Systems biology perspectives on minimal and simpler cells. Microbiol. Mol. Biol. Rev. 78, 487–509. doi: 10.1128/MMBR.00050-13.

Yao, M., Tweddell, R., and Desilets, H. 2002. Effect of two vesicular-arbuscular mycorrhizal fungi on the growth of micropropagated potato plantlets and on the extent of disease caused by *Rhizoctonia solani.* *Mycorriza* 12: 235–242.

Zaher E.A., Abada, K.A., and Zyton, M.A. 2013. Effect of combination between bioagents and solarization on management of crown-and stem-rot of Egyptian clover. *Amer. J. Plant Sci.* 1(3): 43–50.

5 Agronomical, Physiological, and Biochemical Effects as well as the Changes in Mineral Composition of Crops Treated with Microbial-Based Biostimulants

Iqbal Hussain, Muhammad Arslan Ashraf, Shafaqat Ali,
Rizwan Rasheed, Abida Parveen, Muhammad Riaz,
Ali Akbar, Mudassir Iqbal Shad, and Muhammad Iqbal
Government College University Faisalabad

CONTENTS

5.1 INTRODUCTION

The application of plant biostimulants possesses a significant potential for increasing nutritional quality and yield production of crops. The use of chemical fertilizers to improve crop production notably modifies equilibrium in agroecosystem; however, plant biostimulants reestablish this equilibrium and improve plants' yield production capacity (Woo and Pepe 2018). The substances or microorganisms that can improve nutrient acquisition, abiotic stress tolerance, and crop quality are

DOI: 10.1201/9781003188032-5

referred to as plant biostimulants (du Jardin 2015). The commercial plant biostimulants contain a mixture of beneficial bacteria or fungi alongside chemical elements (Si, Se, Na, Co, and Al), natural oligomers and polymers, plant extracts, seaweeds, amino acids, and fulvic and humic acids (Yakhin et al. 2017). Not all the components of plant biostimulants are biological, and this makes the definition a little ambiguous. The "bio" designation may come due to the living components and natural substances in their formulations. Several non-organic factors may mediate critical biological processes such as morphology, metabolism, plant physiology, and behavior within the agroecosystem (Woo and Pepe 2018).

Better crop productivity is a consequence of alleviating the adverse effects of abiotic stress on plants that mediate growth and development during various phenological stages. Plants cannot achieve their yield potential due to several biotic and abiotic stress factors, and our contemporary knowledge of the mechanisms mediating stress responses is rudimentary. Providing plants with optimized growth conditions, adequate nutrition and water, and plant growth regulators (brassinosteroids, strigolactones, gibberellins, cytokinins, and auxins) may depreciate the intensity of abiotic stress effects on plants. Besides, the integration of biostimulants into these traditional strategies modifies plants' physiological responses and improves yield production under abiotic stress. Over the last two decades, commercial enterprises and the scientific community have paid considerable attention to plant biostimulants gleaned from natural substances. Biostimulants give an innovative strategy to modify plant physiological responses to incite growth, circumvent stress effects, and enhance yield production capacity (Yakhin et al. 2017).

5.2 MICROBIAL STIMULANTS

Plant growth-promoting rhizobacteria (PGPR) are among the crucial constituents of microbial stimulants that improve plant growth. PGPR including *Rhizobium, Azotobacter,* and *Azospirillum,* alongside mycorrhizal fungi, arc an encouraging alternative to ensure better yield production under phosphorus- or nitrogen-deficient conditions and serve as an invaluable means to address some environmental limitations (Ruzzi and Aroca 2015). PGPR and mycorrhizal fungi alter the soil ecosystem by tempering the rhizosphere microbial population qualitatively and quantitatively (Fiorentino et al. 2018). The phytostimulation effect of mycorrhizal fungi and PGPR under suboptimal and optimal growth conditions involves many indirect and direct mechanisms, including (i) generation of enzymes, particularly phosphatases, (ii) stimulation of nutrient transporters, (iii) regulation of phytohormone production (gibberellins, ethylene, cytokinins, ABA, and auxins), (iv) strengthened oxidative defense, (v) enhanced photosynthetic capacity and plant–water relations, (vi) more vigorous root system, and (vii) bettered nutrient acquisition, especially P and N and micronutrients (Grobelak et al. 2015). Microbial biostimulants secrete organic compounds of high molecular weight proteins and mucilage, and the production of phenolics, amino acids, sugars, and organic acids lead to low molecular weight organic compounds. Contrary to agrochemicals, microbial stimulants are human- and environment-friendly as the accumulation of microbial stimulants is less toxic in the long term. There are minimal chances of getting resistant strains of pathogens and pests (Hayat et al. 2010). Microbial biostimulants are gaining more interest from the scientific community since plants harbor a plethora of microbes inside plant tissues (endosphere), above-ground plant biomass (phyllosphere), and surrounding soil associated with roots (rhizosphere). Microbial symbiosis is the most common and fundamental process in plants. Microbial symbiosis coevolves with plants and aids plants in nutrient acquisition, improving plant growth and productivity under environmental limitations (Vandenkoornhuyse et al. 2015). The fossil evidence indicated that microbial association with plants is an ancient phenomenon; the symbiotic relationship of arbuscular mycorrhiza with plant roots might have performed critical roles in plants' adjustment to the terrestrial environment (Selosse and Le Tacon 1998). Secondary metabolites produced by microorganisms function as plant biostimulants, insecticides, and

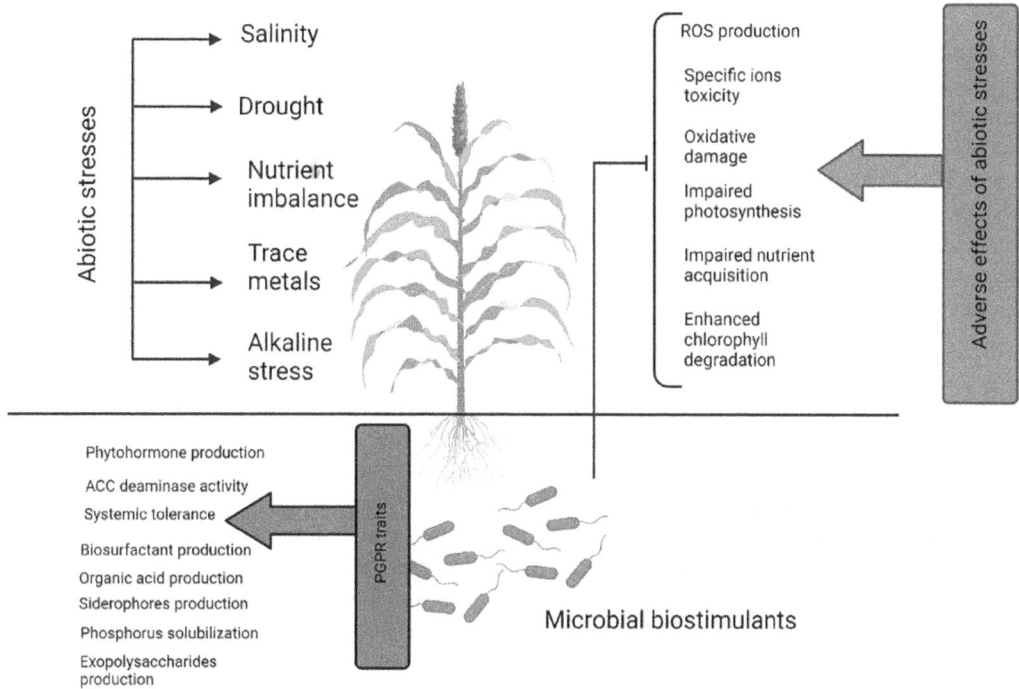

FIGURE 5.1   Effect of microbial bistimulants on photosynthetic pigments, photosynthesis, ROS metabolism, oxidative defense and nutrient acquisition in plants under abiotic stresses.

herbicides. PGPR include endophytic bacteria, free-living bacteria, and root-colonizing bacteria. Soil associated with plant roots shelters free-living bacteria, root-colonizing bacteria exist in the rhizoplane, and endophytic bacteria live within roots. However, these demarcations are not uniform because individual bacteria may adopt all three modes of life depending upon the soil and environmental conditions alongside the host roots (Ruzzi and Aroca 2015). Hence, PGPR are bacteria that inhabit the rhizoplane and the rhizosphere, showing tremendous promise for bettered plant growth and development. There are diverse modes of action shown by PGPR to mediate plant responses under natural environmental conditions. For instance, hormone production, volatile compounds biosynthesis, and bettered nutrient acquisition are how PGPR improve functions as plant biostimulants (Calvo et al. 2014). A schematic representation of microbial biostimulant-mediated stress tolerance is given in Figure 5.1. PGPR functioning as biostimulants improve plant growth via direct or indirect methods. These methods adopted by microbial stimulants are discussed as under.

5.3   PGPR-MEDIATED IMPROVEMENT IN PLANT GROWTH: MECHANISM OF ACTION

Microbial biostimulants can improve plant growth through direct mechanisms that entail bettered nutrient acquisition and phytohormone production. The increase in nutrient acquisition brought by these biostimulants is due to higher nitrogen fixation, mineral nutrients solubilization, and mineralization of organic compounds (Grobelak et al. 2015). Such methods directly impact plant growth and development, and growth-promoting effects vary with microbial strains and plant species. Microbial biostimulants bring a significant increase in nutrient availability around roots surface, consequently improving mineral uptake by plants.

5.4   NUTRIENT ACQUISITION

Microbial biostimulants hold the potential to heighten the nutrient availability and concentration by mediating their supply for plant growth and yield production (Kumar et al. 2016). Plants require nitrogen for important metabolic events in the form of ammonium (NH_4^+) and nitrate (NO_3^-), which are absorbed from the soil. Soil nitrification produces nitrate, which is the predominant form of nitrogen absorbed by plants (Xu et al. 2012). Some microbial biostimulants can solubilize phosphorus, enhancing soil phosphorus contents that plants readily absorb (Widdig et al. 2019). In this context, *Kocuria Turkanensis* 2M4 obtained from soil that adhered to roots surface manifested phosphorus solubilization, phytohormone production, and siderophore production in several plant species (Goswami et al. 2014).

5.5   NITROGEN FIXATION

Biological nitrogen fixation accounts for over two-thirds of all nitrogen fixed globally. The interaction of microbes with plants, either symbiotic or non-symbiotic, carries out the fixation of atmospheric nitrogen to usable forms (nitrates and ammonium). The introduction of a couple of rhizobacterial strains into the soil improves soil quality alongside bettered nodule formation (Ahemad and Kibret 2014). Biological nitrogen fixation is due to the nifH gene that activates other structural genes mediating iron proteins, oxidation, biosynthesis of iron-molybdenum cofactor, and several additional genes essential for enzyme activity (Reed et al. 2011). The growth-promoting activity due to microbial biostimulant inoculation significantly contributed to growth improvement, biocontrol, and bettered soil nitrogen contents (Gouda et al. 2018).

5.6   PHOSPHORUS SOLUBILIZATION

Phosphorus (P) is among the essential nutrients needed by the plants. Soil is a big reservoir of phosphorus. However, only a very small amount of it is available to plants. The limited availability of phosphorus to plants despite the soil being a big phosphorus reservoir is due to the presence of P in insoluble form. The soluble forms of P available to plants are dibasic $\left(HPO_4^{2-}\right)$ and monobasic (H_2PO^{4-}) (Bhattacharyya and Jha 2012). The addition of microbial biostimulants to soil makes P in available soluble forms via secretion of protons and organic acids, acidification (Widdig et al. 2019), chelation, and exchange reactions (Nath et al. 2017). Chelation-mediated phosphorus solubilization is brought by microbial stimulants (fungi and saprophytic bacteria). The secretion of organic acids in root exudates also alters P levels in soil. In some cases, phosphate starvation also induces P solubilization. The microbial stimulants able to solubilize P include *Enterobacter, Burkholderia, Beijerinckia, Bacillus, Azotobacter,* and *Azospirillum* (Liu et al. 2020). Bettered growth and yield due to rhizobacteria hold positive correlation with enhanced solubilization of P sources. Additionally, most widely reported P-solubilizing bacterial strains displaying association with different crop plants include *Azotobacter chroococcum, Rhizobium leguminosarum, P. putida, Pseudomonas chlororaphis, Enterobacter agglomerans, Bradyrhizobium japonicum, Cladosporium herbarum,* and *Bacillus circulans*. Soil microorganisms synthesizing organic acids show significant potential for P solubilization from inorganic phosphorus sources (Bhattacharyya and Jha 2012).

The two most common mechanisms of PGPR for plant growth promotion include the mineralization of phosphorus alongside its solubilization, thereby improving nutrient supply to the host plant (Glick 2012). PGPR were involved in each of the three main components of the soil P cycle: mineralization–immobilization, sorption–desorption, and dissolution–precipitation. Phosphorus-solubilizing bacteria (PSB), for example, dissolve a variety of relatively insoluble phosphorus sources such as zinc phosphate (Saravanan et al. 2007) and calcium phosphate (Rodriguez et al. 2004) by adjusting the soil pH and rendering P obtainable for plants. The presence of PSB in the soil solution improves P availability between the overall soil P and the existing soil P levels through solubilization and mineralization, immobilization of P, and production of rare P types of both organic and inorganic nature. PSB usually convert unavailable P to usable forms through many processes that include acidification, chelation, exchange reactions, proton and organic acid secretion, bettered ACC deaminase activity, siderophore production, and IAA generation. PSB can also mineralize organic phosphates (Nath et al. 2017).

5.7 ORGANIC ACIDS PRODUCTION

The secretion of inorganic acids and organic and proton excretion are established tools of PSB-mediated P solubilization. NH_4^+ assimilation carried out by PSB results in H^+ excretion (Nacoon et al. 2020). The primary process for enhancing P solubilization in some microorganisms appears to be NH_4^+-driven proton release. Microbial metabolism primarily accomplished via fermentation of organic compounds or oxidative phosphorylation produces organic acids, including citric acid, lactic acid, tartaric acid, aspartic acid, and oxalic acid (Wei et al. 2018). PGPR such as IAA producers may also increase the amount of root exudates. Exudates from the roots contain several organic nutrients, including mucilage, nucleosides, amino acids, vitamins, sugars, phytosiderophores, and organic acids. These root exudates function as signals for microorganisms, especially for PSB for metabolism of root exudates, and multiply in this niche (Sharma et al. 2013). PSB- and plant-produced organic acids lower the rhizosphere pH, allowing precipitated P forms to dissolve more easily. PSB-produced organic anions substitute phosphate on the clay soil surface, or PSB may also compete for sites of phosphate fixation (Etesami and Adl 2020). They may also improve cations (Ca^{2+}, Fe^{3+}, and Al^{3+}) chelation or soluble complexes formation with metal ions coupled with the insoluble form of phosphorus, preventing phosphate precipitation, thereby releasing P (Wan et al. 2020). The monovalent anion phosphate $\left(H_2PO_4^-\right)$ is the primary soluble inorganic phosphate type at lower pH levels. Besides, the increase in soil pH produces HPO_4^{2-} and HPO_4^{3-}. PGPR produces organic acids which acidify the environment, consequently releasing P ions from phosphate minerals. This reaction occurs due to the substitution of protons with those of cations associated with phosphate minerals (Etesami and Adl 2020). The introduction of PGPR in alkaline or neutral soils decreases soil pH due to the production of acids, allowing calcium phosphates and apatites to dissolve more easily (Widdig et al. 2019).

5.8 INORGANIC ACIDS PRODUCTION

PSB produce inorganic acids such as HCl, which help to dissolve insoluble phosphates (Li et al. 2019). Bacteria of genera *Thiobacillus* and *Nitrosomonas,* alongside other bacteria, produce sulfuric, nitric, and carbonic acids that decompose organic residues to dissolve phosphate compounds (Etesami and Adl 2020). As a result, *Thiobacillus* inoculation alongside elemental sulfur can increase P solubility and plant biomass (Stamford et al. 2003). On the other hand, inorganic acids are less effective than organic acids at dissolving insoluble PO_4^{3-} (Wei et al. 2018). The other means is the synthesis of H_2S that yields phosphate and ferrous sulfate upon reaction with ferric phosphate. Overall, acidification is not the only mechanism by which phosphate-solubilizing PGPR can solubilize insoluble phosphates, as the potential to abridge pH in certain instances is unrelated to the P solubilization potential.

5.9 BETTERED ACC DEAMINASE ACTIVITY AND INDOLE ACETIC ACID (IAA) GENERATION

Increased root growth through either ACC deaminase activity or IAA production and the extension of existing root systems are all ways by which PGPR can improve the plants' ability to acquire P from the soil (Richardson et al. 2009). ACC deaminase degrades ethylene precursor to impact plant root development. Ethylene production at a high level in a plant can stifle root growth. ACC deaminase enzyme indirectly controls P uptake of roots by modulating the root architecture (Etesami and Adl 2020). Bacterial IAA can boost root exudates and strengthens the roots architecture. Organic acids (citric acid and gluconic acid) in root exudates cause the rhizosphere to become acidic (Dakora and Phillips 2002). Plant growth can be aided by an acidic rhizosphere that mobilizes nutrients such as Fe. IAA mediates the generation of root exudates and strengthens roots. The root exudates contain organic acids that acidify the rhizosphere and offer a reducing environment needed for transforming Fe^{3+} to Fe^{2+}, indicating the possible involvement of IAA-producing PGPR in Fe solubilization and improving plant Fe uptake (Etesami and Adl 2020). The rhizodeposits are rich in phenolics, amino acids, organic acids, secondary metabolites, proteins, and mucilage (Badri and Vivanco 2009). The rhizosphere will experience steep redox gradients due to O_2 consumption in roots and related microflora (Etesami et al. 2014). Similarly, chelating agents such as organic acids found in exudates of microorganisms and roots can chelate Fe^{3+} and allow access to plant roots (Etesami and Adl 2020).

5.10   BETTERED MANGANESE (MN) AND IRON (FE) BIOAVAILABILITY

Manganese (Mn) is a micronutrient needed for several processes in plants. For example, Mn is involved in root development and photosynthesis. Besides, Mn also functions as an enzyme component. Mn is bioavailable as free Mn^{2+}, but has a poor solubility as oxides in the soil. Chemical reactions and microbial activity influence the balance of Mn in different soil forms. Mn solubility and availability in soil is significantly reduced by high soil pH. As a result, Mn deficiency is more probable to appear in limed or alkaline soils. Mn is an important nutrient, and soil Mn availability is influenced by two major factors: pH and redox state. In oxidized soils, Mn is found in the low-soluble mineral pyrolusite in its oxidized form, Mn^{4+}. Some PGPR strains also improve this element's bioavailability (Etesami and Adl 2020). Iron (Fe) is a critical cofactor for several enzymatic reactions and is important nutrient for plants. It is needed for many cellular functions and fundamental physiological processes, including photosynthesis and respiration. Fe is often in Fe^{3+}, which interacts to form highly insoluble oxyhydroxides and hydroxides that are practically inaccessible to microorganisms and plants. High soil pH decreases Fe supply, while acidic soil conditions increase it. The Fe^{3+} activity decreases 1000-fold as pH rises by one unit. Fe is abundant in most soils, but it is mostly in types that are unavailable to plants. Despite the vast amount of Fe in soils, microbes and plants have limited access to it. Microorganisms and plants have established successful Fe absorption strategies. Iron mobilization is needed for Fe uptake by roots, which can be accomplished in two ways in the plant kingdom. Strategy I and strategy II are the names of these two strategies. Except for grasses, rhizosphere in all plant species (dicotyledonous and monocotyledonous) gets acidified due to organic production in strategy I. Also, iron chelate reductase enzyme converts Fe(III) to Fe(II) by transforming Fe^{3+} to Fe^{2+}(II). Fe^{2+} can then be absorbed by Fe(II) transporter in the membrane. On the other hand, in strategy II, grasses deal with Fe deficiency by producing and secreting siderophores, which are then taken up through Fe^{3+} siderophore transporter in root plasmalemma (Hartmann et al. 2009; Altomare and Tringovska 2011; Etesami and Adl 2020).

5.11   SIDEROPHORES PRODUCTION

In reaction to insufficient iron levels in the rhizosphere, most plant-associated bacteria can create iron chelators termed siderophores. The low molecular weight organic compounds with significant affinity to bind certain elements such as metal ions and Fe^{3+}, thereby improving their availability, are referred to as siderophores (Ghazy and El-Nahrawy 2021). Several reports indicate enhanced uptake of Fe by plants alongside contemporaneous plant growth increase (Kalyanasundaram et al. 2021). Numerous investigations have shown that plants can use microbial siderophores as a Fe source using strategies I and II (Robin et al. 2008; Etesami and Adl 2020). According to earlier research, plants cannot uptake/absorb bacterial siderophores, and hence plants acquire Fe via the reduction-based method (Sabet and Mortazaeinezhad 2018). Specific Fe–siderophore membrane receptors transfer ferric siderophores into cells, permitting siderophore release and reuse (Etesami and Adl 2020). The ability of siderophores to deliver Fe^{3+} to root surfaces and intracellular regions of root cells has been discovered to be an essential feature of these chelating agents. As a result, the greater Fe^{3+} ion concentrations available to root phytosiderophores improve their later uptake by plants (Ferreira et al. 2019). Another possibility for the delivery of Fe through siderophores has been demonstrated: ligand exchange (Etesami and Adl 2020). Fe provided through siderophores undergoes a ligand exchange reaction with phytosiderophores before being taken by plants. Therefore, siderophores perform an indirect role in Fe update. This idea confirms the indirect role of siderophore in Fe uptake (Etesami and Adl 2020). Many plants thrive at very low Fe concentrations, especially in contrast to most microorganisms. Several plant species absorb Fe in the presence of microorganisms that have the potential to produce siderophores. Aside from having a strong affinity for Fe(III) ions, siderophores could also form bivalent ion complexes that the plant can absorb (Ferreira et al. 2019).

5.12 DEVELOPMENT OF ABIOTIC STRESS TOLERANCE IN PLANTS BY MICROBIAL STIMULANTS

The rhizosphere is the soil zone surrounding plant roots, where reactions between the root, soil, and biotic and abiotic elements can occur and are mediated by the plant roots. The rhizosphere soil has the highest microbial population due to the organic carbon in the form of root exudates generated by plant roots providing nourishment for microbial growth (Goswami and Deka 2020). These bacteria can interact with plant roots in various ways, including positive, negative, and neutral interactions, successfully impacting plant growth and development (Basu et al. 2021). Actinomycetes, fungi, bacteria, protozoa, and algae are just a few of the organisms that have colonized plant roots. The application of these bacteria aided plant growth and development, which has been well documented in the literature. In the rhizosphere soil, bacteria are the most abundant and dominant microbiota (Oleńska et al. 2020). The number and composition of root exudates and the soil state determine the specificity of microbial interactions. Exudates from plant roots serve as signals for rhizobacteria to approach the root surface and interact via a process known as chemotaxis (Subiramani et al. 2020). PGPR are a wide range of bacterial communities that live in the zonal root region. *Azospirillum, Azotobacter, Enterobacter, Arthrobacter, Bacillus, Serratia, Streptomyces,* and *Pseudomonas,* spp. are among the PGPR. PGPR are classified as free-living (extracellular) or symbiotic (intracellular) bacteria depending on their interaction with plants. PGPR are connected with diverse plants and have several positive effects, including improved plant development and lowered vulnerability to plant pathogens (Noreen et al. 2012). PGPR mechanisms in plant growth promotion include bettered root hairs, greater leaf area, effective seed germination, increased root hairs, phytohormone production, higher nutrient availability, and improved biomass (Kour et al. 2019). Certain investigations have shown that PGPR facilitate phytohormone production, biocontrol agent generation, and improved plant nutrient consumption (Basu et al. 2021). The mechanisms by which PGPR promote plant growth have been elucidated in several studies. PGPR produce phytohormones, nutrient immobilization, organic acids, and siderophores to impact plant growth under environmental constraints (Basu et al. 2021). The modern agriculture often encounters biological (living) and abiotic (non-living) stress factors in the natural environmental conditions. Herbivores (nematodes and insects), parasitic weeds, and plant pathogens (fungi, bacteria, and viruses) are the main biotic stresses that have reduced crop yields significantly. Abiotic stress factors such as cold, heat, drought, and excess water, on the other hand, produce a significant drop in agricultural output, contributing to a yield decrease of more than 50% (Goswami and Deka 2020). Salt stress and drought create oxidative stress, producing biochemical changes and metabolic changes that impair crop output (Abdallah et al. 2016; Abbas et al. 2018). PGPR treatment has been shown to improve plant tolerance to abiotic stresses in several studies, and it can cause "induced systemic resistance" in plants to drought and salt conditions (Goswami and Deka 2020). Furthermore, PGPR increase the absorption of nutrients from the soil, resulting in less fertilizer (nitrates and phosphates) buildup in the cultivated area.

5.13   PGPR-MEDIATED DROUGHT TOLERANCE

Plant growth characteristics and stress-responsive genes are affected by drought stress. To adapt to adverse climatic circumstances, plants change their physiological and morphological features (Ullah et al. 2019). Because of their direct interaction with soil, water, and nutrients, plant roots are critical for perceiving and reacting to various external environmental stimuli. The root system senses stress in the soil and uses abiotic and biotic inputs to regulate and change its genetic program for post-embryonic root growth. Plant roots change their morphology in response to changes in soil surface moisture; that is, water deficits lead to deeper root penetration, whereas increased soil moisture reduces root infiltration (Zhang et al. 2017). Long-term drought causes rhizosphere-colonizing bacteria to release different antioxidants, extracellular polymeric compounds, osmolytes, and phytohormones, which cause root morphological changes and improve stress tolerance. Drought stress relief largely relies on bacterial ABA. PGPR have been shown to be effective in boosting the synthesis of ABA and minimizing the impacts of abiotic stress in the soil in a few studies. PGPR have the potential to help plants cope with drought stress. These microbes colonize the rhizosphere/endorhizosphere of plants and confer drought tolerance by producing phytohormones, exopolysaccharides, volatile compounds, osmolytes accumulation, bettered ACC deaminase activity, strengthened antioxidant system, and altered root morphology (Vurukonda et al. 2016). The term "induced systemic tolerance" (IST) was developed to describe chemical and physical changes in plants caused by microorganisms that result in increased drought tolerance. Plant root systems are home to millions of microorganisms that create a complex ecological community that promotes plant development and production through plant interactions and metabolic activities (Vurukonda et al. 2016). Alterations in the structure of plant-associated microbial communities in root area accumulated in response to different environmental constraints help improve plants' tolerance to drought (Schmidt et al. 2014). Similarly, a significant alteration in the distribution of bacteria was seen in the soil surrounding the root, rhizosphere, and endosphere of pepper plants under drought conditions (Marasco et al. 2012). Likewise, pepper plants inoculated with isolates from deserts manifested a notable tolerance to water scarcity over control plants. The plant potential for uptake was improved due to bettered root system in inoculated plants under drought (Marasco et al. 2013). PGPR have been hypothesized to produce drought resistance in plants by eliciting the so-called rhizobacterial-induced drought endurance and resilience (RIDER) process, which involves various biochemical and physiological changes. It contains changes in the phytohormonal content and antioxidant system. To maintain plant life in the drought-stressed area, PGPR also produce osmolytes and exopolysaccharides (EPS). Drought tolerance is conferred in plants via the formation of heat shock proteins (HPs), volatile organic compounds (VOCs), and dehydrins (Kaushal and Wani 2016). A detailed survey of literature on microbial biostimulant-mediated drought tolerance in plants is given in Table 5.1.

5.14   PGPR ROLE IN SALINITY TOLERANCE

One of the biggest limits on global agricultural productivity is salinity stress. Changes in soil physicochemical qualities lead to irreversible land degradation in salt-stressed areas, resulting in a massive reduction in world food output. Furthermore, prolonged stress causes significant

TABLE 5.1

Effect of Plant Growth-Promoting Rhizobacteria (PGPR) on Plant Growth and Development under Drought Conditions

Plant Species	Drought Conditions	Growth Medium	PGPR Strains	Plant Responses	Reference
Triticum aestivum L.	10% Polyethylene glycol (PEG)	Hydroponics	*Bacillus subtilis* (LDR2), *Dietzia natronolimnaea* (STR1), *Arthrobacter protophormiae* (SA3)	Wheat plants inoculated with PGPR strains manifested significant tolerance to PEG-induced drought under hydroponics conditions. PGPR strains improved photosynthesis and indole acetic acid (IAA) production under drought conditions. The rise in 1-aminocyclopropane-1-carboxylate (ACC) and abscisic acid (ABA) was remarkably counteracted by LDR2 and SA3 PGPR strains, while STR1 did not alter these parameters.	Barnawal et al. (2017)
Mentha piperita L.	100%, 50%, and 35% field capacity	Vermiculite	*Bacillus amyloliquefaciens* GB03 and *Pseudomonas fluorescens* WCS417r	Peppermint plants were inoculated with GB03 and WCS417r PGPR strains to relieve stress injury under mild and severe drought conditions. A notable drop in fresh weight, leaf area, and leaf number was present in plants in response to drought. However, PGPR inoculation remarkably circumvented the negative impacts of drought on growth characteristics. Likewise, PGPR inoculation enhanced phenolics levels, proline and enzymatic antioxidant activities under drought. Further, drought-brought higher lipid peroxidation was also minimal in plants with PGPR inoculation.	Chiappero et al. (2019)
Triticum aestivum L.	Water withholding	Soil	*Klebsiella* sp. (IG 3), *Enterobacter ludwigii* (IG 10), and *Flavobacterium* sp. (IG 5)	These PGPR strains were tested for drought tolerance in nutritional broth supplemented with varying doses (0%–25% PEG) of polyethylene glycol. Under drought stress, the effect of PGPR inoculation on numerous biochemical and physiological parameters and gene expression of stress-sensitive genes was investigated. Drought stress had a substantial impact on growth indices, membrane integrity, water status, expression of stress-responsive genes, and osmolyte accumulation in wheat, all of which were favorably influenced by PGPR inoculation. PGPR strain IG 3 was most effective in impacting the physiological and biochemical processes of seedlings under drought.	Gontia-Mishra et al. (2016)

(Continued)

TABLE 5.1 (*Continued*)

Effect of Plant Growth-Promoting Rhizobacteria (PGPR) on Plant Growth and Development under Drought Conditions

Plant Species	Drought Conditions	Growth Medium	PGPR Strains	Plant Responses	Reference
Pisum sativum L.	Withholding water to 40% moisture content	Soil	*Rhizobium pisi*, DSM 30132 strain	PGPR inoculation significantly alleviated the adverse drought effects on pea by improving growth characteristics and stomatal conduction.	Bashir et al. (2020)
Platycladus orientalis L.	70% Field capacity	Peat, perlite, and vermiculite mixture	*Bacillus subtilis*	*Bacillus subtilis* was studied for its effects on plant growth, hormone production, and drought tolerance under water-stressed conditions. PGPR strain improved leaf water potential and relative water content under drought. Because of PGPR inoculation, root exudates such as organic acids, amino acids, and sugars rose dramatically regardless of water stress levels. Stomatal conduction was also maximal in plants treated with PGPR.	Liu et al. (2013)
Zea mays L.	50 and 30% field capacity	Soil	*Leclercia adecarboxylata, Achromobacter xylosoxidans, Enterobacter cloacae,* and *Pseudomonas aeruginosa*	PGPR strains gave a significant accretion in carotenoids, chlorophyll, and stomatal conductance under drought conditions.	Danish et al. (2020)
Zea mays L.	Limited amount of irrigation	Soil	*Bacillus cereus, B. amyloliquefaciens, B. mojavensis, B. subtilis* subsp. *subtilis, B. safensis, B. altitudinis, Lysinibacillus xylanilyticus,* and *Paenibacillus peoriae*	Drought significantly decreased plant growth characteristics in maize plants. Under drought stress, PGPR greatly enhanced root architecture, chlorophyll, stem diameter, and plant height in maize plants. The findings show that PGPR inoculation can promote plant growth and root morphology through the production of plant growth regulators, bettered water, and nutrient acquisition.	Lin et al. (2020)
Zea mays L.	Withholding water	Soil	*Alcaligenes faecalis* (AF3), *Pseudomonas aeruginosa* (Pa2), and *Proteus penneri* (Pp1)	Drought caused a significant reduction in plant biomass, leaf area, and shoot length. Plants inoculated with PGPR manifested bettered antioxidant enzyme activities, sugar content, proteins, relative water content, and proline under drought.	Naseem and Bano (2014)

(Continued)

TABLE 5.1 (Continued)
Effect of Plant Growth-Promoting Rhizobacteria (PGPR) on Plant Growth and Development under Drought Conditions

Plant Species	Drought Conditions	Growth Medium	PGPR Strains	Plant Responses	Reference
Cicer arietinum L.	Water withholding	Soil	*Bacillus megaterium, Bacillus thuringiensis,* and *Bacillus subtilis*	Drought-sensitive (Punjab Noor-2009) and drought-tolerant (93127) cultivars inoculated with PGPR produced bettered biomass; relative water content; photosystem II efficiency; and accumulation of phenolics, sugar, and proteins.	Khan et al. (2019)
Zea mays L.	75% Field capacity	Soil	*P. monteilii strain WAPP53, P. syringae strain GRFHYTP52, P. putida strain GAP-P45, P. stutzeri strain GRFHAP-P14,* and *P. entomophila strain BV-P13*	PGPR inoculation enhanced antioxidant status, improved osmotic adjustment, and growth in maize under drought. PGPR inoculation reduced electrolyte leakage compared with non-inoculated plants under drought.	Sandhya et al. (2010)
Setaria italica L.	100%, 75%, and 50% field capacity	Soil	*Penicillium sp. EU-FTF-6*	Plants inoculated with drought-tolerant phosphorus-solubilizing strains notably enhanced the accumulation of sugars, proline, and glycine betaine. Further, PGPR inoculation also reduced lipid peroxidation under drought. Drought-tolerant P-solubilizing bacteria may promote plant development and reduce drought stress in crops growing in water-stressed environments.	Kour et al. (2020)
Mentha pulegium L.	70% and 40% field capacity	Soil	*Azospirillum brasilense (Ab)* and *Azotobacter chroococcum (Ac)*	Inoculation with PGPR considerably decreased the adverse effects of water stress on pennyroyal secondary metabolite synthesis and physio-biochemical properties. Besides, PGPR inoculation enhanced the production of oxygenated monoterpenes, flavonoids, phenolics, soluble sugars, and proteins. DPPH radical scavenging activity was maximal in plants given PGPR inoculation.	Asghari et al. (2020)
Sambucus williamsii Hance	70%–75% and 40%–45% field capacity	Soil	*Acinetobacter calcoaceticus X128*	Plants inoculated with PGPR manifested higher photosynthesis, stomatal conductance, and total chlorophyll contents under drought conditions. As a result, inoculating plants with X128 in dry environments can reduce chlorophyll loss, postpone the limitation of photosynthesis in plant leaves caused by non-stomatal factors, and make it easier for leaves to regain their photosynthetic activities if water supply is restored.	Liu et al. (2019)

changes in plant morphological and physiological characteristics, culminating in osmotic imbalances and nutrient cytotoxicity (Abdelaal et al. 2020). It has an impact on nearly every stage of plant development, including reproductive development, vegetative growth, photosynthesis, and germination. Photosynthesis rate is lowered under salt stress due to a fall in water potential, resulting in a considerable reduction in the supply of carbohydrates essential for cell development. When large concentrations of Cl and Na accumulate in chloroplasts, photosynthesis is hindered, resulting in a lower chlorophyll content in plants. Reduced photosynthesis rates can also be caused by a decline in carotenoids, chlorophyll, photosynthetic enzymes, photophosphorylation activity, electron transport, and photosystem II (PSII) activity in addition to the aforementioned reasons. Furthermore, salinity impairs photorespiration, causing alterations in cell homeostasis. Plants' physiological and metabolic activities are affected when ion homeostasis is disrupted (Goswami and Deka 2020). Singh and Jha (2016) studied the impact of CDP-13 bacterial isolate from *Capparis decidua* on alleviating the deleterious effects of salinity on plants. The bacterial isolate was identified as *Serratia marcescens* based on 16S rRNA gene sequence analysis. The bacterial isolate showed significant potential for ammonia production, nitrogen fixation, IAA production, siderophore production, phosphate solubilization, and ACC deaminase activity. CDP-13 bacterial strain was able to survive at 6% NaCl concentration in the growth medium. The inoculation of CDP-13 to wheat plants improved growth under 150–200 mM salinity. The bacterial strain CDP-13 modulated the accumulation of osmoprotectants, indicating the significant role of this bacterial strain in the osmotic adjustment of plants under salinity. The literature review of microbial biostimulant-mediated salinity tolerance in different plant species is given in Table 5.2.

5.15 HEAVY METAL TOLERANCE IN PLANTS DUE TO MICROBIAL BIOSTIMULANTS

Biosphere contamination is a serious environmental issue that has rapidly been expanding since the beginning of the industrialization age and contemporary lifestyle. Various industrial operations continually release a wide range of organic (solvents, VOCs, and hydrocarbons) and inorganic (heavy metals) contaminants into the environment (Manoj et al. 2020). Heavy metals, among the contaminants, play a critical role in the environment owing to their mutagenicity and toxicity. Aluminum (Al), zinc (Zn), mercury (Hg), chromium (Cr), copper (Cu), cadmium (Cd), and lead (Pb) are the most frequent toxic heavy metals (Etesami and Maheshwari 2018). These heavy metals are classified as a "priority contaminants" by the United States Environmental Protection Agency due to their mutagenic and carcinogenic properties. As a result, removing these heavy metal contaminations is a necessary step in maintaining a stable environment. Many physiochemical approaches have been used in the past to clear up metal pollution. The majority of these methods are expensive, produce secondary pollutants, and change the soil's biological character (Ashraf et al. 2017). As a result, adopting a more environmentally friendly strategy is widely preferred. Phytoremediation is the most commonly acknowledged biological technique because of its environment-friendliness and economic feasibility (Jacob et al. 2018). However, various constraints hindered the phytoremediation efficacy of the plants, including low biomass production at elevated metal levels, sensitivity to multi-metals, and a shallow root structure (Kumar Yadav et al. 2018). As a result, increasing the plant's phytoremediation efficacy at greater metal concentrations is a viable strategy, which may be achieved by using PGPR. When PGPR are employed as a bioinoculant, they boosted plant biomass and root development through stabilizing soil structure, nutrient recycling, and modulating heavy metal toxicity and bioavailability (Ashraf et al. 2017). PGPR-mediated improvement in plant metal stress tolerance is given in Table 5.3.

TABLE 5.2

PGPR-Mediated Salinity Tolerance in Plants

Plant Species	Salinity Level	PGPR Strain	Plant Responses	Reference
Phaseolus vulgaris L.	25 mM NaCl	Paenibacillus sp. and Aneurinibacillus aneurinilyticus	The bacterial isolates showed significant ACC deaminase activity; indole acetic acid production capacity; and hydrogen cyanide, ammonia, and siderophore production. Besides, these PGPR strains also manifested bettered $ZnSO_4$ solubilization. PGPR strains significantly enhanced growth characteristics under saline conditions.	Gupta and Pandey (2019)
Glycine max L.	50 and 75 mM NaCl	Bradyrhizobium japonicum USDA 110 and Pseudomonas putida TSAU1	The authors performed hydroponics experiment to examine the effect of PGPR on root architecture, proteins, growth, and phosphorus and nitrogen uptake under salinity. The maximal values for soluble proteins, phosphorus and nitrogen contents were seen in plants inoculated with both PGPR strains under salinity.	Egamberdieva et al. (2017)
Vigna radiata L.	9 dS m^{-1} NaCl	Bacillus cereus	Inoculation of plants with PGPR strain resulted in a significant growth improvement in plants under salinity. Stress-induced chlorophyll degradation and oxidative damage were mitigated by PGPR treatment. The inoculated plants showed higher phosphorus, nitrogen, potassium, and proline accumulation alongside a significant decline in sodium contents under salinity. The activities of antioxidant enzymes were also higher in plants inoculated with PGPR under salinity. PGPR inoculation also altered soil biological activity, which was reflected in the form of increased dehydrogenase activity, microbial biomass carbon, alkaline phosphate, total organic carbon, and available phosphorus content.	Islam et al. (2016a)
Zea mays L.	100 mM NaCl	Pseudomonas and Rhizobium species	The effect of PGPR inoculation on stress tolerance alleviation in maize was evaluated. Two maize cultivars, namely Av 4001 and Agaiti-2002, were used in the experiment. Plants were given PGPR treatments at the seedling stage. Maize plants manifested detrimental salinity effects on growth and development. However, microorganisms inoculation improved plant growth by decreasing electrolyte leakage, regulating osmotic adjustment, and improving relative water content and potassium uptake under saline conditions.	Bano and Fatima (2009)

(Continued)

TABLE 5.2 (Continued)
PGPR-Mediated Salinity Tolerance in Plants

Plant Species	Salinity Level	PGPR Strain	Plant Responses	Reference
Mentha arvensis L.	0, 100, 300, and 500 mM NaCl	*Exiguobacterium oxidotolerans, Halomonas desiderata,* and *Bacillus pumilus*	Salinity induced a notable drop in growth characteristics, yield, and oil content. However, inoculation of bacterial strains alleviated the adverse effects of salinity.	Bharti et al. (2014)
Triticum aestivum L.	150 mM NaCl	*Dietzia natronolimnaea* STR1	The authors reported salinity tolerance in inoculated plants due to modulation in SOS pathway, antioxidant machinery, and ABA signaling.	Bharti et al. (2016)
Oryza sativa L.	150 mM	*Bacillus megaterium* and *Pseudomonas aeruginosa*	When inoculated with PGPR strains, salt-tolerant Dhandi and salt-sensitive GR 11 rice cultivars manifested a significant increase in salinity tolerance.	Jha and Subramanian (2015)
Oryza sativa L.	0.5, 1.0, 1.5, 2.0 and 2.5 g NaCl kg⁻¹ soil	*Pseudomonas pseudoalcaligenes* and *Bacillus pumilus*	Inoculation of rice cultivar (GJ-17) with *Pseudomonas pseudoalcaligenes* resulted in a significant increase in glycine betaine accumulation and bettered biomass production under mild salinity levels. However, inoculation with both PGPR strains gave better stress tolerance at higher salinity levels. However, proline accumulation was lower in inoculated plants compared with non-inoculated plants under saline conditions.	Jha et al. (2011)
Hordeum maritimum L.	200 mM NaCl	*Bacillus mojavensis, B. pumilus,* and *Pseudomonas fluorescens*	Salt-tolerant (Kerkna) and salt-sensitive (Rihane) suffered a significant reduction in growth characteristics, chlorophyll, and shoot water content. The inoculation of plants with PGPR resulted in bettered salinity tolerance. Besides, improvement in salinity tolerance due to PGPR was better in sensitive barley cultivar than susceptible cultivar under stress or non-stress conditions.	Mahmoud et al. (2017)
Zea mays L.	−0.2, −0.6, −1.0, and −1.6 MPa NaCl	*Azospirillum spp.*	Two maize cultivars with contrasting salinity tolerance, namely cv. 323 (salt-sensitive) and cv.324 (salt-tolerant), were inoculated with *Azospirillum spp.* under salinity. The authors recorded an improvement in salinity tolerance due to regulation in ions homeostasis, osmotic adjustment, and nitrogen metabolism under saline conditions in plants treated with PGPR.	Hamdia et al. (2004)

TABLE 5.3

Microbial Stimulants-Mediated Improvement in Phytoremediation through the Reduction in Metal Phytotoxic Effects on Plants

Plant Species	Type of Heavy Metals	Microbial Strain	Plant Response	Reference
Triticum aestivum L.	Pb, Mn, Cd, Co, Cr, and Cu	*Pseudomonas moraviensis* and *Cenchrus ciliaris*	The PGPR inoculation reduced the harmful effects of heavy metals, and the inclusion of carriers further aided the PGPR.	Hassan et al. (2017)
Zea mays L.	Cr	*Leclercia adecarboxylata* and *Agrobacterium fabrum*	The inoculated plants had a notable improvement in plant leaves, shoot length, roots, plant height, and shoot and root dry weights. Besides, PGPR treatment significantly increased chlorophyll contents under Cr stress. The acquisition of nutrients was also bettered in plants inoculated with PGPR.	Danish et al. (2019)
Zea mays L.	Cr	*P. mirabilis*	The effects of Cr toxicity on maize plant growth and development were considerable. Inoculation with certain PGPR reduced Cr toxicity. Plant performance under Cr toxicity was increased by PGPR inoculation. By increasing enzymatic antioxidant activity, PGPR lowered oxidative stress.	Islam et al. (2016b)
Cicer arietinum L.	Cr	*Bacillus* species PSB10	The plant growth-promoting and chromium-reducing PGPR strain significantly increased chickpea crop growth, grain protein, seed production, leghaemoglobin, chlorophyll, and nodulation compared to plants cultivated in the absence of bioinoculation in the presence of various chromium concentrations.	Wani and Khan (2010)
Lycopersicon esculentum L.	Cd	*Enterobacter ludwigii* DJ3, *Enterobacter sp.* EG16, and *Funneliformis mosseae*	Mycorrhizal treatment alone or in combination with bacteria significantly improved plant biomass under Cd toxicity. Further, microbial stimulants dropped Cd aerial translocation.	Li et al. (2020)
Zea mays L.	Cd	*Enterobacter* (CIK-521R), *Leifsonia* (CIK-521), *Klebsiella* (CIK-518), and *Bacillus* (CIK-517 and CIK-519)	Exogenous administration of Cd and PGPR, respectively, resulted in the detrimental and favorable control of maize growth. Cd exposure (0–80 mg Cd L^{-1}) resulted in a considerable decrease in relative seedling growth of maize cultivars in seed germination experiments. When compared to their respective controls, bacterial strains dramatically boosted shoot/root growth and dry biomass in normal and Cd-contaminated soil.	Ahmad et al. (2016)
Eruca sativa L.	Cd	*Pseudomonas putida* (ATCC 39213)	*E. sativa* plants infected with *P. putida* had the greatest increase in Cd uptake compared to uninoculated plants. Plants that were not infected with *P. putida* had poor growth and high Cd residual levels, which were used to determine Cd toxicity. Despite the hyperaccumulation of Cd in the whole *E. sativa* plant, *P. putida* increased plant development at all amounts of Cd input compared to non-inoculated plants. Inoculation with *P. putida* increased plants' Cd uptake capability and favored healthy growth under Cd stress, according to the findings.	Kamran et al. (2015)
Zea mays L., *Helianthus annuus* L., and pumpkin: *Cucurbita pepo* L.	Cu	*Pseudomonas cedrina* K4 and *Stenotrophomonas sp.* A22	The findings of the research showed that the EDTA+PGPR treatment might increase Cu bioavailability and selective absorption by plants, enhancing plant phytoremediation efficacy in Cu-contaminated environments.	Abbaszadeh-Dahaji et al. (2019)

REFERENCES

Abbas, T., M. Rizwan, S. Ali, M. Adrees, A. Mahmood, M. Zia-ur-Rehman, M. Ibrahim, M. Arshad, and M.F. Qayyum. 2018. Biochar application increased the growth and yield and reduced cadmium in drought stressed wheat grown in an aged contaminated soil. *Ecotoxicol. Environ. Saf.* 148:825–833. doi:10.1016/j.ecoenv.2017.11.063.

Abbaszadeh-Dahaji, P., A. Baniasad-Asgari, and M. Hamidpour. 2019. The effect of Cu-resistant plant growth-promoting rhizobacteria and EDTA on phytoremediation efficiency of plants in a Cu-contaminated soil. *Environ. Sci. Pollut. Res.* 26(31):31822–31833. doi:10.1007/s11356-019-06334-0.

Abdallah, M.M.S., Z.A. Abdelgawad, and H.M.S. El-Bassiouny. 2016. Alleviation of the adverse effects of salinity stress using trehalose in two rice varieties. *S. Afr. J. Bot.* 103:275–282. doi:10.1016/j.sajb.2015.09.019.

Abdelaal, K.A., L.M. EL-Maghraby, H. Elansary, Y.M. Hafez, E.I. Ibrahim, M. El-Banna, M. El-Esawi, and A. Elkelish. 2020. Treatment of sweet pepper with stress tolerance-inducing compounds alleviates salinity stress oxidative damage by mediating the physio-biochemical activities and antioxidant systems. *Agronomy* 10(1):26.

Ahemad, M., and M. Kibret. 2014. Mechanisms and applications of plant growth promoting rhizobacteria: Current perspective. *J. King Saud Univ. Sci.* 26(1):1–20. doi:10.1016/j.jksus.2013.05.001.

Ahmad, I., M.J. Akhtar, H.N. Asghar, U. Ghafoor, and M. Shahid. 2016. Differential effects of plant growth-promoting rhizobacteria on maize growth and cadmium uptake. *J. Plant Growth Regul.* 35(2):303–315. doi:10.1007/s00344-015-9534-5.

Altomare, C., and I. Tringovska. 2011. Beneficial soil microorganisms, an ecological alternative for soil fertility management. In: Lichtfouse, E. (ed.) *Genetics, Biofuels and Local Farming Systems.* Springer, Dordrecht, pp. 161–214. doi:10.1007/978-94-007-1521-9_6.

Asghari, B., R. Khademian, and B. Sedaghati. 2020. Plant growth promoting rhizobacteria (PGPR) confer drought resistance and stimulate biosynthesis of secondary metabolites in pennyroyal (*Mentha pulegium* L.) under water shortage condition. *Sci. Hortic.* 263:109132. doi:10.1016/j.scienta.2019.109132.

Ashraf, M.A., I. Hussain, R. Rasheed, M. Iqbal, M. Riaz, and M.S. Arif. 2017. Advances in microbe-assisted reclamation of heavy metal contaminated soils over the last decade: A review. *J. Environ. Manage.* 198:132–143. doi:10.1016/j.jenvman.2017.04.060.

Badri, D.V., and J.M. Vivanco. 2009. Regulation and function of root exudates. *Plant Cell Environ.* 32(6):666–681. doi:10.1111/j.1365-3040.2009.01926.x.

Bano, A., and M. Fatima. 2009. Salt tolerance in *Zea mays* (L). following inoculation with Rhizobium and Pseudomonas. *Biol. Fertil. Soils* 45(4):405–413. doi:10.1007/s00374-008-0344-9.

Barnawal, D., N. Bharti, S.S. Pandey, A. Pandey, C.S. Chanotiya, and A. Kalra. 2017. Plant growth-promoting rhizobacteria enhance wheat salt and drought stress tolerance by altering endogenous phytohormone levels and TaCTR1/TaDREB2 expression. *Physiol. Plant.* 161(4):502–514. doi:10.1111/ppl.12614.

Bashir, T., S. Naz, and A. Bano. 2020. Plant growth promoting rhizobacteria in combination with plant growth regulators attenuate the effect of drought stress. *Pak. J. Bot.* 52(3): 783–792.

Basu, A., P. Prasad, S.N. Das, S. Kalam, R.Z. Sayyed, M.S. Reddy, and H. El Enshasy. 2021. Plant Growth Promoting Rhizobacteria (PGPR) as green bioinoculants: Recent developments, constraints, and prospects. *Sustainability* 13(3):1140.

Bharti, N., D. Barnawal, A. Awasthi, A. Yadav, and A. Kalra. 2014. Plant growth promoting rhizobacteria alleviate salinity induced negative effects on growth, oil content and physiological status in Mentha arvensis. *Acta Physiol. Plant.* 36(1):45–60.

Bharti, N., S.S. Pandey, D. Barnawal, V.K. Patel, and A. Kalra. 2016. Plant growth promoting rhizobacteria Dietzia natronolimnaea modulates the expression of stress responsive genes providing protection of wheat from salinity stress. *Sci. Rep.* 6(1):34768. doi:10.1038/srep34768.

Bhattacharyya, P. N., & Jha, D. K. (2012). Plant growth-promoting rhizobacteria (PGPR): emergence in agriculture. World Journal of Microbiology and Biotechnology, 28(4), 1327–1350.

Calvo, P., L. Nelson, and J.W. Kloepper. 2014. Agricultural uses of plant biostimulants. *Plant Soil* 383(1):3–41. doi:10.1007/s11104-014-2131-8.

Chiappero, J., L.d.R. Cappellari, L.G. Sosa Alderete, T.B. Palermo, and E. Banchio. 2019. Plant growth promoting rhizobacteria improve the antioxidant status in Mentha piperita grown under drought stress leading to an enhancement of plant growth and total phenolic content. *Ind. Crops Prod.* 139:111553. doi:10.1016/j.indcrop.2019.111553.

Dakora, F.D., and D.A. Phillips. 2002. Root exudates as mediators of mineral acquisition in low-nutrient environments. In: Adu-Gyamfi, J.J. (ed.). *Food Security in Nutrient-Stressed Environments: Exploiting Plants' Genetic Capabilities.* Springer Netherlands, Dordrecht, pp. 201–213. doi:10.1007/978-94-017-1570-6_23.

Danish, S., S. Kiran, S. Fahad, N. Ahmad, M.A. Ali, F.A. Tahir, M.K. Rasheed, K. Shahzad, X. Li, D. Wang, M. Mubeen, S. Abbas, T.M. Munir, M.Z. Hashmi, M. Adnan, B. Saeed, S. Saud, M.N. Khan, A. Ullah, and W. Nasim. 2019. Alleviation of chromium toxicity in maize by Fe fortification and chromium tolerant ACC deaminase producing plant growth promoting rhizobacteria. *Ecotoxicol. Environ. Saf.* 185:109706. doi:10.1016/j.ecoenv.2019.109706.

Danish, S., M. Zafar-ul-Hye, F. Mohsin, and M. Hussain. 2020. ACC-deaminase producing plant growth promoting rhizobacteria and biochar mitigate adverse effects of drought stress on maize growth. *PLOS One.* 15(4):e0230615. doi:10.1371/journal.pone.0230615.

Du Jardin, P., 2015. Plant biostimulants: Definition, concept, main categories and regulation. *Sci. Hortic.* 196:3–14 doi:10.1016/j.scienta.2015.09.021.

Egamberdieva, D., S. Wirth, D. Jabborova, L.A. Räsänen, and H. Liao. 2017. Coordination between Bradyrhizobium and Pseudomonas alleviates salt stress in soybean through altering root system architecture. *J. Plant Interact* 12(1):100–107. doi:10.1080/17429145.2017.1294212.

Etesami, H., and S.M. Adl. 2020. Plant growth-promoting rhizobacteria (PGPR) and their action mechanisms in availability of nutrients to plants. In: Kumar, M., Kumar, V., Prasad, R. (eds.) *Phyto-Microbiome in Stress Regulation.* Springer, Singapore, pp. 147–203. doi:10.1007/978-981-15-2576-6_9.

Etesami, H., H.M. Hosseini, H.A. Alikhani, and L. Mohammadi. 2014. Bacterial Biosynthesis of 1-Aminocyclopropane-1-Carboxylate (ACC) Deaminase and Indole-3-Acetic Acid (IAA) as Endophytic Preferential Selection Traits by Rice Plant Seedlings. *J. Plant Growth Regul.* 33(3):654–670. doi:10.1007/s00344-014-9415-3.

Etesami, H., and D.K. Maheshwari. 2018. Use of plant growth promoting rhizobacteria (PGPRs) with multiple plant growth promoting traits in stress agriculture: Action mechanisms and future prospects. *Ecotoxicol. Environ. Saf.* 156:225–246. doi:10.1016/j.ecoenv.2018.03.013.

Ferreira, M.J., H. Silva, and A. Cunha. 2019. Siderophore-producing rhizobacteria as a promising tool for empowering plants to cope with iron limitation in saline soils: A review. *Pedosphere* 29(4):409–420. doi:10.1016/S1002-0160(19)60810-6.

Fiorentino, N., V. Ventorino, S.L. Woo, O. Pepe, A. De-Rosa, L. Gioia, I. Romano, N. Lombardi, M. Napolitano, G. Colla, and Y. Rouphael. 2018. Trichoderma-based biostimulants modulate rhizosphere microbial populations and improve N uptake efficiency, yield, and nutritional quality of leafy vegetables. *Front. Plant Sci.* 9(743). doi:10.3389/fpls.2018.00743.

Ghazy, N., and S. El-Nahrawy. 2021. Siderophore production by Bacillus subtilis MF497446 and Pseudomonas koreensis MG209738 and their efficacy in controlling Cephalosporium maydis in maize plant. *Arch. Microbiol.* 203(3):1195–1209. doi:10.1007/s00203-020-02113-5.

Glick, B.R., 2012. Plant growth-promoting bacteria: Mechanisms and applications. *Scientifica* 2012:963401.

Grobelak, A., A. Napora, and M. Kacprzak. 2015. Using plant growth-promoting rhizobacteria (PGPR) to improve plant growth. *Ecol. Eng.* 84:22–28.

Gontia-Mishra, I., S. Sapre, A. Sharma, and S. Tiwari. 2016. Amelioration of drought tolerance in wheat by the interaction of plant growth-promoting rhizobacteria. *Plant Biol.* 18(6):992–1000. doi:10.1111/plb.12505.

Goswami, D., S. Pithwa, P. Dhandhukia, and J.N. Thakker. 2014. Delineating Kocuria turfanensis 2M4 as a credible PGPR: A novel IAA-producing bacteria isolated from saline desert. *J. Plant Interact* 9(1): 566–576. doi:10.1080/17429145.2013.871650.

Goswami, M., and S. Deka. 2020. Plant growth-promoting rhizobacteria—alleviators of abiotic stresses in soil: A review. *Pedosphere* 30(1):40–61. doi:10.1016/S1002-0160(19)60839-8.

Gouda, S., R.G. Kerry, G. Das, S. Paramithiotis, H.S. Shin, and J.K. Patra. 2018. Revitalization of plant growth promoting rhizobacteria for sustainable development in agriculture. *Microbiol. Res.* 206:131–140. doi:10.1016/j.micres.2017.08.016.

Gupta, S., and S. Pandey. 2019. ACC deaminase producing bacteria with multifarious plant growth promoting traits alleviates salinity stress in French bean (*Phaseolus vulgaris*) plants. *Front. Microbiol.* 10(1506). doi:10.3389/fmicb.2019.01506.

Hamdia, M.A.E.S., M.A.K. Shaddad, and M.M. Doaa. 2004. Mechanisms of salt tolerance and interactive effects of Azospirillum brasilense inoculation on maize cultivars grown under salt stress conditions. *Plant Growth Regul.* 44(2):165–174. doi:10.1023/B:GROW.0000049414.03099.9b.

Hartmann, A., M. Schmid, D.V. Tuinen, and G. Berg. 2009. Plant-driven selection of microbes. *Plant Soil* 321(1):235–257. doi:10.1007/s11104-008-9814-y.

Hassan, T.U., A. Bano, and I. Naz. 2017. Alleviation of heavy metals toxicity by the application of plant growth promoting rhizobacteria and effects on wheat grown in saline sodic field. *Int. J. Phytoremediation* 19(6):522–529. doi:10.1080/15226514.2016.1267696.

Hayat, R., S. Ali, U. Amara, R. Khalid, and I. Ahmed. 2010. Soil beneficial bacteria and their role in plant growth promotion: A review. *Ann. Microbiol.* 60(4):579–598. doi:10.1007/s13213-010-0117-1.

Islam, F., T. Yasmeen, M.S. Arif, S. Ali, B. Ali, S. Hameed, and W. Zhou. 2016a. Plant growth promoting bacteria confer salt tolerance in Vigna radiata by up-regulating antioxidant defense and biological soil fertility. *Plant Growth Regul.* 80(1):23–36. doi:10.1007/s10725-015-0142-y.

Islam, F., T. Yasmeen, M.S. Arif, M. Riaz, S.M. Shahzad, Q. Imran, and I. Ali. 2016b. Combined ability of chromium (Cr) tolerant plant growth promoting bacteria (PGPB) and salicylic acid (SA) in attenuation of chromium stress in maize plants. *Plant Physiol. Biochem.* 108:456–467. doi:10.1016/j.plaphy.2016.08.014.

Jacob, J.M., C. Karthik, R.G. Saratale, S.S. Kumar, D. Prabakar, K. Kadirvelu, and A. Pugazhendhi. 2018. Biological approaches to tackle heavy metal pollution: A survey of literature. *J. Environ. Manage.* 217:56–70. doi:10.1016/j.jenvman.2018.03.077.

Jha, Y., R.B. Subramanian, and S. Patel. 2011. Combination of endophytic and rhizospheric plant growth promoting rhizobacteria in Oryza sativa shows higher accumulation of osmoprotectant against saline stress. *Acta Physiol. Plant.* 33(3):797–802.

Jha, Y., and R.B. Subramanian. 2015. Reduced cell death and improved cell membrane integrity in rice under salinity by root associated bacteria. *Theor. Exp. Plant Physiol.* 27 (3–4):227–235.

Kalyanasundaram, G.T., N. Syed, and K. Subburamu. 2021. Chapter 17- Recent developments in plant growth-promoting rhizobacteria (PGPR) for sustainable agriculture. In: Viswanath, B. (ed.) *Recent Developments in Applied Microbiology and Biochemistry.* Academic Press, Cambridge, MA, pp. 181–192. doi:10.1016/B978-0-12-821406-0.00017-5.

Kamran, M.A., J.H. Syed, S.A.M.A.S. Eqani, M.F.H. Munis, and H.J. Chaudhary. 2015. Effect of plant growth-promoting rhizobacteria inoculation on cadmium (Cd) uptake by *Eruca sativa*. *Environ. Sci. Pollut. Res.* 22(12):9275–9283. doi:10.1007/s11356-015-4074-x.

Kaushal, M., and S.P. Wani. 2016 Plant-growth-promoting rhizobacteria: Drought stress alleviators to ameliorate crop production in drylands. *Ann. Microbiol.* 66(1):35–42. doi:10.1007/s13213-015-1112-3.

Khan, N., A. Bano, and M.D.A. Babar. 2019. Metabolic and physiological changes induced by plant growth regulators and plant growth promoting rhizobacteria and their impact on drought tolerance in *Cicer arietinum* L. *PLOS One.* 14(3):e0213040. doi:10.1371/journal.pone.0213040.

Kour, D., K.L. Rana, A.N. Yadav, I. Sheikh, V. Kumar, H.S. Dhaliwal, and A.K. Saxena. 2020. Amelioration of drought stress in Foxtail millet (*Setaria italica* L.) by P-solubilizing drought-tolerant microbes with multifarious plant growth promoting attributes. *Environ. Sustain.* 3(1):23–34. doi:10.1007/s42398-020-00094-1.

Kour, D., K.L. Rana, N. Yadav, A.N. Yadav, A. Kumar, V.S. Meena, B. Singh, V.S. Chauhan, H.S. Dhaliwal, and A.K. Saxena. 2019. Rhizospheric microbiomes: Biodiversity, mechanisms of plant growth promotion, and biotechnological applications for sustainable agriculture. In: Kumar, A., Meena, V.S. (eds.) *Plant Growth Promoting Rhizobacteria for Agricultural Sustainability: From Theory to Practices.* Springer, Singapore, pp. 19–65. doi:10.1007/978-981-13-7553-8_2.

Kumar, P., Pandey, P., Dubey, R. C., & Maheshwari, D. K. (2016). Bacteria consortium optimization improves nutrient uptake, nodulation, disease suppression and growth of the common bean (Phaseolus vulgaris) in both pot and field studies. Rhizosphere, 2, 13–23.

Kumar Yadav, K., N. Gupta, A. Kumar, L.M. Reece, N. Singh, S. Rezania, and S.A. Khan. 2018. Mechanistic understanding and holistic approach of phytoremediation: A review on application and future prospects. *Ecol. Eng.* 120:274–298. doi:10.1016/j.ecoleng.2018.05.039.

Li, Y., J. Zeng, S. Wang, Q. Lin, D. Ruan, H. Chi, M. Zheng, Y. Chao, R. Qiu, and Y. Yang, 2020. Effects of cadmium-resistant plant growth-promoting rhizobacteria and Funneliformis mosseae on the cadmium tolerance of tomato (*Lycopersicon esculentum* L.). *Int. J. Phytoremediation.* 22(5):451–458. doi:10.1080/15226514.2019.1671796.

Li, Y., J. Zhang, J. Zhang, W. Xu, and Z. Mou. 2019. Characteristics of inorganic phosphate-solubilizing bacteria from the sediments of a eutrophic lake. *Int. J. Environ. Res. Public Health.* 16(12):2141.

Lin, Y., D.B. Watts, J.W. Kloepper, Y. Feng, and H.A. Torbert. 2020. Influence of plant growth-promoting rhizobacteria on corn growth under drought stress. *Commun. Soil Sci. Plant Anal.* 51(2):250–264. doi:10.1080/00103624.2019.1705329.

Liu, F., H. Ma, L. Peng, Z. Du, B. Ma, and X. Liu. 2019. Effect of the inoculation of plant growth-promoting rhizobacteria on the photosynthetic characteristics of Sambucus williamsii Hance container seedlings under drought stress. *AMB Express* 9(1):169. doi:10.1186/s13568-019-0899-x.

Liu, F., S. Xing, H. Ma, Z. Du, and B. Ma 2013. Cytokinin-producing, plant growth-promoting rhizobacteria that confer resistance to drought stress in Platycladus orientalis container seedlings. *Appl. Microbiol. Biotechnol.* 97(20):9155–9164. doi:10.1007/s00253-013-5193-2.

Liu, J., W. Qi, Q. Li, S.G. Wang, C. Song, and X.Z. Yuan. 2020. Exogenous phosphorus-solubilizing bacteria changed the rhizosphere microbial community indirectly. *3 Biotech* 10(4):164. doi:10.1007/s13205-020-2099-4.

Mahmoud, O.M.B., I.B. Slimene, O.T. Zribi, C. Abdelly, and N. Djébali. 2017. Response to salt stress is modulated by growth-promoting rhizobacteria inoculation in two contrasting barley cultivars. *Acta Physiol. Plant* 39(6):120.

Manoj, S.R., C. Karthik, K. Kadirvelu, P.I. Arulselvi, T. Shanmugasundaram, B. Bruno, and M. Rajkumar. 2020. Understanding the molecular mechanisms for the enhanced phytoremediation of heavy metals through plant growth promoting rhizobacteria: A review. *J. Environ. Manage.* 254:109779. doi:10.1016/j.jenvman.2019.109779.

Marasco, R., E. Rolli, B. Ettoumi, G. Vigani, F. Mapelli, S. Borin, A.F. Abou-Hadid, U.A. El-Behairy, C. Sorlini, and A. Cherif. 2012. A drought resistance-promoting microbiome is selected by root system under desert farming. *PLOS One.* 7(10):e48479.

Marasco, R., E. Rolli, G. Vigani, S. Borin, C. Sorlini, H. Ouzari, G. Zocchi, and D. Daffonchio. 2013. Are drought-resistance promoting bacteria cross-compatible with different plant models? *Plant Signal. Behav.* 8(10):e26741. doi:10.4161/psb.26741.

Nacoon, S., S. Jogloy, N. Riddech, W. Mongkolthanaruk, T.W. Kuyper, and S. Boonlue. 2020. Interaction between phosphate solubilizing bacteria and arbuscular mycorrhizal fungi on growth promotion and tuber inulin content of *Helianthus tuberosus* L. *Sci. Rep.* 10(1):4916. doi:10.1038/s41598-020-61846-x.

Naseem, H., and A. Bano. 2014. Role of plant growth-promoting rhizobacteria and their exopolysaccharide in drought tolerance of maize. *J. Plant Interact.* 9(1):689–701. doi:10.1080/17429145.2014.902125.

Nath, D., B.R. Maurya, and V.S. Meena. 2017. Documentation of five potassium-and phosphorus-solubilizing bacteria for their K and P-solubilization ability from various minerals. *Biocatal. Agric. Biotechnol.* 10: 174–181.

Noreen, S., B. Ali, and S. Hasnain. 2012. Growth promotion of *Vigna mungo* (L.) by Pseudomonas spp. exhibiting auxin production and ACC-deaminase activity. *Ann. Microbiol.* 62(1):411–417. doi:10.1007/s13213-011-0277-7.

Oleńska, E., W. Małek, M. Wójcik, I. Swiecicka, S. Thijs, and J. Vangronsveld. 2020. Beneficial features of plant growth-promoting rhizobacteria for improving plant growth and health in challenging conditions: A methodical review. *Sci. Total Environ.* 743:140682. doi:10.1016/j.scitotenv.2020.140682.

Reed, S.C., C.C. Cleveland, and A.R. Townsend. 2011. Functional ecology of free-living nitrogen fixation: A contemporary perspective. *Annu. Rev. Ecol. Evol. Syst.* 42(1):489–512. doi:10.1146/annurev-ecolsys-102710-145034.

Richardson, A.E., J.M. Barea, A.M. McNeill, and C. Prigent-Combaret. 2009. Acquisition of phosphorus and nitrogen in the rhizosphere and plant growth promotion by microorganisms. *Plant Soil* 321(1):305–339. doi:10.1007/s11104-009-9895-2.

Robin, A., G. Vansuyt, P. Hinsinger, J.M. Meyer, J.F. Briat, and P. Lemanceau. 2008. Chapter 4 iron dynamics in the rhizosphere: Consequences for plant health and nutrition. In: *Advances in Agronomy*, vol. 99. Academic Press, pp. 183–225. doi:10.1016/S0065-2113(08)00404-5.

Rodriguez, H., T. Gonzalez, I. Goire, and Y.J.N. Bashan. 2004. Gluconic acid production and phosphate solubilization by the plant growth-promoting bacterium Azospirillum spp. *Naturwissenschaften* 91(11):552–555.

Ruzzi, M., and R. Aroca. 2015. Plant growth-promoting rhizobacteria act as biostimulants in horticulture. *Sci. Hortic.* 196:124–134. doi:10.1016/j.scienta.2015.08.042.

Sabet, H., and F. Mortazaeinezhad. 2018. Yield, growth and Fe uptake of cumin (*Cuminum cyminum* L.) affected by Fe-nano, Fe-chelated and Fe-siderophore fertilization in the calcareous soils. *J. Trace Elem. Med. Biol.* 50:154–160.

Sandhya, V., S.Z. Ali, M. Grover, G. Reddy, and B. Venkateswarlu. 2010. Effect of plant growth promoting Pseudomonas spp. on compatible solutes, antioxidant status and plant growth of maize under drought stress. *Plant Growth Regul.* 62(1):21–30. doi:10.1007/s10725-010-9479-4.

Saravanan, V.S., M. Madhaiyan, and M. Thangaraju. 2007. Solubilization of zinc compounds by the diazotrophic, plant growth promoting bacterium Gluconacetobacter diazotrophicus. *Chemosphere* 66(9):1794–1798. doi:10.1016/j.chemosphere.2006.07.067.

Schmidt, R., M. Köberl, A. Mostafa, E. Ramadan, M. Monschein, K. Jensen, R. Bauer, and G. Berg. 2014. Effects of bacterial inoculants on the indigenous microbiome and secondary metabolites of chamomile plants. *Front. Microbiol.* 5(64). doi:10.3389/fmicb.2014.00064.

Selosse, M.A., and F. Le Tacon. 1998. The land flora: A phototroph-fungus partnership? *Trends Ecol. Evol.* 13(1):15–20. doi:10.1016/S0169-5347(97)01230-5.

Sharma, S.B., R.Z. Sayyed, M.H. Trivedi, and T.A. Gobi. 2013. Phosphate solubilizing microbes: Sustainable approach for managing phosphorus deficiency in agricultural soils. *Springer Plus* 2(1):587. doi:10.1186/2193-1801-2-587.

Singh, R.P., and P.N. Jha, 2016. The multifarious PGPR serratia marcescens CDP-13 augments induced systemic resistance and enhanced salinity tolerance of wheat (*Triticum aestivum* L.). *PLOS One* 11(6):e0155026. doi:10.1371/journal.pone.0155026.

Stamford, N.P., P.R.D. Santos, A.M.M.F.D. Moura, C.E.D.R. Santos, and A.D.S.D. Freitas. 2003. Biofertilzers with natural phosphate, sulphur and Acidithiobacillus in a siol with low available-P. *Sci. Agric.* 60(4):767–773.

Subiramani, S., S. Ramalingam, T. Muthu, S.H. Nile, and B. Venkidasamy. 2020. Development of abiotic stress tolerance in crops by plant growth-promoting Rhizobacteria (PGPR). In: Kumar, M., Kumar, V., Prasad, R. (eds.), *Phyto-Microbiome in Stress Regulation*. Springer, Singapore, pp. 125–145. doi:10.1007/978-981-15-2576-6_8.

Ullah, A., M. Nisar, H. Ali, A. Hazrat, K. Hayat, A.A. Keerio, M. Ihsan, M. Laiq, S. Ullah, S. Fahad, A. Khan, A.H. Khan, A. Akbar, and X. Yang. 2019. Drought tolerance improvement in plants: An endophytic bacterial approach. *Appl. Microbiol. Biotechnol.* 103(18):7385–7397. doi:10.1007/s00253-019-10045-4.

Vandenkoornhuyse, P., A. Quaiser, M. Duhamel, A. Le Van, and A. Dufresne. 2015. The importance of the microbiome of the plant holobiont. *New Phytol.* 206(4):1196–1206. doi:10.1111/nph.13312.

Vurukonda, S.S.K.P., S. Vardharajula, M. Shrivastava, and A. SkZ. 2016. Enhancement of drought stress tolerance in crops by plant growth promoting rhizobacteria. *Microbiol. Res.* 184:13–24. doi:10.1016/j.micres.2015.12.003.

Wan, W., Y. Qin, H. Wu, W. Zuo, H. He, J. Tan, Y. Wang, and D. He. 2020. Isolation and characterization of phosphorus solubilizing bacteria with multiple phosphorus sources utilizing capability and their potential for lead immobilization in soil. *Front. Microbiol.* 11:752–752. doi:10.3389/fmicb.2020.00752.

Wani, P.A., and M.S. Khan. 2010. Bacillus species enhance growth parameters of chickpea (*Cicer arietinum* L.) in chromium stressed soils. *Food Chem. Toxicol.* 48(11):3262–3267. doi:10.1016/j.fct.2010.08.035.

Wei, Y., Y. Zhao, M. Shi, Z. Cao, Q. Lu, T. Yang, Y. Fan, and Z. Wei. 2018. Effect of organic acids production and bacterial community on the possible mechanism of phosphorus solubilization during composting with enriched phosphate-solubilizing bacteria inoculation. *Bioresour. Technol.* 247:190–199. doi:10.1016/j.biortech.2017.09.092.

Widdig, M., P.M. Schleuss, A.R. Weig, A. Guhr, L.A. Biederman, E.T. Borer, M.J. Crawley, K.P. Kirkman, E.W. Seabloom, P.D. Wragg, and M. Spohn. 2019. Nitrogen and phosphorus additions alter the abundance of phosphorus-solubilizing bacteria and phosphatase activity in grassland soils. *Front. Environ. Sci.* 7(185). doi:10.3389/fenvs.2019.00185.

Woo, S.L., and O. Pepe. 2018. Microbial consortia: Promising probiotics as plant biostimulants for sustainable agriculture. *Front. Plant Sci.* 9(1801). doi:10.3389/fpls.2018.01801.

Xu, M., Schnorr, J., Keibler, B., & Simon, H. M. (2012). Comparative analysis of 16S rRNA and amoA genes from archaea selected with organic and inorganic amendments in enrichment culture. Applied and environmental microbiology, 78(7), 2137–2146.

Yakhin, O.I., A.A. Lubyanov, I.A. Yakhin, and P.H. Brown. 2017. Biostimulants in plant science: A global perspective. *Front. Plant Sci.* 7(2049). doi:10.3389/fpls.2016.02049.

Zhang, H., A. Khan, D.K.Y. Tan, and H. Luo. 2017. Rational water and nitrogen management improves root growth, increases yield and maintains water use efficiency of cotton under mulch drip irrigation. *Front. Plant Sci.* 8(912). doi:10.3389/fpls.2017.00912.

6 Microbial Biostimulants in Protecting against Nematodes

Muhammad Raheel, Waqas Ashraf,
Amir Riaz, Qaiser Shakeel and Sajjad Ali
The Islamia University of Bahawalpur

Hafiz Muhammad Aatif and Muhammad Zeeshan Mansha
Bahauddin Zakariya University, Bahadur Sub Campus, Layyah

CONTENTS

6.1 INTRODUCTION

Nematodes are usually microscopic, non-segmented invertebrates that reside marine, freshwater and terrestrial environments. They are wormlike in appearance, which is the most abundant animal life form in the world. Nematodes residing in soil are either saprophytes or parasites and are generally organized as bacterivores, fungivores, omnivores, plant parasites and predators. There are more than 4,300 species of plant parasitic nematodes that belong to 197 genera (Decraemer and Hunt, 2006). The most damaging plant parasitic nematodes include *Meloidogyne* spp. (root-knot nematodes), *Heterodera* and *Globodera* spp. (cyst nematodes), *Tylenchulus semipenetrans* (a citrus nematode), *Radopholus similis* (a burrowing nematode) and *Pratylenchus* spp. (root lesion nematodes) (Ebone et al., 2019; Gamalero and Glick, 2020). The life cycle of most of these nematodes is similar. Their eggs hatch into juveniles (immature stages of nematodes similar in appearance to mature ones). These immature nematodes grow in size, and the end of each juvenile stage is marked by ecdysis or molting. There are four juvenile stages in nematodes. After final molt, nematodes are distinguished into adult females and males. The mature female nematodes may lay fertile eggs with or without fusion with male nuclei by plasmogamy and pseudogamy or through parthenogenesis. The infection cycle of nematodes starts with the help of stylet and knobs. The hollow stylet is pushed by knobs into the plant tissue to suck cell sap. This activity is responsible for causing symptoms on roots and above-ground plant parts. The root symptoms of infected plants include root

DOI: 10.1201/9781003188032-6

tips injury, excessive branching of roots, size reduction in taproots and abnormal enlargement and shape of roots. The combined damage of nematodes and other pathogens may result in root rots. The above-ground symptoms due to these nematodes are often similar to wilting or those due to mineral stress. The other above-ground symptoms are twisting and distortion of stems and leaves, replacement of seeds with nematode galls, etc. (Agrios, 2005).

The plant parasitic nematodes attacking plants are worldwide in occurrence and are causing huge crop losses. They are responsible for annual crop losses of about US\$ 173 billion globally (Gamalero and Glick, 2020). They cause 12%–25% reduction in total production of most economically important crops. Furthermore, it is supposed that in future, the crop damage by phyto-nematodes will probably increase due to climate change. They are also involved in association with some other plant pathogens such as bacteria and fungus as these pathogens rely on the injuries caused by nematodes to enter the plants for causing diseases in plants (Migunova and Sasanelli, 2021). Moreover, some of these nematodes such as *Xiphinema* spp. and *Longidorus* spp. also serve as a virus vector (Hao et al., 2012). To manage these plant parasites and their disease complexes, it is obligatory to reduce their numbers in soil.

For the effective management of plant parasitic nematodes, different approaches are used, including cultural control, physical control, biological control and chemical control. Chemical control through nematicides has exclusively been used since the 1950s. These nematicides can act on respiration, transmission of nerve impulse and steroid metabolism of nematodes. These nematicides may be grouped into fumigants and non-fumigants depending upon their mode of action (Agrios, 2005; Gamalero and Glick, 2020). Fumigants are effective against a broad range of pathogenic bacteria, fungi, nematodes and other organisms living in soil. However, their application is extremely dangerous to human and animal health. They are also toxic to non-target earthworms, mites and beneficial insects. Furthermore, their efficiency is only against nematodes that have active host searching. They have limited success against endoparasitic nematodes and lead to the development of resistance in nematodes against these chemicals. Moreover, agrochemicals usage is presently limited in many developed countries of world. These limitations of chemical nematicides have stimulated the search for alternative and eco-friendly approaches against nematodes. Among these approaches, the use of microbial biostimulants is a vital option for integrated pest management.

6.2 BACTERIAL BIOSTIMULANTS IN PROTECTING PLANTS AGAINST NEMATODES

6.2.1 Plant Growth-Promoting Rhizobacteria

Nitrogen fixers, sporulating *Bacillus* and fluorescent *Pseudomonas* are examples of plant growth-promoting rhizobacteria (PGPR) that can be utilized for their positive attributes toward plant growth and suppression of their pathogens. These PGPR can also be found in association with roots of plants or sometimes on leaves and flowers. Treatment of plants with these bacteria offers enhancement in biomass and vital mineral contents of plants, increase in seed germination, improvement in plant nutrition, greater production of secondary metabolites and increased resistance against various biotic and abiotic stresses (Glick, 2020). These also have the potential to protect plants against nematodes. There are about 10% of these bacteria that exhibit antagonistic potential against plant parasitic nematodes. Different species of PGPR such as *Agrobacterium, Arthrobacter, Azotobacter, Azospirillum, Bacillus, Burkholderia, Chromobacterium, Clostridium, Corynebacterium* and *Serratia* have been used against plant parasitic nematodes (Khabbaz et al., 2019).

The plant parasitic nematodes are suppressed by PGPR through direct and indirect ways. Direct antagonism is established by the production of lytic enzymes and toxic proteins, parasitism or volatile organic compounds (VOCs). On the contrary, the indirect ways of expression of antagonism is through competition with nematodes for minerals and space, or molecular discharge, which modulate nematode behavior (toward host searching, nourishing and gender ratio) and induction of systemic resistance. These rhizobacteria release lytic enzymes such as chitinases, proteases and peptidyl peptide hydrolases that can damage both chitinous egg shells and nematode cuticle by cleavage (Gamalero and Glick, 2020). Other important enzymes include collagenases, lipases, glucanases, cellulases and pectinases. Collagenases are synthesized by *B. cereus*, which are harmful to juveniles of *M. javanica*. Various species of rhizobacteria such as *B. thuringiensis* FB833T, *B. thuringiensis* FS213P and *B. amyloliquefaciens* FR203A synthesize lipases against *Xiphinema index*. Furthermore, Pseudomonas spp. is revealed for the synthesis of glucanases, cellulases and pectinases against juveniles of *M. incognita* (Kohl et al., 2019). Another PGPR strain *B. cereus* BCM2 was shown to be completely fatal to J_2 stage of root-knot nematodes through the synthesis of chitosanase and alkaline serine protease. These enzymes induced leakage of nematode contents at cuticle level (Hu et al., 2020).

Several studies have revealed that rhizobacteria *B. thuringiensis* produce crystal toxins such as Cry5, Cry6, Cry12, Cry13, Cry14, Cry21 and Cry55 against a wide range of nematodes. These toxins may enter the nematodes through their feeding structures, i.e., stylets (Migunova and Sasanelli, 2021). The toxin Cry5B produced by *B. thuringiensis* is also being proposed to manage some species of root-knot nematodes and pan hookworms. Similarly, the protoxin named Cry6Aa2 is proved lethal to *M. hapla* as it decreases egg hatching, shows toxicity to infective juveniles (J_2 stage) and reduces their motility and root penetration. Furthermore, species of *Xenorhabdus* and *Photorhabdus* bacteria release toxins and antibiotics that suppress endoparasitic nematodes (Gamalero and Glick, 2020; Migunova and Sasanelli, 2021).

The third mechanism of suppressing plant parasitic nematodes is through the production of VOCs. These compounds have high vapor pressure and low molecular weight. They may diffuse an extended area over air, water and soil. These compounds not only enhance the growth of plants, but also suppress plant pathogens and nematodes. The major classes of these compounds that are reported to suppress plant parasitic nematodes are alcohols, alkenes, alkanes, ketones, esters, terpenoids and sulfur families. Different bacterial species and strains such as *Arthrobacter nicotianae*, *Achromobacter xylosoxidans*, *B. amyloliquefaciens* FZB42 and *Pseudochrobactrum saccharolyticum* are known to produce these compounds having nematicidal effects against *M. incognita*, while *Bacillus* spp., *Xanthomonas* spp. and *Paenibacillus* spp. show such activity against *M. graminicola* due to the production of VOCs (Xiang et al., 2018; Aeron et al., 2020; Gamalero and Glick, 2020). Many commercial products of PGPR are also widely manufactured and distributed in many countries, and their results against different plant parasitic nematodes are also significant (Table 6.1).

6.2.2 GENUS *PASTEURIA*

The members of the genus *Pasteuria* are mycelial, endospore-forming and gram-positive bacteria. These bacteria have septate mycelium and are dichotomously branched. These bacteria are obligate parasites and persist in soil or benthic community as durable and resting endospores. These are worldwide in occurrence and are reported from all continents. These are dispersed by the decaying host. These have a great biocontrol potential against plant parasitic nematodes. Their endospores serve as resting propagules that are highly resistant to desiccation and temperature extremes. The endospores also behave as an infective stage responsible for the horizontal transmission of parasite. They remain in non-motile form in soil matrix. When an appropriate nematode host arrives in its territory, these endospores become attached to its cuticle. One to several hundred endospores may become attached to the cuticle of the nematode, but a nematode can be infected by a single endospore attached (Chen and Dickson, 1998).

TABLE 6.1

List of PGPR-Based Commercial Products against Plant Parasitic Nematodes

Product Name	Microbial Origin	Company or Institution	Target
Avid or Abamectin	*Bacillus thuringiensis*	Syngenta	Root-knot and other nematodes
RhizoVital	*B. amyloliquefaciens* FZB42	ABiTEP GmbH, Germany	*M. incognita*
Deny	*Burkholderia cepacia*	Stine Microbial Products, the USA	*M. incognita*
Stanes Sting	*Bacillus subtilis*	Introduced by Gaara Company, Egypt, from India (T. Stanes and Company Limited)	*M. incognita*
Poncho	*B. firmus* I-1582	Votivo Crop Science, Raleigh, NC	*Meloidogyne* spp., *H. glycines* and *R. reniformis*
VOTiVO	*B. firmus* GB-126	Bayer, Germany	*R. reniformis*
BioNem-WP	*B. firmus*	AgroGreen, Israel	*M. incognita*
Biosafe-WP	*B. firmus*	AgroGreen, Israel	
BioNemaGon	*B. firmus*	AgriLife, India	*Meloidogyne* spp., *Heterodera* spp. and *Helicotylenchus* spp.
Onix	*B. methylotrophicus*	Laboratorio de Bio Controle Farroupilha S.A., Brazil	*M. javanica*
Sheathgua (Sudozone)	*Pseudomonas fluorescens*	Agriland Biotech, India	Root-knot and cyst nematodes
Nemaless	*Serratia marcescens*	Agricultural Research Center, Egypt	*M. incognita*
BioYield	*B. amyloliquefaciens* IN937a and *B. subtilis* GB03	Gustafson LLC, the USA	*M. incognita*
Nemix C	*B. licheniformis* and *B. subtilis*		Root-knot nematodes
Presense	*B. subtilis* strain FMCH002 and *B. licheniformis* strain FMCH001	FMC, Brazil	Plant parasitic nematodes
Pathway Consortia	*B. coagulans*, *B. megaterium*, *B. licheniformis*, *B. subtilis*, *P. fluorescens* and *Streptomyces* spp.	Pathway Holdings, the USA	Plant parasitic nematodes
BioStart	*B. licheniformis*, *B. laterosporus* and *B. chitinosporus*	BIO-CAT, the USA	Root-knot nematodes
BioStartL	*B. laterosporus* and *B. licheniformis* (mixture)	Rhcon-Vltova	
Xlan Mile	*B. cereus*	XinYI Zhongkai Agro-Chemical Industry Co., Ltd, China	Root-knot nematodes
Micronema	*Serratia* spp., *Pseudomonas* spp., *Azotobacter* spp., *B. circulans* and *B. thuringiensis*	Agricultural Research Center, Egypt	*M. incognita*
Equity	47 Strains of Bacilli	Naturize Biosciences LLC, Jacksonville, FL, the USA	*M. incognita*
Ag Blend	Rhizobacteria- and microbe-based metabolites formed by a microbial population during anaerobic fermentation	Advanced Microbial Solutions LLC, Pilot Point, TX, the USA	Root-knot nematodes

Modified from Migunova and Sasanelli (2021).

TABLE 6.2

List of *Pasteuria*-Based Commercial Products against Plant Parasitic Nematodes

Product name	Bacterial spp.	Company	Target
Econem	*Pasteuria penetrans*	Syngenta	*Belonolaimus longicaudatus*
Econem	*P. penetrans*	Pasteuria Bioscience, the USA	*Meloidogyne incognita* and *M. arenaria*
Econem	*P. penetrans*	Nematech, Japan	*M. incognita*
Clariva pn	*P. nishizawae* Pn1	Syngenta, Brazil	*Heterodera glycines*
Naviva ST	*Pasteuria* sp. *Ph3*	Syngenta	*R. reniformis*
NewPro	*Pasteuria* sp. *Ph3P+ usage* Bl1	Syngenta	*B. longicaudatus*

Modified from Migunova and Sasanelli (2021).

The attachment of endospores to cuticle depends upon the age of plant parasitic nematodes, while the age of nematode cuticle is affected by root exudates (Mohan et al., 2020). After attachment with cuticle, these endospores form a germ tube that penetrates into the nematode body cavity. After the penetration of the germ tube, primary colonies are formed in pseudocoelom. The shapes of these colonies vary from grape clusters to cauliflower florets. Upon fragmentation of mother colony, daughter colonies are produced, which in turn form clusters of sporangia. Later on, these sporangia give rise to endospores. This sporulation is completed within the nematode body usually after complete or partial consumption of its contents. Resultantly, a remarkable reduction in reproductive capacity and fecundity of nematodes occurs. Moreover, these *Pasteuria* may kill the female nematodes and transform its body into a bag of endospores. The life cycle of these *Pasteuria* spp. can be completed in juvenile, adult or both stages of nematodes. For the successful use of *Pasteuria* as agents of biocontrol of nematodes, their density should be 104–105 endospores/g of soil. So far, the members of the genus *Pasteuria* have been known to show parasitic activities against 323 nematode species of 116 genera. These nematode species include many economic important nematodes such as *M. incognita, M. arenaria, M. hapla, M. graminicola, Aphelenchoides besseyi, R. similis, Pratylenchus penetrans, Globodera rostochiensis* and *H. cajani* (Tiran et al., 2007; Mohan et al., 2012; Abd-Elgawad and Askary, 2018; Ciancio, 2018). These bacteria exhibit diverse levels of host specificity that may differ from species to populations. The commercial products of these bacteria are also available in many countries of the world and are providing effective control of plant parasitic nematodes (Table 6.2).

6.3 FUNGAL BIOSTIMULANTS IN PROTECTING PLANTS AGAINST NEMATODES

6.3.1 MYCORRHIZAL FUNGI

There are two types of mycorrhizal fungi, i.e., ectomycorrhizal fungi (ECMF) and arbuscular mycorrhizal fungi (AMF). The ECMF are capable of colonizing the intercellular space of roots of trees and forming a Hartig network. These fungi are members of the phyla Ascomycota, Basidiomycota and few Zygomycota. On the other hand, AMF belong to Glomeromycota and many of the fungal-based microbial biostimulants used against plant parasitic nematodes are formulated from these AMF (Berruti et al., 2016; Dominguez-Nunez and Albanesi, 2019). These fungi are ubiquitous and form obligate mutualistic relations with roots of several species of plants. These were mainly known to supply plants with additional nutrients such as nitrogen, phosphorus and some microelements. These are also identified for their role in the enhancement of plant photosynthetic activity and resistance to soilborne root pathogens. Therefore, these are applied to many crops to minimize crop losses due to soilborne plant pathogens. Since AMF and many plant parasitic nematodes are all native soil organisms, they coexist in plant roots. Therefore, several researches have revealed their biocontrol potential against nematodes. Their applications to various host plants induce tolerance against different species of root-knot nematodes such as *Meloidogyne incognita*, *M. javanica*, *M. hapla* and *M. arenaria* (Affokpon et al., 2011). Some chief species of AMF having potential against nematodes include *Acaulospora* spp., *Glomus intraradices*, *G. manihotis*, *G. mosseae*, *Funneliformis mosseae*, *Kuklospora* spp. and *Rhizophagus irregularis* (Zhang et al., 2008; Jefwa et al., 2010). Many of the mycorrhizal fungi are also available commercially for their use against nematodes, such as Aegis™, Ekoprop Nemax™, Micofort™, Micosat F™, Micosat Jolly™ and Mychodeep™ (D'Addabbo et al., 2019). The AMF species are reported to reduce root-knot galling index, the number of females, the number of eggs per root and the number of eggs per egg mass. The products of *Funneliformis mosseae-* and *Rhizophagus irregularis*-based microbial biostimulants were also reported to be suppressive to *Meloidogyne* spp. These can be used alone or in combination with other plant extracts or other microorganisms in field and greenhouse trials. The mechanism of AMF to counter plant parasitic nematodes is still to be explained because their action is through host plant. The suggested mechanism for the AMF outcome against phytoparasitic nematodes includes the rivalry for nutrients and space, biological and chemical changes in rhizosphere, modification of root metabolism, with altered molecular compositions of discharged secretions, and induction of systemic resistance in host plants (Flor-Peregrín et al., 2014; Marro et al., 2018). These biostimulants are also reported to suppress some other plant parasitic nematodes such as *Helicotylenchus multicinctus* (the banana spiral nematode), *Nacobbus aberrans* (the false root-knot nematode) (Marro et al., 2014), *Radopholus similis* and *Scutellonema bradys*. A significant reduction in the number of cysts, eggs inside cysts and population of *Heterodera cajani* was reported after the inoculation of plants with *Funneliformis mosseae* and *Rhizophagus fasciculatus*, respectively (Marro et al., 2018). In tomato plants, these mycorrhizal fungi are reported to induce systemic defense against *Pratylenchus penetrans* and *M. incognita* by activating chitinases encoding genes, pathogenesis-related proteins, enzymatic engagement in biosynthesis of lignin and reactive oxygen species detoxification through the involvement of enzymes as well as in the shikimate pathway that in response yields substances that form different aromatic secondary metabolites to counter plant parasitic nematodes. *R. irregularis* treatment in tomato plants produces systemic resistance against plant parasitic nematodes because of activities of phenolics and defensive plant enzymes such as peroxidases, polyphenol oxidase and superoxide dismutase (SOD). These enzymes and phenolics greatly decrease hydrogen peroxide (H_2O_2) and malondialdehyde (MDA) contents in roots and also increase the crop growth (Sharma and Sharma, 2017).

6.3.2 ENDOPHYTIC FUNGI

Plant endophytes are organisms that reside within the tissues of host without causing any visual symptoms. Endophytic fungi are ubiquitous to plants. These are members of phyla Ascomycota, Basidiomycota, Zygomycota and Oomycota. Their relationship with plants ranges from symbiotic to pathogenic. These are known for manufacturing a group of bioactive molecules having anti-inflammatory, antioxidant, antimicrobial, cytotoxic and immunosuppressive activities. Many industrially important enzymes are also produced from these fungi. This group of fungi have an additional role in ecosystems by providing protection against stresses due to abiotic as well as biotic factors (Lugtenberg et al., 2016; Yan et al., 2019). These are able to induce systemic acquired resistance (SAR) in plants and induced systemic resistance (ISR) against the invasion of pathogens and pests. These fungi can be utilized against plant parasitic nematodes as they can directly attack, repel, immobilize, kill or reduce their host searching ability (Schouten, 2016). These can compete with plant parasitic nematodes and can also hinder the development of nurse cell formation. For example, the role of one endophytic fungus named *Acremonium implicatum* is reported for parasitism and destruction of eggs of root-knot nematodes. Another fungus *Chaetomium globosum* is reported to produce secondary metabolites having nematicidal activity against root-knot nematodes (Poveda et al., 2020). These metabolites include 3-methoxyepicoccone, 4,5,6-trihydroxy-7-methylphthalide, chaetoglobosin A, chaetoglobosin B and flavipin. *Daldinia cf. concentrica* is observed to produce VOCs against root-knot nematodes. The activity of *Bursaphelenchus xylophilus* (pinewood nematode) is badly affected due to alternariol 9-methyl ether production by *Alternaria* spp. Moreover, the endophytic fungus *Fusarium oxysporum* is reported to induce systemic resistance against root-knot and burrowing nematodes. Similarly, *Fusarium moniliforme* is reported to develop ISR in rice against *M. graminicola*. This development occurs because of the production of 4-hydroxybenzoicacid, indole-3-acetic acid and gibepyrone D that not only are directly nematotoxic, but are also known for inducing defense response in plants (Poveda et al., 2020).

6.3.3 Genus *Trichoderma*

Trichoderma (teleomorph Hypocrea, Ascomycota and Dikarya) is a well-studied genus of fungi that is comprised of more than 200 species. *Trichoderma* spp. grow naturally in various habitats and are omnipresent colonialists of cellulose-based material; consequently, they are present in degenerating material and rhizosphere of plants (Lopez-Bucio et al., 2015). They assist plants to resist environmental stresses such as drought and salinity through supporting their growth and reprograming gene expression. The fungal mycelium produces various compounds that increase the branching capability of plant root system. These are potential biocontrol agents against plant pathogens because of antibiosis, mycoparasitism and competition with pathogens and defense stimulation in host against diseases (Hermosa et al., 2012). Various studies have proved their role in controlling plant parasitic nematodes; for example, *Trichoderma harzianum* is reported to reduce diseases caused by root-knot nematodes by reducing their egg hatching, the number of galls on plants, the number of egg masses and the number of egg masses per egg mass. The genus *Trichoderma* shows direct parasitism of eggs and juveniles of cyst nematodes, so it is also effective in controlling cyst nematodes. The fungus increases the levels of chitinase and protease enzymes, which allow its direct penetration into eggs because eggshell is mainly composed of chitin. Therefore, the number of viable eggs is reduced and so the number of second-stage juveniles. Spores of fungal species named *T. longibrachiatum* completely surround and destroy the cyst of *Heterodera avenae* by producing their enzymes. This fungus also shows inhibitory effects on *H. avenae* by hindering the development of eggs, juveniles and females (Zhang et al., 2014; Poveda et al., 2020).

Several investigations have been carried out to find the capability of *Trichoderma* spp. to induce resistance in numerous species of plants against plant parasitic nematodes, leading to metabolomic, proteomic and transcriptomic modifications in plants. This stimulation in systemic defense permits faster response of plant after pathogen attack; therefore, the chances of disease increase are reduced. This resistance is mostly controlled through jasmonic acid (JA) or ethylene signaling. Induction of defense in tomato against *M. incognita* is mediated by JA/ET and JA pathways (Martinez-Medina et al., 2017a, b).

TABLE 6.3

List of Fungal-Based Commercial Products against Plant Parasitic Nematodes

Product Name	Fungal Source	Company	Target	Reference
BioAct/MeloCon	*Paecilomyces lilacinus* strain 251	Bayer CropScience Biologics	Root-knot and cyst nematodes, *R. similis* and *Pratylenchus* spp.	Tranier et al. (2014)
Mycotrol O	*Beauveria bassiana* strain GHA	ARLA	Root-knot nematodes	Tranier et al. (2014)
Biostat	*Purpureocillium lilacinum*		Root-knot nematodes, *Ditylenchus* spp., *Tylenchulus* spp., *Pratylenchus* spp., *Radopholus* spp. and *Rotylenchulus* spp.	Tranier et al. (2014)
Mycobac	*Trichoderma lignorum*	Laboratorios Laverlam, Columbia	Plant parasitic nematodes	Woo et al. (2014)
Mycotal	*Verticillium lecanii*	Koppert	*M. incognita*	Tranier et al. (2014)
Micostat Ffito	*T. harzianum* TH01 and *Glomus* spp.	CCS Aosta S.r.l., Italy	Plant parasitic nematodes	Woo et al. (2014)
Trianum	*T. harzianum* T-22	Koppert	Plant parasitic nematodes	Tranier et al. (2014)
Met52	*Metarhizium anisopliae* F52		Plant parasitic nematodes	Tranier et al. (2014)
Nemaxxion Biol	*Trichoderma* spp. and *Paecilomyces* spp.	GreenCorp, Mexico	Root-knot nematodes	Tranier et al. (2014)
Rem G	*Arthrobotrys* spp., *Dactylella* spp., *Glomus* spp. and *Paecilomyces* spp.	Green Solutions, Italy	Plant parasitic nematodes	Tranier et al. (2014)

Commercial products of many fungal-based microbial biostimulants are available and are effective against a variety of soilborne plant pathogens. Some of them also have the potential to be used as nematicides against plant parasitic nematodes (Table 6.3).

6.4 CONCLUSIONS

Nematodes pose a severe danger to global food output. Microbial biostimulants, for example, are excellent biological control techniques for the large-scale management of these worms to boost agricultural output. These microbial biostimulants are effective in a completely different way in the field compared to traditional chemical-based nematicides as these trigger a chain of activities that result in nematode population suppression, stimulation of ISR and changes in biochemical processes in plants. These are eco-friendly and long-lasting in the environment and are also available in commercial formulations. To ensure more effective and environment-friendly agriculture, more exploration into the isolation of local strains is required.

REFERENCES

Abd-Elgawad, M.M.M., and T.H. Askary. 2018. Fungal and bacterial nematicides in integrated nematode management strategies, *Egypt. J. Biol. Pest Control.* 28: 74.

Aeron, A., E. Khare, C.K. Jha, V.S. Meena, S.M.A. Aziz, M.T. Islam, K. Kim, S.K. Meena, A. Pattanayak, and H. Rajashekara. 2020. Revisiting the plant growth promoting rhizobacteria: Lessons from the past and objectives for the future. *Arch. Microbiol.* 202: 665–676.

Affokpon, A., D.L. Coyne, L. Lawouin, C. Tossou, R.D. Agbèdè, and J. Coosemans. 2011. Effectiveness of native West African arbuscular mycorrhizal fungi in protecting vegetable crops against root-knot nematodes. *Biol. Fertil. Soils* 47: 207–217.

Agrios, G.N., ed. 2005. *Plant Pathology.* New York: Elsevier Academic Press.

Berruti, A., E. Lumini, R. Balestrini, and V. Bianciotto. 2016. Arbuscular mycorrhizal fungus as natural biofertilizers: let´s benefit from past successes. *Front. Microbiol.* 6:1559. doi:10.3389/fmicb.2015.01559.

Chen, Z.X., and D.W. Dickson. 1998. Review of *Pasteuria penetrans*: Biology, ecology, and biological control potential. *J. Nematol.* 30(3): 313–340.

Ciancio, A. 2018. Biocontrol potential of *Pasteuria* spp. for the management of plant parasitic nematodes. *CAB Rev.* 13: 1–13.

D'Addabbo, T., S. Laquale, M. Perniola, and V. Candido. 2019. Biostimulants for plant growth promotion and sustainable management of phytoparasitic nematodes in vegetable crops. *Agronomy* 9: 616.

Decraemer, W., and D.J. Hunt. 2006. Structure and classification. In: Perry, R.N., M. Moens. (eds.). *Plant Nematology.* Wallingford: CABI Publishing. doi:10.1079/9781845930561.0003.

Dominguez-Nunez, J.A., and A.S. Albanesi. 2019. Ectomycorrhizal fungi as biofertilizers in forestry. In: A. Beck, S. Uhac, and A.A. DE Marco (eds.). *Biostimulants in Plant Science.* Rijeka: IntechOpen. doi: 10.5772/intechopen.88585.

Ebone, L.A., M. Kovaleski, and C.C. Deuner. 2019. Review article nematicides: History, mode and mechanism action. *Plant Sci. Today* 6: 91–97.

Flor-Peregrín, E., R. Azcón, V. Martos, S. Verdejo-Lucas, and M. Talavera. 2014. Effects of dual inoculation of mycorrhiza and endophytic, rhizospheric or parasitic bacteria on the root-knot nematode disease of tomato. *Biocontr. Sci. Technol.* 24: 1122–1136.

Gamalero, E., and B.R. Glick. 2020. The use of plant growth-promoting bacteria to prevent nematode damage to plants. *Biology* 9: 381. doi: 10.3390/biology9110381.

Glick, B.R. 2020. Introduction to plant growth bacteria. In: *Beneficial Plant-Bacterial Interactions.* Cham. Springer International Publishing, pp. 1–28. doi: 10.1007/978-3-030-44368-9_1.

Hao, Z., L. Fayolle, D. Van Tuinen, O. Chatagnier, X. Li, and S. Gianinazzi. 2012. Local and systemic mycorrhiza-induced protection against the ectoparasitic nematode *Xiphinema index* involves priming of defence gene responses in grapevine. *J. Exp. Bot.* 63: 3657–3672. doi:10.1093/jxb/ ers046.

Hermosa, R., A. Viterbo, I. Chet, and E. Monte. 2012. Plant-beneficial effects of *Trichoderma* and of its genes. *Microbiology* 158: 17–25. doi:10.1099/mic.0.52274-0.

Hu, H.J., Y. Gao, X. Li, S.L. Chen, S.Z. Yan, and X.J. Tian. 2020. Identification and nematicidal characterization of proteases secreted by endophytic bacteria *Bacillus cereus* BCM2. *Phytopathology* 110: 336–344.

Jefwa, J.M., B. Vanlauwe, D. Coyne, P. van Asten, S. Gaidashova, E. Rurangwa, M. Mwashasha, and A. Elsen. 2010. Benefits and potential use of Arbuscular Mycorrhizal Fungi (AMF) in banana and Plantain (Musa spp.) systems in Africa. *Acta Hortic.* 879: 479–486.

Khabbaz, S.E., D. Ladhalakshmi, M. Babu, A. Kandan, V. Ramamoorthy, D. Saravanakumar, T. Al-Mughrabi, and S. Kandasamy. 2019. Plant Growth Promoting Bacteria (PGPB)- a versatile tool for plant health management. *Can. J. Pestic. Pest Manag.* 1: 1–25.

Kohl, J., R. Kolnaar, and W.J. Ravensberg. 2019. Mode of action of microbial biological control agents against plant diseases: Relevance beyond efficacy. *Front. Plant Sci.* 10: 845.

Lopez-Bucio, J., R. Pelagio-Flores, and A. Herrera-Estrella. 2015. *Trichoderma* as biostimulant: exploiting the multilevel properties of a plant beneficial fungus. *Sci. Hortic.* 196: 109–123. doi: 10.1016/j. scienta.2015.08.043.

Lugtenberg, B.J., J.R. Caradus, and L.J. Johnson. 2016. Fungal endophytes for sustainable crop production. *FEMS Microbiol. Ecol.* 92: 194. doi: 10.1093/femsec/fiw194.

Marro, N., P. Lax, M. Cabello, M.E. Doucet, and A.G. Becerra. 2014. Use of the arbuscular mycorrhizal fungus *Glomus intraradices* as biological control agent of the nematode *Nacobbus aberrans* parasitizing tomato. *Braz. Arch. Biol. Technol.* 57: 668–674.

Marro, N., M. Caccia, M.E. Doucet, M. Cabello, and A. Becerra. 2018. Mycorrhizas reduce tomato root penetration by false root knot nematode *Nacobbus aberrans. Appl. Soil Ecol.* 124: 262–265.

Martinez-Medina, A., F.V. Appels, and S.C. van Wees. 2017a. Impact of salicylic acid-and jasmonic acid-regulated defenses on root colonization by *Trichoderma harzianum* T-78. *Plant Signal. Behav.* 12: e1345404. doi: 10.1080/15592324.2017.

Martinez-Medina, A., I. Fernandez, G.B. Lok, M.J. Pozo, C.M. Pieterse, and S.C. van Wees. 2017b. Shifting from priming of salicylic acid-to jasmonic acid-regulated defenses by *Trichoderma protects* tomato against the root knot nematode *Meloidogyne incognita*. *New Phytol.* 213: 1363–1377. doi: 10.1111/nph.14251.

Migunova, V.D., and N. Sasanelli. 2021. Bacteria as biological tool against phytoparasitic nematodes. *Plants* 10: 389.

Mohan, S., T.H. Mauchline, J. Rowe, P.R. Hirsch, and K.G. Davies. 2012. *Pasteuria* endospores from *Heterodera cajani* (Nematoda: Heteroderidae) exhibit inverted attachment and altered germination in cross-infection studies with *Globodera pallida* (Nematoda: Heteroderidae). *FEMS Microbiol. Ecol.* 79: 675–684.

Mohan, S., K.K. Kumar, V. Sutar, S. Saha, J. Rowe, and K.G. Davies. 2020. Plant root-exudes recruit hyper-parasitic bacteria of phytonematodes by altered cuticle aging: Implications for biological control strategies. *Front. Plant Sci.* 11: 763.

Poveda, J., P. Abril-Urias, and C. Escobar. 2020. Biological control of plant-parasitic nematodes by filamentous fungi inducers of resistance: *Trichoderma*, Mycorrhizal and Endophytic Fungi. *Front. Microbiol.* 11: 992. doi:10.3389/fmicb.2020.00992.

Schouten, A. 2016. Mechanism involved in nematode control by endophytic fungi. *Annu. Rev. Phytopathol.* 54: 121–142. doi: 10.1146/annurev-phyto-080615-100114.

Sharma, I.P., and A.K. Sharma. 2017. Co-inoculation of tomato with an arbuscular mycorrhizal fungus improves plant immunity and reduce root-knot nematode infection. *Rhizosphere* 4: 25–28. doi: 10.1016/j.rhisph.2017.

Tiran, B., J. Yang, and K.Q. Zhang. 2007. Bacteria in the biological control of plant-parasitic nematodes: Populations, mechanisms of action, and future aspects. *FEMS Microbiol. Ecol.* 61: 197–213.

Tranier, M-S., J. Pognant-Gros, R.D.C. Quiroz, C.N.A. Gonzalez, T. Mateille and S. Roussos. 2014. Commercial biological control agents targeted against plant-parasitic root-knot nematodes. *Braz. Arch. Biol. Technol.* 57: 831–841.

Woo, S.L., M. Ruocco, F. Vinale, M. Nigro, R. Marra, N. Lombardi, A. Pascale, S. Lanzuise, G. Manganiello, and M. Lorito. 2014. *Trichoderma*-based products and their widespread use in agriculture. *Open Mycol. J.* 8: 71–126.

Xiang, N., K.S. Lawrence, and P.A. Donald. 2018. Biological control potential of plant growth-promoting rhizobacteria suppression of *Meloidogyne incognita* on cotton and *Heterodera glycines* on soybean: A review. *J. Phytopthol.* 66: 449–458.

Yan, L., J. Zhu, X. Zhao, J. Shi, C. Jiang, and D. Shao. 2019. Beneficial effects of endophytic fungi colonization on plants. *Appl. Microbiol. Biotechnol.* 103: 3327–3340. doi:10.1007/s0025.

Zhang, L., J. Zhang, P. Christie, and X. Li. 2008. Pre-inoculation with arbuscular mycorrhizal fungi suppresses root knot nematode (*Meloidogyne incognita*) on cucumber (*Cucumis sativus*). *Biol. Fertil. Soils.* 45: 205.

Zhang, S., Y. Gan, B. Yu, and Y. Xue. 2014. The parasitic and lethal effects of *Trichoderma longibrachiatum* against *Heterodera avenae*. *Biol. Control.* 72: 1–8. doi: 10.1016/j.biocontrol.2014.01.009.

7 Nitrogen-Fixing Biofertilizers and Biostimulants

Gulab Khan Rohela and Pawan Saini
Central Sericultural Research and Training Institute,
Central Silk Board, Pampore

CONTENTS

7.1 INTRODUCTION

Plants need nutrient elements such as nitrogen, phosphorous, potassium, sulfur and magnesium for normal growth and development. Among all the nutrient elements, 16 elements are regarded as essential for the normal growth of plants. Among them, 13 mineral elements are taken up by the plants from the soil; whereas carbon, hydrogen and oxygen are from atmosphere. Plants take up most of the nutrient elements in soluble form from the soil in a continuous manner for growth and crop productivity (Kauffman et al. 2007). This leads to the depletion of soluble nutrient elements from the soil and causes a threat to sustainability in crop production. Even though certain mineral elements are present in plenty, they are in insoluble form. So plants cannot uptake them in insoluble form and they need to be converted into soluble form for making them absorbable by the plants (Du Jardin 2012). For example, phosphorous, nitrogen and other elements are present in large amounts in the soil, but they are usually found in insoluble form. Further, the plants will utilize only part of the applied chemical fertilizers and rest of them will be deposited in the soil and converted to insoluble form.

DOI: 10.1201/9781003188032-7

It is understood that by the application of chemical fertilizers, the load of insoluble nutrient elements in the soil is increasing year by year. The increased load of insoluble form of chemical fertilizers is not only decreasing the fertility rate of the soil, but also having its detrimental effects on the count of useful soil microbiota. To convert and reuse the insoluble form of nutrient elements, the concept of biofertilizers has come into the limelight. The term biofertilizers includes the application of selective soil microorganisms such as algae, bacteria and fungi that are capable of either fixing the nitrogen or converting the insoluble form of mineral elements into soluble and absorbable form to plants (Yakhin et al. 2017). Biofertilizers have several advantages over the chemical fertilizers as they are eco-friendly, cost-effective and easy to multiply; improve the fertility of soil; and create rapid transformation of soil constituents. They are considered as renewable form of fertilizers and provide a sustainable agricultural production system.

Besides biofertilizers, there is another important group designated as biostimulants, which are very much useful for plant growth and development. Generally, biostimulants are comprised of diverse compounds and microorganisms. This is the most effective approach to making essential nutrients available to plants for the enhancement of plant growth. Several workers (Traon et al. 2014; Calvo et al. 2014; Yakhin et al. 2017) have defined biostimulants differently, but all the definitions point toward the enhancement in the plant growth and development. Of these definitions, Du Jardin (2012) provides a comprehensive definition as follows: Biostimulants can be a substance or microorganism or any other material that enhances the overall phenological performance with respect to growth, development, cellular differentiation, efficient uptake of nutrients and water, tolerance toward abiotic stresses and production of quality harvest. It is also to mention that biostimulants do not have any effect on insect pests; therefore, these cannot be categorized under pesticide (European Biostimulants Industry, 2012).

The most important task in most of the countries is to produce and provide sufficient food for the growing population. The utilization of biofertilizers and biostimulants is beneficial and provides shifting fortunes in agriculture. Among all these types, nitrogen-fixing biofertilizers and biostimulants are of most importance as they are supplying the most critical and majorly required element, *i.e.*, nitrogen, to the plants. In this book chapter, nitrogen-fixing biofertilizers (bacterial, algal and fungal organisms) and biostimulants are discussed in detail.

7.2 CLASSIFICATION OF BIOFERTILIZERS AND BIOSTIMULANTS

These two important classes which include useful substances or microorganisms are less expensive and environment-friendly for sustainable agriculture. Generally, biofertilizers are comprised beneficial microbes that not only are helpful in the growth of crop plants, but also improve the soil health status. Based upon the type of microbe, the biofertilizers can also be classified as bacterial (*Rhizobium, Azospirillum, Azotobacter, Phosphobacteria*), fungal (Mycorrhiza), algal (blue green algae and *Azolla*) and actinomycetes biofertilizers (*Frankia*) (Sharma 2004). On the other hand, biostimulants also improve the overall growth and development in trace amount. The very first classification of biostimulants was given by Filatov (1951b), where four classes were formed. After five decades, Karnok (2000) grouped 59 materials into 15 biostimulants. Ikrinaww and Kobin (2004) prepared a list of nine biostimulants that consist of natural raw substances. Biostimulant components involved in the formulations, active ingredients and mode of action were proposed as some of the criteria by Basak (2008) for grouping the biostimulants. Du Jardin (2012, 2015) has categorized the biostimulants into eight and seven categories. The classification of biostimulants into different categories is presented in Table 7.1.

7.3 ROLE OF BIOFERTILIZERS AND BIOSTIMULANTS IN SUSTAINABLE AGRICULTURE

7.3.1 Biofertilizers

Chemical-based fertilizers provide essential nutrients such as nitrogen, phosphorous and potassium, which are important for plant growth and development and lead toward enhancement in

TABLE 7.1

Classification of Biostimulants Proposed by Several Workers (Yakhin et al. 2017)

Filatov (1951b)	Ikrina and Kolbin (2004)	Kauffman et al. (2007)	Du Jardin (2012)	Calvo et al. (2014)	Halpern et al. (2015)	Du Jardin (2015)	Torre et al. (2016)
Carboxylic fatty acids (oxalic and succinic acid)	Microorganisms (bacteria and fungi)	Humic substances	Humic substances	Microbial inoculants	Humic substances	Humic and fulvic acids	Humic substances
Carboxylic fatty hydroxy acids (malic and tartaric acid)	Plant materials (land, freshwater and marine)	Hormone-containing products (seaweed extracts)	Complex organic materials	Humic acids	Protein hydrolysates and amino acid formulations	Protein hydrolysates and other N-containing compounds	Seaweed extracts
Unsaturated fatty acids, aromatic and phenolic acids (cinnamic and hydroxycinnamic acids, and coumarin)	Sea shellfish, animals and bees	Amino acid-containing products	Beneficial chemical elements	Fulvic acids	Seaweed extract	Seaweed extracts and botanicals	Hydrolyzed proteins and amino acids
Phenolic aromatic acids containing several benzene rings linked via carbon atoms (humic acids)	Humate- and humus-containing substances		Inorganic acids	Protein hydrolysates and amino acids	Plant growth-promoting microorganisms	Chitosan and other biopolymers	Inorganic salts
	Vegetable oils		Chitin and chitosan derivatives	Seaweed extracts		Inorganic compounds	Microorganisms
	Natural minerals		Antitranspirants			Beneficial fungi	
	Water (activated, degassed and thermal)		Free amino acids and other N-containing substances			Beneficial bacteria	
	Resins						
	Other raw materials (oil and petroleum fractions, and shale substances)						

overall crop productivity. But on the contrary, the excessive utilization of chemical-based fertilizers produces adverse effects on soil rhizosphere, which is the native of several agriculturally important microorganisms. Among the essential nutrient elements, nitrogen is the major nutrient element as it is involves in cellular differentiation, enzymatic activities, biosynthesis of several pigments and compounds, etc. Biofertilizers are an alternate way for achieving the goal of enhanced crop productivity including small and marginal farmers in an economic way. A number of nitrogen-fixing microorganisms are known, of which few such as *Rhizobium, Azospirillum, Azotobacter* and *Azolla* are commercially exploited. Biofertilizers are inputs containing microorganisms, which are capable of mobilizing nutritive elements from non-usable to usable form through biological processes such as biological nitrogen fixation (BNF), phytohormone and siderophore production, nutrient solubilization, ACC deaminase and antifungal activity (Bashan and Holguin 1997). These nitrogen-fixing biofertilizers are helpful in the establishment and growth of crop plants and enhance the biomass production and grain yield. Rhizobium is generally utilized for biofertilizers. These biofertilizers are also known as plant growth-promoting rhizobacteria (PGPR), which not only enhance the plant growth, but also improve the soil health status, prevent soil erosion, acidification and nutrient leaching, and increase water-holding capacity (Bhatt and Bhatt 2021). In this way, biofertilizers play an important role in several agro-horti-forestry-based cropping systems and make agriculture more sustainable.

7.3.2 Biostimulants

Biostimulants are very diverse in nature as these include natural or biological substances/ microorganisms and promote plant growth and development. They also have diverse physiological functions such as increase in photosynthesis, nutrient uptake and their mobilization, and improvisation of harvest quality with respect to enhancement in protein, micronutrients and tolerance to abiotic stresses (Du Jardin 2015). Biostimulants do not act as fertilizers as these do not serve any mineral nutrients to plants though they improve the mineral composition in different plant tissues. It has been found that biostimulants are helpful in imparting the resistance/ tolerance toward salinity and drought stress and enhancing the yield, quality and nutrient uptake in various crop plants as reviewed by Calvo et al. (2014), Halpern et al. (2015), Du Jardin (2015) and Paradikovic et al. (2019) and very much helpful in sustainable agriculture.

7.4   NITROGEN-FIXING BIOFERTILIZERS

Among the major nutrient elements, nitrogen element is required in large amounts to the plants for growth and development as nitrogen is a key element in the biosynthesis of amino acids, purines, pyrimidines, proteins, etc. The nitrogen element is fixed in plants either through industrial nitrogen fixation by the application of chemical fertilizers, or through BNF by the usage of nitrogen biofertilizers such as bacterial (*Azotobacter, Azospirillum, Beijerinckia, Nitrosomonas, Nitrobacter* and *Rhizobium*), algal (*Cyanobacteria* and *Anabaena*) and fungal (*Alternaria, Cladosporium, Phoma* and *Rhodotorula*) organisms. These nitrogen-fixing microorganisms possess *nif* genes coding nitrogenase enzyme, which can convert the inert nitrogen into plant-absorbing form of ammonia.

7.4.1 Bacterial Biofertilizers

7.4.1.1   Azotobacter

The genus *Azotobacter* is a bacterial genus with several species which can fix the nitrogen into plants in a non-symbiotic way; hence, the bacteria belonging to this genus are called non-symbiotic nitrogen fixers. Beijerinck (1901) was the first person to isolate and culture the species of *Azotobacter*, which are of oval shape and quite large in size (up to 3μm wide and 10μm long) when compared to other bacteria. *Azotobacter* group of bacteria are gram negative, and some of them produce colored pigments. This bacterium is usually found in alkaline and neutral soils, in the rhizosphere region and also in freshwater bodies with slime-like layer around it. Beijerinck (1901) discovered that *Azotobacter* can fix the atmospheric nitrogen by without associating or without any symbiotic relationship with plant; i.e., they can fix the nitrogen in a free-living state; hence, they are considered as non-symbiotic nitrogen-fixing microorganisms.

This genus is in use as biofertilizers since more than a century (Gerlach and Vogel 1902). Among the different species, the most important in terms of nitrogen fixation are *Azotobacter agilis*, *A. armeniacus*, *A. brasilense*, *A. beijerinckii*, *A. chroococcum*, *A. insignis*, *A. macrocytogenes*, *A. nigricans*, *A. paspali*, *A. salinestris*, *A. tropicalis* and *A. vinelandii* (Mulder and Brontonegoro 1974). Among the listed species, at genome level, most of the work is carried upon the species of *A. vinelandii* (Setubal et al. 2009). Because of its plant growth-promoting activity, *Azotobacter* is considered as a beneficial microorganism in promoting the sustainable agricultural crop production (Jimenez et al. 2011).

Azotobacter is considered as a PGPR as it has various advantageous effects on plant growth through the release of biologically active substances, by promoting the growth of rhizospheric microorganisms, by improving plant nutrition, by non-symbiotic BNF, by improving the soil fertility and by the production of inhibitor substances, which act against plant pathogens found in the soil (Lenart 2012; Kurrey et al. 2018). Azotobacter utilizes the atmospheric nitrogen for its cellular protein synthesis, which is later mineralized and released into the soil. In this way, Azotobacter fixes the atmospheric nitrogen in a non-symbiotic manner into the soil, which will be available as a considerable amount of nitrogen for utilization by crop plants (Gothandapani et al. 2017). This group of bacteria also produces growth hormones and releases siderophores, which in turn help in improving the plant nutrition and promoting the plant growth (Ansari et al. 2017).

Among the various species of Azotobacter, *A. chroococcum* and *A. vinelandii* are commonly found in the rhizosphere soils. *A. chroococcum* is reported as the first nitrogen fixer, which is a free-living and aerobic bacterium (Beijerinck 1901). *A. chroococcum* is also investigated as a microbial inoculant in agriculture; the usage of this bacterium as an inoculant has resulted in improved crop productivity through the production of plant growth-promoting substances (Gothandapani et al. 2017). *A. chroococcum is* reported to produce plant growth hormones such as indole-3-acetic acid (IAA), cytokinins and gibberellins-like compounds, which promote the growth of plants and also impose positive effects on the soil microflora (Hennequin and Blachere 1966; Brown et al. 1968; Wani et al. 2013).

A. chroococcum and *A. vinelandii* are reported to produce B vitamins such as biotin, pantothenic acid, niacin and riboflavin (Revillas et al. 2000). *A. chroococcum* and *A. vinelandii* are also reported to produce amino acids such as alanine, arginine, aspartic acid, cysteine, glutamic acid, glycine, histidine, isoleucine, leucine, lysine, methionine, phenyl alanine, proline, serine, threonine, tryptophan, tyrosine and valine (Lopez et al. 1981; Revillas et al. 2000). *A. vinelandii* is reported to excrete different types of siderophores in low-iron-containing soils. This will help the plants to uptake iron from the produced siderophores. *Azotobacter* also has the ability of nutrient cycling; it makes the nutrient elements such as sulfur, phosphorous and nitrogen available to plants by accelerating the mineralization process of organic residues present in the soil (Levai et al. 2008). Azotobacter also produces different antibiotics that will help in combating the plant pathogens (Kraepiel et al. 2009).

7.4.1.2 Azospirillum

Azospirillum is a genus of bacteria which is known to influence the plant growth positively through its growth-promoting substances and by the associative type of nitrogen fixation in plants. These are gram-negative, free-living nitrogen-fixing bacteria that can adhesively attach the root hairs of plants and can fix the atmospheric nitrogen associatively (Bashan et al. 2004). They are present in large numbers in the rhizospheric region of cereal crops and grasses across the tropical and temperate climates all over the world (Patriquin et al. 1983). Initially, the genus *Spirillum* was reported first by Beijernick (1925), and later, it was renamed as *Azospirillum* due to its ability to fix atmospheric nitrogen. *Azospirillum* species were mainly studied in association with *Saccharum* spp., *Oryza sativa*, *Triticum aestivum*, *Zea mays*, etc. (Fukami et al. 2016).

Until now, 22 species of *Azospirillum* have been reported; among them, the important species are *Azospirillum amazonense*, *A. brasilense*, *A. halopraeferens*, *A. irakense* and *A. lipoferum* (Tarrand et al. 1978; Reinhold et al. 1987; Khammas et al., 1989). Among the listed species of *Azospirillum*, *A. brasilense*, *A. amazonense* and *A. lipoferum* have widely been studied for their physiological and genetic characteristics using whole genome sequencing (Baldani and Baldani 2005; Sant Anna et al. 2011; Fibach-Paldi et al. 2012). Colonization of *A. lipoferum* in rice (*Oryza sativa*) and *A. brasilense* in wheat (*Triticum aestivum*), *Arabidopsis* and maize (*Zea mays*) plants are well studied (Spaepen et al. 2014; Fukami et al. 2018).

Apart from fixing the nitrogen, *Azospirillum* also implies other positive effects on the plant growth and development. *Azospirillum* produces plant growth hormones such as abscisic acid (Cohen et al. 2008), auxins (Spaepen and Vanderleyden 2015), cytokinins (Tien et al. 1979), ethylene (Strzelczyk et al. 1994), gibberellins (Bottini et al. 1989), nitric oxide (Fibach-Paldi et al. 2012) and salicylic acid (Sahoo et al. 2014). Phytohormones which are produced by *Azospirillum* can greatly affect the growth of root system and help in the uptake of mineral nutrients from soil (Ardakani and Mafakheri 2011). *Azospirillum* also helps the plants in combating the abiotic stresses of drought and salinity (Creus et al. 2004; Kim et al. 2012). *Azospirillum* helps in the biological control of plant pathogens by limiting the iron availability to the pathogens (Khan et al. 2002; Romero et al. 2003). Reports are there regarding the solubilization of insoluble phosphates by *Azospirillum* and making available the utilizable form of phosphorous to the plants (Turan et al. 2012).

The ability of *Azospirillum* to form strong associations with plant root system was first discovered by Bulow and Dobereiner (1975). Due to the beneficial effects of *Azospirillum* on plant growth, improved tolerance to stress conditions and improved soil fertility, this bacterial genus is commercialized in several countries such as Argentina, Australia, Brazil, China, France, India, Mexico, Paraguay, South Africa and Uruguay (Cassan and Diaz Zorita 2016). *Azospirillum* when used as inoculants in agricultural fields causes changes in the roots' architecture. The change in roots' architecture is induced by the production of phytohormones such as indole-3-acetic acid by the bacteria (Cassan et al. 2014). Based on the specific association of *Azospirillum* species with root system of crop plants, they are named as *A. oryzae* (rhizosphere of *Oryza sativa*) (Xie and Yokota 2005), *A. melinis* (association with tropical molasses grass *Melinis minutiflora*) (Peng et al. 2006) and *A. zeae* (rhizosphere of *Zea mays*) (Mehnaz et al. 2007).

Among all the species of *Azospirillum*, *A. brasilense*, *A. halopraeferens*, *A. lipoferum* and *A. oryzae* were widely used as biofertilizers, especially in cereal crops for better growth of plants and for increased productivity. *Azospirillum* has multiple positive effects on the growth of plants, which is best explained by the "multiple mechanism theory" (Bashan and Levanony 1990). The multiple positive effects induced by *Azospirillum* include associative type of nitrogen fixation, production of different phytohormones, biosynthesis and release of amino acids, and mineralization of phosphorous and other elements. These multiple factors lead to successful *Azospirillum*–plant interaction resulting in better nutrient uptake, favoring the root growth and countering the plant pathogens and withstanding the stressful conditions with better tolerance (Mehnaz 2015; Vejan et al. 2016). The accelerated growth of plants is better in the presence of *Azospirillum*-based biologically fixed nitrogen in comparison with chemical fertilizer-based nitrogen source (Dobereiner and Day 1975).

Azospirillum as microbial inoculants in agricultural fields is reported to increase the plant height, number of leaves, stem diameter, length of the roots, number of root hairs, plant dry weight and finally the yield in cereal crops (Bashan and de-Bashan 2010). Based on the revelation of positive results of *Azospirillum* as microbial inoculants in rice fields, later several researchers used *Azospirillum* as an inoculant on other crop plants such as tomato (Bashan et al. 1989; Ribaudo et al. 2006), wheat (Saubidet et al. 2002), maize (Dobbelaere et al. 2001), sorghum (Baldani et al. 1986), pearl millet (Tien et al. 1979) and cucumber (Pereyra et al. 2010).

7.4.1.3 Rhizobium

Plants require nitrogen in huge amounts as a macronutrient for the biosynthesis of amino acids, proteins and nitrogen bases and for the synthesis of photosynthetic chlorophyll pigment (Buren and Rubio 2017). Nitrogen element plays a critical role in plants' active growth and development. Even though there is plenty of nitrogen available in the atmosphere, it is found in unutilizable gaseous form, *i.e.*, the dinitrogen gas (N_2). Plants can assimilate the nitrogen either in reduced form (ammonia), or in oxidized form (nitrates and nitrates) (Wagner 2012). The biological conversion of dinitrogen (N_2) into ammonia by organisms such as bacteria, archaea and algae is of utmost importance for providing nitrogen to the plants.

Rhizobia are a group of bacteria that interact symbiotically with plants, most commonly with leguminous crops (Sprent and James 2007). In this symbiotic relationship, bacteria induce the formation of nodules on roots and stem regions of plants and provide nitrogen to plants in reduced form (ammonia) by converting it from dinitrogen, *i.e.*, atmospheric nitrogen gas, with the help of nitrogenase enzyme and, in turn, bacteria uptake other nutrient elements from the plants (Vance 1998). This group of bacteria in association with leguminous crops fixes the atmospheric nitrogen and thereby increases the soil fertility, improves the plants growth and limits the usage of chemical fertilizers (Ouma et al. 2016). In agricultural areas, the symbiotic relationships between rhizobia and legumes alone are estimated to fix approximately 80% of biological nitrogen (O'Hara 1998).

Rhizobia are gram-negative bacteria belonging to the group of either alphaproteobacteria or betaproteobacteria. The Rhizobia bacteria are represented with eight families and 18 genera. The eight families of Rhizobia are *Allorhizobium, Bradirhizobiaceae, Brucelaceae, Burkholderiaceae, Methylobacteriaceae, Pararhizobium, Phyllobacteriaceae* and *Rhizobiaceae* (Lindstrom and Mousavi 2020). Among the 18 genera of Rhizobia, the genus *Rhizobium* belonging to *Rhizobiaceae* family is the largest one with 112 species (Mousavi 2016). *Rhizobium* is the first reported genus of Rhizobia bacteria that can interact with leguminous plants resulting in the formation of root nodules and fixing of nitrogen. The rhizobium–legume interaction starts with a specific signal exchange between the plants and bacteria (Oldroyd 2013); upon recognition of bacteria, the plant will induce cell divisions in cortical cells to form a primordial of root nodule, which leads to the formation of infection thread. In the infection thread, the bacteria get invaded and make a symbiotic association with plants (Jones et al. 2007).

The successful legume–*Rhizobium* interactions and development of root nodules will help in the incorporation of atmospheric nitrogen into soil ecosystems and account for more than 50% of BNF on land (Tate 1995). Moreover, *Rhizobium*-based BNF is compared as a renewable source of nitrogen in agricultural fields (Peoples et al. 1995a). In legume crops, *Rhizobium* can assimilate approximately about 300 Kg of nitrogen per hectare of agricultural land in a year (Peoples et al. 1995b). However, the amount of nitrogen fixed differs according to the way and method of application system (Sellstedt et al. 1993). The symbiotic relation between leguminous crops and *Rhizobium* can adequately supply and fix nitrogen into agricultural fields. In other research studies, it is revealed that intercropping of non-leguminous crops (*Zea mays*) with leguminous crops (*Arachis hypogaea*) can adequately supply nitrogen in an eco-friendly manner, which can improve the growth and productivity of non-leguminous crops (Mandimba 1995).

Along with the role of BNF, Rhizobia bacteria also involve in the phosphorous solubilization, siderophore formation and its secretion into the soil and production of phytohormones such as abscisic acid, cytokinins, gibberellins and indole-3-acetic acid. Rhizobia species such as *Bradyrhizobium japonicum, Rhizobium leguminosarum* and *R. meliloti* can solubilize the phosphorus from insoluble phosphate form and can provide it to plants (Marra et al. 2011; Hemissi et al. 2015). Rhizobia bacteria can produce siderophores that can sequester iron from the soil and provide it to plants, which will help the plants to uptake iron from the siderophores (Chabot et al. 1996). For easier and faster growth of root nodules, Rhizobia bacteria synthesize indole-3-acetic acid, an auxin responsible for the active growth of plant roots and root nodules. This will enable the bacteria to fix atmospheric nitrogen more efficiently (Mishra et al. 2009).

Certain strains of *Rhizobium* bacterium is also reported for the synthesis of cytokinins and gibberellins, which will induce active cell division and are responsible for root growth (Boiero et al. 2007; Senthilkumar et al. 2009). *Rhizobium* and *Bradyrhizobium* were also reported for the synthesis of abscisic acid, a phytohormone which protects plants from combating abiotic stresses such as drought and low-temperature stress (Dobbelaere et al. 2003). Rhizobium is also involved in providing resistance in plants against plant pests and pathogens (Huang et al. 2007; Siddiqui et al. 2007). In conclusion, the rhizobium provides multiple benefits to the plants for their active growth, defense against pathogens and survival under stress conditions.

7.4.2 Algal Biofertilizers

7.4.2.1 Blue Green Algae (Cyanobacteria)

Blue green algae (BGA) are generally known as the paddy microorganism as they are usually found in paddy fields. This group of biofertilizers include several algae such as *Aulosira, Anabaena, Cylindrospermum, Nostoc, Plectonema, Tolypothrix, Schizothrix and Calothrix* (Prasad and Prasad 2001). Most of the nitrogen-fixing BGA are filamentous and consist of chain of vegetative cells including specialized cells called heterocysts that function as a micro-nodule for the synthesis and fixation of nitrogen. BGA also synthesize and liberate growth-promoting substances such as auxins and amino acids that stimulate rice growth (Sharma 2004). The technique of BGA production in abundant amount is developed in Japan through algalization technique. Algalization is the application of BGA culture in field as a biofertilizer, and it was proposed by Venkataraman in the year 1961. It is suggested that nitrogen-fixing BGA enhance the fertility of tropical rice field. The BGA act as photosynthetic nitrogen fixers and utilize the energy derived from the photosynthesis process. They are known as cyanobacteria and are free-living organisms. These are usually found in both marine and freshwater and occur abundantly in marine water (Pisciotta et al. 2010; Williams and Laurens 2010; Milledge 2011; Hoekman et al. 2012). These are the most suitable class of biofertilizer for rice. The cyanobacteria have the ability to convert inorganic nitrogen into organic nitrogen for its utilization by plants (Kumar 2016).

The major functions of these cyanobacteria are (i) maintenance of soil fertility by increasing soil porosity and production of adhesive substances; (ii) phytohormones (auxin, gibberellin, cytokinin), amino acids and vitamins excretion (Roger and Reynaud 1982; Rodriguez et al. 2006); (iii) improvement in the soil water-holding capacity (Roger and Reynaud 1982); (iv) enhancement in soil biomass through death and decomposition (Saadatnia and Riahi 2009); (v) reduction in soil salinity (Saadatnia and Riahi 2009); (vi) reduction in weed growth (Saadatnia and Riahi 2009); (vii) increase in soil phosphate (Wilson 2006); and (viii) bioremediation (Ibraheem 2007). In different crop plants, different types of algal biofertilizer combinations have been reported, which leads to improvement in plant growth and yield (Thajuddin and Subramanian 2005).

Azolla, a small floating water fern, is another useful alga that is widely utilized as a biofertilizer in rice. It is widely distributed in rice-growing tracts of tropics and temperate zones and grows on the irrigated rice fields. There are six species of Azolla, viz. A. caroliniana, A. nilotica, A. filiculoides, A. mexicana, A. microphylla and A. pinnata. Of these, A. pinnata is commonly present in India. A. microphylla which is heat tolerant is introduced from Latin America in India. A. filiculoides is a temperate species received from Europe. In rice, Azolla has been used with Anabaena, which is free-living cyanobacteria as Azolla–Anabaena complex (Vaishampayan et al. 2001).

7.4.3 Fungal Biofertilizers

The microorganisms that influence the plants' health through plant–microbe interactions, which can lead to the active growth and development of plants as well as improvement of soil fertility, are called biofertilizers (Malusa et al. 2012). Along with bacteria, several soil fungi are also regarded as biofertilizers due to their active role in providing nutrients to plants either symbiotically or non-symbiotically.

7.4.3.1 Mycorrhizae

The term mycorrhizae includes the association of fungi with the root system of plants (Frank 1885). This mycorrhizal association benefits the terrestrial ecosystems and tree species for the uptake of nutrients from the soil (Brundrett 2009). The mycorrhizal fungi are considered as the biofertilizers as they will help in improving the plants' health through the uptake of mineral elements. The mycorrhizae are broadly classified into two subdivisions known as ectomycorrhizae and endomycorrhizae based on the association of fungi either externally or internally.

7.4.3.1.1 Ectomycorrhizal Fungi

Ectomycorrhizal fungi mostly belong to *Ascomycetes*, *Basidiomycetes* and *Zygomycetes* – very few members – where they will form a Hartig network and mantle structure in association with the plant root system. The Hartig network acts as an interface between the plant root and fungi for the exchange of nutrients and metabolites (Frank 1885). The mantle formed by the ectomycorrhizal fungi is connected to mycelia filaments of fungi which extend into the soil, because of which they will carry out the mobilization, solubilization, absorption and translocation of mineral nutrient elements and water from the soil to the root system of plants. Some of the beneficial ectomycorrhizal fungal associations with crops were studied in *Tricholoma matsutake*, *Cantharellus* spp., *Boletus edulis*, etc.

7.4.3.1.2 Endomycorrhizal Fungi

An endomycorrhizal fungus makes a strong symbiotic association with the root system of higher plant species. They invade the plant root system and form specialized structures inside the root cells such as arbutoids, arbuscles, ericoids, vesicles and monotropoids (Smith and Read 2008). Endomycorrhizae also extend their hyphae into the intercellular spaces of root cells. Based on the specialized structures formed, they are classified as arbutoid mycorrhizae, arbuscular mycorrhizae, ericoid mycorrhizae and monotropoid mycorrhizae (Peterson et al. 2004).

7.4.3.2 Arbuscular Mycorrhizal Fungi

Arbuscular mycorrhizal fungi (AMF) are a small group of endomycorrhizal fungi that play an important role in plants' active growth through symbiotic association with root system. It enables the plants to uptake the nutrient elements from soil, supplies water content and also provides protection against plant pathogens, and in return, AMF uptake photosynthetic products from the plants (Berruti et al. 2016). AMF can interact with almost 90% of plants available on the land as they lack the specificity in making the symbiotic associations (Brachmann and Parniske 2006). The AMF act as an interface between soil and plants as the mycelium of AMF can grow inside the root system and soil; through the mycelia, plants can uptake nutrients and water from the soil (Smith and Read 2008). The AMF can assimilate nutrients from distant areas that are inaccessible to the roots of plants (Smith and Smith 2012). AMF are most markedly recognized for the mobilization of phosphorous and providing it to the plants through symbiotic association; hence, AMF are most commonly regarded as phosphate biofertilizers (Smith et al. 2011). Some of the important genera of AMF that can mobilize the phosphorous from the soil are *Acaulospora*, *Gigaspora*, *Glomus* and *Scutellospora*.

7.4.3.3 Fungal Organisms as Phosphate and Potassium Solubilizers

Various species of *Cladosporium* spp. have the ability to solubilize the potassium from the soil (Bahadur et al. 2016). Thus, *Cladosporium* spp. can be used as a potassium solubilizer to provide potassium to plants. Other fungi which have the ability to solubilize potassium are *Aspergillus* sp. and *Penicillium* sp. *Penicillium* and *Aspergillus* fungi were also used as phosphate solubilizers to provide phosphorous to plants (Anand et al. 2016). *Aspergillus niger* fungal organism when used as a phosphate solubilizer increased the height of the plant, size of leaves, number of fruits and fruit size in treated plants in comparison with untreated plants (Din et al. 2019). Mycorrhizal fungi were reported to be utilized as zinc solubilizers for providing zinc microelement to plants (Raj 2007). The nematofungus *Arthrobotrys oligospora* was also reported as a phosphate solubilizer (Duponnois et al.

2006). Among the *Aspergillus* spp., *A. awamori*, *A. candidus*, *A. clavatus*, *A. flavus, A. foetidus*, *A. fumigatus*, *A. niger*, *A. nidulans*, *A. ochraceus*, *A. parasiticus*, *A. rugulosus*, *A. sydowii*, *A. terreus*, *A. tubingensis, A. versicolor* and *A. wentii* were reported as phosphate solubilizers (Tarafdar et al. 2003; Pradhan and Sukla 2005; Mittal et al. 2008).

7.5   NITROGEN-FIXING BIOSTIMULANTS

Recently, plant biostimulants and their applications in agro-horti-forestry crop plants have been reviewed by several workers, and they provide comprehensive information regarding plant biostimulants and their beneficial effects. The role of three important biostimulants, *i.e.*, humic substances, protein-based products and seaweed extracts, is briefly discussed here.

7.5.1   Humic Substances

Humic substances are organic materials that are derived from the decomposition of flora and fauna, which include soil microbes, plants and animal dead material existing in soil acting as a substrate for the formation of humic substances (Nardi et al. 2017; Du Jardin 2015). The humic substances consist of organic acids such as humic and fulvic acids. The humic substances form more than 60 percent fraction of global soil organic materials (Muscolo et al. 2007). These organic heterogeneous substances have been classified based on their molecular weight and solubility in organic acids such as humic acid and fulvic acid (Du Jardin 2015). Humic acid is soluble in basic media and requires alkali for its extraction from soil. These are high molecular weight substances. On the contrary, the fulvic acid is soluble in both acid and alkali media and has low molecular weight (Stevenson 1994; Berbara and Gracia 2014; Nardi et al. 2009). The soluble humic substances generally occur in freshwater (Steinberg et al. 2008). There are several ways of humic substance application in plants, i.e., foliar (Yildirim 2007; Katkat et al. 2009), irrigation (Salman et al. 2005) and direct application to soil (Katkat et al. 2009). Humic substances play a diverse role in the phenological stages of plants by positively influencing the growth, physiology and metabolism and by ameliorating abiotic stress tolerance. These can successfully be extracted naturally from organic materials such as volcanic eruption-derived soils and organic vermicompost or other organic residues such as silkworm litter or earthworm casts (Du Jardin 2015). Besides these, agricultural by-products/wastes can also be used for humic substances extraction.

   The effects of humic substances (humic acid and fulvic acid) are well documented in different crop plants during different growth stages. The application of humic substances to plants leads to enhancement in growth *via* increase in biomass and yield (Yildirim 2007; Costa et al. 2008), fruit and flower number (Kirn et al. 2010; Mazhar et al. 2012), root and shoot number (Adani et al. 1998; Canellas et al. 2011; Dobbss et al. 2010; Canellas et al. 2010; Peng et al. 2001; Tahir et al. 2011; Canellas et al. 2002, 2009) and the quality of harvest (Karakurt et al. 2009; Morard et al. 2011; Befrozfar et al. 2013). Besides this, humic substances also exert a positive influence on nutrient uptake and utilization of nitrogen, phosphorous, zinc, copper, iron, etc. (Bocanegra et al. 2006; Pandeya et al. 1998; Cimrin et al. 2010; Murillo et al. 2005; Sanchez-Sanchez et al. 2002). Also, it is very much useful in imparting resistance or tolerance toward drought and salinity (Garcia et al. 2013; Moghaddam and Soleimani 2012; Shahid et al. 2012). Thus, it is found that the components of humic substances, *i.e.*, humic and fulvic acids, both enhance the plant growth by interacting with soil and physiological factors.

7.5.2   Protein Hydrolysates and Amino Acids

Proteins are polymers of amino acids, and about 20 different types of amino acids are required for protein synthesis. It has been found that the protein base formulations have a stimulatory effect on plant growth by involving in physiological and biochemical pathways. The protein-derived products have been grouped into two major groups: protein hydrolysates containing mixture of peptides and plant/animal amino acids (Calvo et al. 2014). The protein hydrolysates preparation involves enzymatic, chemical or thermal hydrolysis process involving connective or epithelial tissues of animals, plants or animal residues (Morales-Payan and Stall 2003; Cavani et al. 2006; Kauffman et al. 2007; Ertani et al. 2009, 2013; Grabowska et al. 2012), collagen/elastin tissue (Cavani et al. 2006), carob germ protein (Parrado et al. 2008) and other tissues or residues. The commercial forms of protein hydrolysates are also available in the market, for example Siapton®, Amino16®, ILSATOP and Macro-Sorb (Calvo et al. 2014). Combinations of different amino acids are also available as biostimulants commercially. These protein hydrolysates or amino acid-based biostimulants enhance the nitrogen uptake and its assimilation in plants with the aid of enzymes or genes involved in nitrogen metabolism. They also improve the soil fertility by increasing the microbial activity and soil respiration. The amino acids accelerate the enzymatic activities in roots by chelating with different metal ions such as Mn, Fe, Cu and Zn and also increase the bioavailability of micronutrients (Du Jardin 2015).

Plant height and the number of flowers per plant increased with Siapton treatment, and the number of fruits per plant increased with carob germ hydrolysates in tomato (Parrado et al. 2008). Similarly, in papaya, fruit yield is increased when fertilization occurs with Siapton (Morales-Payan and Stall 2003). The application of Amino16® enhances fruit yield by increasing the number of fruits or weight of fruits in tomato (Koukounararas et al. 2013). Positive effects have been observed in nitrogen assimilation pathways by using protein hydrolysates, which resulted in the enhancement in antioxidant activity in maize (Maini 2006). In combination with amino acids such as glycine, betaine and other derivatives, protein hydrolysates induce the plant defense system and impart tolerance to abiotic stresses (Ashraf and Foolad 2007; Chen and Murata 2008; Ertani et al. 2013). Among the amino acids, proline and glycine betaine are already recognized as a well-known abiotic stress indicator. In several crop plants, these two amino acids impart tolerance to a variety of abiotic stresses. Similarly, other amino acids such as glutamic acid, aspartic acid and arginine also provide tolerance to salt stress (Chang et al. 2010; dos Reis et al. 2012).

7.5.3 Seaweed Extract

In ancient times, seaweed was used as an organic fertilizer to improve the nutrient status of soil and crop productivity. It is used as a biostimulant in plants as it is useful in enhancing the plant growth and development by increasing seed germination, flowering, fruiting, plant height, root and shoot growth, quality of produce and yield. The liquefiable form of seaweed is also available, which are derived from *Ascophyllum nodosum, Durvillaea antarctica, D. potatorum, Macrocystis pyrifera, Ecklonia maxima* (Khan et al. 2009), *Laminaria, Sargassum* and *Turbinaria* spp. (Hong et al. 2007; Sharma et al. 2012). Seaweed is applicable to both soil and plants. Seaweed extracts are associated with enhancement in root and shoot growth, improved nutrient uptake and increase in the nutrient content, which improves the physiological and biochemical activities in different plants. These are also involved in alleviating a wide range of abiotic stresses.

7.6 CONCLUSIONS AND FUTURE PROSPECTS

The increased usage of chemical fertilizers, weedicides and pesticides has increased the agricultural productivity, but at the same time, it also causes several environmental stresses leading to increased load of complex and highly aromatic chemical compounds in the soil, which has led to soil infertility, decreased trend of natural recycling of nutrients in the soil and decline in the count of beneficial microbes and microfauna in the soil. In the context of increased trend of chemical fertilizer-based soil pollution, the modern agricultural practices should concentrate more intensively on the usage of biofertilizers and biostimulants to restore the sustainability status of soil for eco-friendly recycling of nutrient elements and for stable food production to feed the growing population of the world. Further research has to be carried out on the phytomicrobiome-based interactions at molecular level to better understand the role of biofertilizers and biostimulants in promoting the growth and development of plants. Future studies should also concentrate on crop-specific biofertilizers and biostimulants or consortia of microbes for better yield and sustainable agricultural production.

REFERENCES

Adani, F., P. Genevini, P. Zaccheo, and G. Zocchi. 1998. The effect of commercial humic acid on tomato plant growth and mineral nutrition. *J. Plant. Nutr.* 21:561–575.

Anand, K., B. Kumari, and M. Mallick. 2016. Phosphate solubilizing microbes: An effective and alternative approach as biofertilizers. *J. Pharm. Pharm. Sci.* 8:37–40.

Ansari, R.A., R. Rizvi, A. Sumbul, and I. Mahmood. 2017. PGPR: Current vogue in sustainable crop production. In: V. Kumar, M. Kumar, S. Sharma, and R. Prasad (eds.), *Probiotics and Plant Health*. Springer, Singapore, pp. 455–472.

Ardakani, M., and S. Mafakheri. 2011. Designing a sustainable agroecosystem for wheat (*Triticum aestivum* L.) production. *J. Appl. Environ. Biol. Sci.* 1:401–413 https://doi.org/10.1016/S1573-5214(07)80001-7.

Ashraf, M., and M.R. Foolad. 2007. Roles of glycine betaine and proline in improving plant abiotic stress resistance. *Environ. Exp. Bot.* 59:206–216.

Bahadur, I., B.R. Maurya, A. Kumar, V.S. Meena, and R. Raghuwanshi. 2016. Towards the soil sustainability and potassium-solubilizing microorganisms. In: V. Meena, B. Maurya, J. Verma, and R. Meena (eds.), *Potassium Solubilizing Microorganisms for Sustainable Agriculture*. Springer, Heidelberg, pp. 255–266.

Baldani, J.I., and V.L.D. Baldani. 2005. History on the biological nitrogen fixation research in graminaceous plants: Special emphasis on the Brazilian experience. *An Acad. Bras. Cienc.* 77:549–579. https://doi.org/10.1590/S0001-37652005000300014.

Baldani, V.L.D., M.A.B. Alvarez, J. Baldani, and J. Döbereiner. 1986. Establishment of inoculated *Azospirillum* spp. in the rhizosphere and in roots of field grown wheat and sorghum. *Plant Soil* 90:35–46.

Basak, A. 2008. Biostimulators–definitions, classification and legislation. In: H. Gawrónska (ed.), *Monographs Series: Biostimulators in Modern Agriculture. General Aspects*. Wié Jutra, Warsaw, pp. 7–17.

Bashan, Y., and G. Holguin. 1997. *Azospirillum*–plant relationships: Environmental and physiological advances (1990–1996). *Can. J. Microbiol.* 43(2):103–121.

Bashan, Y., and L.E. de-Bashan. 2010. How the plant growth-promoting bacterium *Azospirillum* promotes plant growth: A critical assessment. *Adv. Agro.* 108:77–136.

Bashan, Y., G. Holguin, and L.E. de-Bashan. 2004. *Azospirillum*-plant relationships: Physiological, molecular, agricultural, and environmental advances (1997–2003). *Can. J. Microb.* 50(8):521–577.

Bashan, Y., M. Singh, and H. Levanony. 1989. Contribution of *Azospirillum brasilense* cd to growth of tomato seedlings is not through nitrogen fixation. *J. Bot.* 67:2429–2434.

Bashan, Y., and H. Levanony. 1990. Current status of *Azospirillum* inoculation technology: *Azospirillum* as a challenge for agriculture. *Can. J. Micro.* 36:591–608. https://doi.org/10.1139/m90-105.

Befrozfar, M.R., D. Habibi, A. Asgharzadeh, M. Sadeghi-Shoae, and M.R. Tookallo (2013) Vermicompost, plant growth promoting bacteria and humic acid can affect the growth and essence of basil (*Ocimum basilicum* L.). *Ann. Biol. Res.* 4:8–12.

Beijerinck, M.W. 1901. UberoligonitrophileMikroben.Zentralblatt fur Bakteriologie, Parasitenkunde, Infektionskrankheiten und Hygiene. *Abteilung II* 7:561–582.

Beijerinck, M.W. 1925. Uberein*Spirillum*, welches freienStickstoffbindenkann. *Zentral bl Bakteriol Parasitenkd Infekt Abt* 63:353.

Berbara, R.L.L., and A.C. García. 2014. Humic substances and plant defense metabolism. In: P. Ahmad, and M.R. Wani (eds.), *Physiological Mechanisms and Adaptation Strategies in Plants under Changing Environment*, vol. 1. Springer Science+Business Media, New York, pp. 297–319.

Berruti, A., E. Lumini, R. Balestrini, and V. Bianciotto. 2016. Arbuscular mycorrhizal fungi as natural biofertilizers: Let's benefit from past successes. *Front. Microbiol.* 6:1559. https://doi.org/10.3389/fmicb.2015.01559.

Bhatt, K., and P. Bhatt. 2021. Rhizospheric biology: Alternate tactics for enhancing sustainable agriculture. In: A. Verma, J.K. Saini, A.E.-L. Hesham, H.B. Singh (eds.), *Phytomicrobiome Interactions and Sustainable Agriculture*. Wiley-Blackwell. pp. 164–186.

Bocanegra, M.P., J.C. Lobartini, and G.A. Orioli. 2006. Plant uptake of iron chelated by humic acids of different molecular weights. *Commun. Soil Sci. Plant Anal.* 37:1–2.

Boiero, L., D. Perrig, O. Masciarelli, C. Penna, F. Cassan, and V. Luna. 2007. Phytohormone production by three strains of *Bradyrhizobium japonicum* and possible physiological and technological implications. *Appl. Micro. Biot.* 74:874–880.

Bottini, R., M. Fulchieri, D. Pearce, and R.P. Pharis. 1989. Identification of gibberellins A1, A3, and iso-A3 in cultures of *Azospirillum lipoferum*. *Plant Physiol.* 90:45–47. https://doi.org/10.1104/pp.90.1.45.

Brachmann, A., and M. Parniske. 2006. The most widespread symbiosis on Earth. *PLoS Biol.* 4(7):e239. https://doi.org/10.1371/journal.pbio.0040239.

Brown, M.E., R.M. Jackson, and S.K. Burlingham. 1968. Growth and effects of bacteria introduced into soil. *Ecol. Soil Bact.* 531–551.

Brundrett, M.C. 2009. Mycorrhizal associations and other means of nutrition of vascular plants: Understanding the global diversity of host plants by resolving conflicting information and developing reliable means of diagnosis. *Plant Soil.* 320:37–77.

Bulow, V.J.F., and J. Dobereiner. 1975. Potential for nitrogen fixation in maize genotypes in Brazil. *Pro. Natl. Acad. Sci.* 72:2389–2393. https://doi.org/10.1073/pnas.72.6.2389.

Buren, S., and L.M. Rubio. 2017. State of the art in eukaryotic nitrogenase engineering. *FEMS Microbiol. Lett.* 365:fnx274.

Calvo, P., L. Nelson, and J.W. Kloepper. 2014. Agricultural uses of plant biostimulants. *Plant Soil* 383(1):3–41.

Canellas, L.P., A. Piccolo, L.B. Dobbss, et al. 2010. Chemical composition and bioactivity properties of size-fractions separated from a vermin compost humic acid. *Chemosphere* 78:457–466.

Canellas, L.P., D.J. Dantas, N.O. Aguiar, et al. 2011.Probing the hormonal activity of fractionated molecular humic components in tomato auxin mutants. *Ann. Appl. Biol.* 159:202–211.

Canellas, L.P., F.L. Olivares, A.L. Okorokaova-Façanha, and A.R. Façanha. 2002. Humic acids isolated from earthworm compost enhance root elongation, lateral root emergence, and plasma membrane H$^+$-ATPase activity in maize roots. *Plant Physiol.* 130:1951–1957.

Canellas, L.P., R. Spaccini, A. Piccolo, et al. 2009. Relationships between chemical characteristics and root growth promotion of humic acids isolated from Brazilian oxisols. *Soil Sci.* 174:611–620.

Cassan, F., and M. Diaz-Zorita. 2016. *Azospirillum* sp. in current agriculture: From the laboratory to the field. *Soil Biol. Biochem.* 103:117–130. https://doi.org/10.1016/j.soilbio.2016.08.020.

Cassan, F., J. Vanderleyden, and S. Spaepen. 2014. Physiological and agronomical aspects of phytohormone production by model plant growth-promoting rhizobacteria (PGPR) belonging to the genus *Azospirillum*. *J. Plant Growth Regul.* 33:440–459. https://doi.org/10.1007/s00344-013-9362-4.

Cavani, L., A.T. Halle, C. Richard, and C. Ciavatta. 2006. Photosensitizing properties of protein hydrolysate-based fertilizers. *J. Agric. Food Chem.* 54:9160–9167.

Chabot, R., H. Antoun, and M.C. Cescas. 1996 Growth promotion of maize and lettuce by phosphate solubilizing *Rhizobium leguminosarum* bv. phaseoli. *Plant Soil* 184:311–321.

Chang, C., B. Wang, L. Shi, et al. 2010. Alleviation of salt stress induced inhibition of seed germination in cucumber (*Cucumis sativus* L.) by ethylene and glutamate. *J. Plant Physiol.* 167:1152–1156.

Chen, T.H.H., and N. Murata. 2008. Glycinebetaine: An effective protectant against abiotic stress in plants. *Trends Plant Sci.* 13:499–505.

Cimrin, K.M., T. Onder, M. Turan, and T. Burcu. 2010. Phosphorus and humic acid application alleviate salinity stress of pepper seedling. *Afr. J. Biotechnol.* 9:5845–5851.

Cohen, A.C., R. Bottini, and P.N. Piccol. 2008. *Azospirillum brasilense* Sp. 245 produces ABA in chemically-defined culture medium and increases ABA content in *Arabidopsis* plants. *Plant Growth Regul.* 54:97–103. https://doi.org/10.1007/s10725-007-9232-9.

Costa, G., P. Labrousse, C. Bodin, et al. 2008. Effects of humic substances on the rooting and development of woody plant cuttings. *Acta Hortic.* 779:255–261.

Creus, C.M., R.J. Sueldo, and C.A. Barassi. 2004. Water relations and yield in *Azospirillum* inoculated wheat exposed to drought in the field. *Can. J. Bot.* 82:273–281. https://doi.org/10.1139/b03-119.

Din, M., R. Nelofer, M. Salman, F.H. Khan, A. Khan, M. Ahmad, F. Jalil, J.U. Din, and M. Khan. 2019. Production of nitrogen fixing *Azotobacter* (SR-4) and phosphorus solubilizing *Aspergillus niger* and their evaluation on *Lagenaria siceraria* and *Abelmoschus esculentus*. *Biotechnol. Rep.* 22:e00323.

Dobbelaere, S., A. Croonenborghs, A. Thys, D. Ptacek, J. Vanderleyden, P. Dutto, C. Labandera-Gonzalez, J. Caballero-Mellado, J.F. Aguirre, and Y. Kapulnik. 2001. Responses of agronomically important crops to inoculation with *Azospirillum* Funct. *Plant. Biol.* 28:871–879.

Dobbelaere, S., J. Vanderleyden, and Y. Okon. 2003. Plant growth promoting effects of diazotrophs in the rhizosphere. *Crit. Rev. Plant Sci.* 22:107–149.

Dobbss, L.B., L.P. Canellas, F.L. Olivares, et al. 2010. Bioactivity of chemically transformed humic matter from vermicompost on plant root growth. *J. Agric. Food Chem.* 58:3681–3688.

Dobereiner, J., and J.M. Day. 1975. Associative symbioses in tropical grasses: Characterization of microorganisms and dinitrogen-fixing sites. In W.E. Newton and E.J. Nyman (eds.), *Proceedings of the 1st International Symposium on N2-Fixation*, vol. 2. Washington State University Press, Pullman, WA, pp. 518–588.

dos Reis, S.P., A.M. Lima, and C.R.B. de Souza. 2012. Recent molecular advances on downstream plant responses to abiotic stress. *Int. J. Mol. Sci.* 13:8628–8647.

Du Jardin, P. 2012. The Science of Plant Biostimulants–A bibliographic analysis, Ad hoc study report. European Commission.

Du Jardin, P. 2015. Plant biostimulants: Definition, concept, main categories and regulation. *Sci. Hortic.* 196:3–14.

Duponnois, R., M. Kisa, and C. Plenchette. 2006. Phosphate solubilizing potential of the nematophagous fungus *Arthrobotrys oligospora. J. Plant Nutr. Soil Sci.* 169(2):280–282.

Ertani, A., D. Pizzeghelio, A. Altissimo, and S. Nardi. 2013. Use of meat hydrolyzate derived from tanning residues as plant biostimulant for hydroponically grown maize. *J. Plant. Nutr. Soil. Sci.* 176:287–296.

Ertani, A., L. Cavani, D. Pizzeghello, et al. 2009. Biostimulant activity of two protein hydrolyzates in the growth and nitrogen metabolism of maize seedlings. *J. Plant. Nutri. Soil Sci.* 172:237–244.

European Biostimulants Industry Council. 2012. EBIC and biostimulants in brief. http://www.biostimulants.eu/.

Fibach-Paldi, S., S. Burdman, and Y. Okon. 2012. Key physiological properties contributing to rhizosphere adaptation and plant growth promotion abilities of *Azospirillum brasilense. FEMS Microbiol. Lett.* 326:99–108. https://doi.org/10.1111/j.1574-6968.2011.02407.x.

Filatov, V.P. 1951b. Tissue treatment (Doctrine on biogenic stimulators). II. Hypothesis of tissue therapy, or the doctrine on biogenic stimulators. *Priroda* 12:20–28.

Frank, B. 1885. Ueber die auf Wurzelsymbiose beruhende Ernährung gewiser Bäume durch unterirdishe Pilze. *Berichte der Deutschen Botanischen Gesellschaft* 3:128–145.

Fukami, J., C. de la Osa, F.J. Ollero, M. Megias, and M. Hungria. 2018. Co-inoculation of maize with *Azospirillum brasilense* and *Rhizobium tropici* as a strategy to mitigate salinity stress. *Funct. Plant Biol.* 45:328–339. https://doi.org/10.1071/FP17167.

Fukami, J., M.A. Nogueira, R.S. Araujo, and M. Hungria. 2016. Accessing inoculation methods of maize and wheat with *Azospirillum brasilense. AMB Express* 6:1. https://doi.org/10.1186/s13568-015-0171-y.

Garcia, A.C., L.A. Santos, F.G. Izquierdo, et al. 2013. Potentialities of vermicompost humic acids to alleviate water stress in rice plants (*Oryza sativa* L.). *J. Geochem. Explor.* 136:48–54.

Gerlach, M., and J. Vogel. 1902. Nitrogen fixing bacteria. *Zentralblatt fur Bakteriologie* 2:817.

Gothandapani, S., S. Sekar, and J.C. Padaria. 2017. *Azotobacter chroococcum*: Utilization and potential use for agricultural crop production: An overview. *Int. J. Adv. Res. Biol. Sci.* 4(3):35–42.

Grabowska, A., E. Kunicki, A. Sekara, A. Kalisz, and R. Wojciechowska. 2012. The effect of cultivar and biostimulant treatment on the carrot yield and its quality. *Veg. Crops Res. Bull.* 77:37–48.

Halpern, M., A. Bar-Tal, M. Ofek, D. Minz, T. Muller, and U. Yermiyahu. 2015. The use of biostimulants for enhancing nutrient uptake. *Adv. Agron.* 130:141–174.

Hara, G.W.O. 1998. The role of nitrogen fixation in crop production. *J. Crop Prod.* 1:115–138.

Hemissi, I., N. Abdi, A. Bargaz, M. Bouraoui, Y. Mabrouk, M. Saidi, and B. Sifi. 2015. Inoculation with phosphate solubilizing Mesorhizobium strains improves the performance of chickpea (*Cicer aritenium* L.) under phosphorus deficiency. *J. Plant Nutr.* 38(11):1656–1671.

Hennequin, J.R., and H. Blachere. 1966. Research on the synthesis of phytohormones and phenolic compounds by Azotobacter and bacteria of the rhizosphere. *Annales de l'Institut Pasteur* 111(3):89.

Hoekman, S.K., A. Broch, C. Robbins, E. Ceniceros, and M. Natarajan. 2012. Review of biodiesel composition, properties, and specifications, *Renew. Sustain. Energy. Rev.* 16(1):143–169. https://doi.org/10.1016/j.rser.2011.07.143.

Hong, D.D., H.M. Hien, and P.N. Son. 2007. Seaweeds from Vietnam used for functional food, medicine and biofertilizer. *J. Appl. Phycol.* 19:817–826.

Huang, H.C., R.S. Erickson, and T.F. Hsieh. 2007. Control of bacterial wilt of bean (*Curtobacterium flaccumfaciens* pv. *Flaccumfaciens*) by seed treatment with *Rhizobium leguminosarum. Crop Prot.* 26:1055–1061.

Ibraheem, I.B. 2007. Cyanobacteria as alternative biological conditioners for bioremediation of barren soil. *Egypt. J. Phycol.* 8(100):99–116.

Ikrina, M.A., and A.M. Kolbin. 2004. *Regulators of Plant Growth and Development*, Vol. 1. Chimia, Moscow.

Jimenez, D.J., J.S. Montana, and M.M. Martinez. 2011. Characterization of free nitrogen fixing bacteria of the genus *Azotobacter* in organic vegetable grown Colombian soils. *Braz. J. Microbiol.* 42:846–858.

Jones, K.M., H. Kobayashi, B.W. Davies, M.E. Taga, and G.C. Walker. 2007. How rhizobial symbionts invade plants: The Sinorhizobium–Medicago model. *Nat. Rev. Microbiol.* 5:619–633. https://doi.org/10.1038/nrmicro1705.

Karakurt, Y., H. Unlu, H. Unlu, and H. Padem. 2009. The influence of foliar and soil fertilization of humic acid on yield and quality of pepper. *Acta Agric. Scand. Sect. B.* 59:233–237.

Karnok, K.J. 2000. Promises, promises: Can biostimulants deliver? *Golf Course Manage.* 68:67–71.

Katkat, A.V., H. Çelik, M.A. Turan, and B.B. Asik. 2009. Effects of soil and foliar applications of humic substances on dry weight and mineral nutrients uptake of wheat under calcareous soil conditions. *Aust. J. Basic Appl. Sci.* 3(2):1266–1273.

Kauffman, G.L., D.P. Kneivel, and T.L. Watschke. 2007. Effects of a biostimulant on the heat tolerance associated with photosynthetic capacity, membrane thermostability, and polyphenol production of perennial ryegrass. *Crop. Sci.* 47:261–267. https://doi.org/10.2135/cropsci2006.03.0171.

Khammas, K.M., E. Ageron, P.A.D. Grimont, and P. Kaiser. 1989. *Azospirillum irakense* sp. nov., a nitrogen fixing bacterium associated with rice roots and rhizosphere soil. *Res. Microbiol.* 140:679–693.

Khan, M.R., K. Kounsar, and A. Hamid. 2002. Effect of certain rhizobacteria and antagonistic fungi on root-nodulation and root-knot nematode disease of green gram. *Nematol. Mediterr.* 30:85–89.

Khan, W., U.P. Rayirath, S. Subramanian, M.N., Jithesh, P. Rayorath, D.M. Hodges, A.T. Critchley, J.S. Craigie, J. Norrie, and B. Prithiviraj. 2009. Seaweed extracts as biostimulants of plant growth and development. *Plant Growth Regul.* 28:386–399.

Kim, Y.C., B.R. Glick, Y. Bashan, and C.M. Ryu. 2012. Enhancement of plant drought tolerance by microbes. In: R. Aroca (ed.), *Plant Responses to Drought Stress: From Morphological to Molecular Features.* Springer Verlag, Berlin, pp. 383–413. ISBN 978-3-642-32653-0.

Kirn, A., S.R. Kashif, and M. Yaseen. 2010. Using indigenous humic acid from lignite to increase growth and yield of okra (*Abelmoschus esculentus* L.). *Soil Environ.* 29:187–191.

Koukounararas, A., P. Tsouvaltzis, and A.S. Siomos. 2013. Effect of root and foliar application of amino acids on the growth and yield of greenhouse tomato in different fertilization levels. *J. Food. Agric. Environ.* 11:644–648.

Kraepiel, A.M.L., J.P. Bellenger, T. Wichard, and F.M. Morel. 2009. Multiple roles of siderophores in free-living nitrogen-fixing bacteria. *Biometals* 22(4):573.

Kumar, N. 2016. Effect of algal bio-fertilizer on the *Vigna radiata*: A critical review. *Int. J. Eng. Res. Appl.* 6(2):85–94.

Kurrey, D.K., R. Sharma, M.K. Lahre, and R.L. Kurrey. 2018. Effect of *Azotobacter* on physio-chemical characteristics of soil in onion field. *Pharma Inn. J.* 7(2):108–113.

Lenart, A. 2012. Occurrence, characteristics, and genetic diversity of *Azotobacter chroococcum* in various soils of Southern Poland. *Pol. J. Environ. Stud.* 21(2):415–424.

Levai, L., V. Szilvia, B. Nora, et al. 2008. Can wood ash and biofertilizer play a role in organic agriculture? *Agronomski Glasnic.* 3:263–271.

Lindstrom, K., and S.A. Mousavi. 2020. Effectiveness of nitrogen fixation in rhizobia. *Microbial. Biotechnol.* 13(5):1314–1335. https://doi.org/10.1111/1751-7915.13517.

Lopez, J.G., M.V. Toledo, S. Reina, and V. Salmeron. 1981. Root exudates of maize on production of auxins, gibberellins, cytokinins, amino acids and vitamins by *Azotobacter chroococcum* chemically defined media and dialysed soil media. *Toxicol. Environ. Chem.* 33:69–78.

Maini, P. 2006. The experience of the first biostimulant, based on amino acids and peptides: A short retrospective review on the laboratory researches and the practical results. *Fertilitas. Agrorum.* 1:29–43.

Malusa, E., L. Sas-Paszt, and J. Ciesielska. 2012. Technologies for beneficial microorganisms inocula used as biofertilizers. *Sci. World J.* 2012:1–12. https://doi.org/10.1100/2012/491206.

Mandimba, G.R. 1995. Contribution of nodulated legumes on the growth of *Zea mays* L. under various cropping systems. *Symbiosis* 19:213–222.

Marra, L.M., S.M. de Oliveira II, C.R.F. de Sousa Soares, and F.M. de Souza Moreira. 2011. Solubilisation of inorganic phosphates by inoculant strains from tropical legumes. *Sci. Agric.* 68(5):603–609.

Mazhar, A.A.M., S.I. Shedeed, N.G. Abdel-Aziz, and M.H. Mahgoub. 2012. Growth, flowering and chemical constituents of *Chrysanthemum indicum* L. plant in response to different levels of humic acid and salinity. *J. Appl. Sci. Res.* 8:3697–3706.

Mehnaz, S. 2015. *Azospirillum*: A biofertilizer for every crop. In: N.K. Arora (ed.), *Plant Microbes Symbiosis: Applied Facets.* Springer, New Delhi, pp. 297–314. https://doi.org/10.1007/978-81-322-2068-8_15.

Mehnaz, S., B. Weselowski, and G. Lazarovits. 2007. *Azospirillum zeae* sp.nov.adiazotrophic bacterium isolated from rhizosphere soil of *Zea mays*. *Int. J. Syst. Evol. Microbiol.* 57:2805–2809. https://doi.org/10.1099/ijs.0.65128-0.

Milledge, J.J. 2011. Commercial application of microalgae other than as biofuels: A brief review. *Rev. Environ. Sci. Biotechnol.* 10(1):31–41. https://doi.org/10.1007/s11157-010-9214-7.

Mishra, P.K., S. Mishra, G. Selvakumar, J.K. Bisht, S. Kundu, and H.S. Gupta. 2009. Co-inoculation of *Bacillus thuringeinsis* -KR1 with *Rhizobium leguminosarum* enhances plant growth and nodulation of pea (*Pisum sativum* L.) and lentil (*Lens culinaris* L.). *World J. Micro. Biot.* 25:753–761.

Mittal, V.O., H. Singh, J. Nayyar, J. Kaur, and R. Tewari. 2008. Stimulatory effect of phosphate-solubilizing fungal strains (*Aspergillus awamori* and *Penicillium citrinum*) on the yield of chickpea (*Cicer arietinum* L. cv. GPF2). *Soil Biol. Biochem.* 40(3):718–727.

Moghaddam, A.R.L., and A. Soleimani. 2012. Compensatory effects of humic acid on physiological characteristics of pistachio seedlings under salinity stress. *Acta Hortic.* 940:252–255.

Morales-Payan, J.P., and W.M. Stall. 2003. Papaya (*Carica papaya*) response to foliar treatments with organic complexes of peptides and amino acids. *Proc. Fla. State. Hortic. Soc.* 116:30–32.

Morard, P., B. Eyheraguibel, M. Morard, and J. Silvestre. 2011. Direct effects of humic-like substance on growth, water, and mineral nutrition of various species. *J. Plant Nutr.* 34:46–59.

Mousavi, S.A. 2016. Revised taxonomy of the family *Rhizobiaceae*, and phylogeny of mesorhizobia nodulating *Glycyrrhiza* spp. Dissertationes Schola Doctoralis Scientiae Circumiectalis, Alimentariae, Biologicae.

Mulder, E.G., and S. Brontonegoro. 1974. Free living heterotrophic nitrogen fixing bacteria. *Bio. Nitro. Fix.* 57:205–222.

Murillo, J.M., E. Madejon, P. Madejon, and F. Cabrera. 2005. The response of wild olive to the addition of a fulvic acid-rich amendment to soils polluted by trace elements (SW Spain). *J. Arid Environ.* 63:284–303.

Muscolo, A., M. Sidari, O. Francioso, V. Tugnoli, and S. Nardi. 2007. The auxin-like activity of humic substances is related to membrane interactions in carrot cell cultures. *J. Chem. Ecol.* 33(1):115–129.

Nardi, S., A. Ertani, and O. Francioso. 2017. Soil–root cross-talking: The role of humic substances. *J. Plant. Nutr. Soil Sci.* 180(1):5–13.

Nardi, S., P. Carletti, D. Pizzeghello, and A. Muscolo. 2009. Biological activities of humic substances. In: N. Senesi, B. Xing, P.M. Huang (eds.), *Biophysico-Chemical Processes Involving Natural Nonliving Organic Matter in Environmental Systems*. Wiley, Hoboken, NJ, pp. 305–339.

Oldroyd, G.E. 2013. Speak, friend, and enter: Signalling systems that promote beneficial symbiotic associations in plants. *Nat. Rev. Microbiol.* 11:252–264. https://doi.org/10.1038/nrmicro2990.

Ouma, E.W., A.M. Asango, J. Maingi, and E.M. Njeru. 2016. Elucidating the potential of native rhizobial isolates to improve biological nitrogen fixation and growth of common bean and soybean in smallholder farming systems of Kenya. *Intern. J. Agron.* 2016:1–7.

Pandeya, S.B., A.K. Singh, and P. Dhar. 1998. Influence of fulvic acid on transport of iron in soils and uptake by paddy seedlings. *Plant. Soil* 198:117–125.

Paradikovic, N., T. Teklic, S. Zeljkovic, M. Lisjak, and M. Spoljarevic. 2019. Biostimulants research in some horticultural plant species—A review. *Food Energy Secur.* 8(2):e00162.

Parrado, J., J. Bautista, E.F. Romero, et al. 2008. Production of a carob enzymatic extract: Potential use as a biofertilizer. *Bioresour. Technol.* 99:2312–2318.

Patriquin, D.G., J. Do bereiner, and D.K. Jain. 1983. Sites and processes of association between diazotrophs and grasses. *Can. J. Microbiol.* 29:900–915.

Peng, A., Y. Xu, and Z.J. Wang. 2001. The effect of fulvic acid on the dose effect of selenite on the growth of wheat. *Biol. Trace Elem. Res.* 83:275–279.

Peng, G., H. Wang, G. Zhang, W. Hou, Y. Liu, E.T. Wang, and Z. Tan. 2006. *Azospirillum melinis* sp. nov., a group of diazotrophs isolated from tropical molasses grass. *Int. J. Syst. Evol. Microbiol.* 56:1263–1271. https://doi.org/10.1099/ijs.0.64025-0.

Peoples, M.B., D.F. Herridge, and J.K. Ladha. 1995a. Biological nitrogen fixation: An efficient source of nitrogen for sustainable agricultural production. *Plant Soil* 174:3–28.

Peoples, M.B., J.K. Ladha, and D.F. Herridge. 1995b. Enhancing legume N2 fixation through plant and soil management. *Plant Soil.* 174:83–101.

Pereyra, C.M., N.A. Ramella, M.A. Pereyra, C.A. Barassi, and C.M. Creus. 2010. Changes in cucumber hypocotyl cell wall dynamics caused by *Azospirillum brasilense* inoculation. *Plant Physiol. Biochem.* 48:62–69.

Peterson, R.L., H.B. Massicotte, and L.H. Melville. 2004. *Mycorrhizas: Anatomy and Cell Biology*. CABI Publishing, CAB International, Wallingford, Oxon.

Pisciotta, J.M., Y. Zou, and I.V. Baskakov. 2010. Light-dependent electrogenic activity of cyanobacteria. *PloS One* 5(5):e10821. https://doi.org/10.1371/journal.pone.0010821.

Pradhan, N., and L.B. Sukla. 2005. Solubilization of inorganic phosphate by fungi isolated from agriculture soil. *Afr. J. Biotech.* 5:850–854.

Prasad, R.C., and B.N. Prasad. 2001. Cyanobacteria as a source biofertilizer for sustainable agriculture in Nepal. *J. Plant Sci. Bot. Orient.* 1:127–133.

Raj, S. 2007. Bio-fertilizers for micronutrients. *Biofertil. Newsl.* 1–8.

Reinhold, B., T. Hurek, I. Fendrik, B. Pot, M. Gillis, K. Kersters, S. Thielemans, and J. De Ley. 1987. *Azospirillum halopraeferens* sp. nov., a nitrogen-fixing organism associated with roots of Kallar Grass (*Leptochloafusca* (L.) Kunth). *Int. J. Syst. Bacteriol.* 37:43–51.

Revillas, J.J., B. Rodelas, C. Pozo, M.V. Toledo, and J. Gonalez-Lopez. 2000. Production of B-group vitamins by two *Azotobacter* strains with phenolic compounds as sole carbon source under diazotrophic and adiazotrophic conditions. *J. Appl. Microbiol.* 89:486–493.

Ribaudo, C., E. Krumpholz, F. Cassán, R. Bottini, M. Cantore, and J. Cura. 2006. *Azospirillum* sp. promotes root hair development in tomato plants through a mechanism that involves ethylene. *J. Plant Growth Regulat.* 25:175–185.

Rodriguez, A.A., A.M. Stella, M.M. Storni, G. Zulpa, and M.C. Zaccaro. 2006. Effects of cyanobacterial extracellular products and gibberellic acid on salinity tolerance in *Oryza sativa* L. *Saline Syst.* 2(1):7. https://doi.org/10.1186/1756-1448-2-7.

Roger, P.A., and P.A. Reynaud. 1982. *Free—Living Blue—Green Algae in Tropical Soils, Microbiology of Tropical Soils and Plant Productivity.* Springer, Dordrecht, pp. 147–168.

Romero, A.M., O.S. Correa, S. Moccia, and J.G. Rivas. 2003. Effect of *Azospirillum*-mediated plant growth promotion on the development of bacterial diseases on fresh-market and cherry tomato. *J. Appl. Microbiol.* 95:832–838. https://doi.org/10.1046/j.1365-2672.2003.02053.x.

Saadatnia, H., and H. Riahi. 2009. Cyanobacteria from paddy fields in Iran as a biofertilizer in rice plants. *Plant Soil Environ.* 55(5):207–212.

Sahoo, R.K., M.W. Ansari, M. Pradhan, T.K. Dangar, S. Mohanty, and N. Tuteja. 2014. Phenotypic and molecular characterization of native *Azospirillum* strains from rice fields to improve crop productivity. *Protoplasma* 251:943–953. https://doi.org/10.1007/s00709-013-0607-7.

Salman, S.R., S.D. Abou-Hussein, A.M.R. Abdel-Mawgoud, and M.A. El-Nemr. 2005. Fruit yield and quality of watermelon as affected by hybrids and humic acid application. *J. Appl. Sci. Res.* 1(1):51–58.

Sanchez-Sanchez, A., J. Sanchez-Andreu, M. Juarez, J. Jorda, and D. Bermudez. 2002. Humic substances and amino acids improve effectiveness of chelate FeEDDHA in lemon trees. *J. Plant. Nutr.* 25:2433–2442.

Sant Anna, F.H., L.G.P. Almeida, R. Cecagno, et al. 2011. Genomic insights into the versatility of the plant growth-promoting bacterium *Azospirillum amazonense. BMC Genomics* 12:409.

Saubidet, M.I., N. Fatta, and A.J. Barneix. 2002. The effect of inoculation with *Azospirillum brasilense* on growth and nitrogen utilization by wheat plants. *Plant Soil* 245:215–222.

Sellstedt, A., L. Staahl, M. Mattsson, K. Jonsson, and P. Hoegberg. 1993. Can the 15N dilution technique be used to study N2 fixation in tropical tree symbioses as affected by water deficit? *J. Exp. Bot.* 44:1749–1755.

Senthilkumar, M., M. Madhaiyan, S.P. Sundaram, and S. Kannaiyan. 2009. Intercellular colonization and growth promoting effects of Methylobacterium sp. with plant-growth regulators on rice (*Oryza sativa* L. CvCO-43). *Micro. Res.* 164:92–104.

Setubal, J.C., et al. 2009. Genome sequence of *Azotobacter vinelandii*, an obligate aerobe specialized to support diverse anaerobic metabolic processes. *J. Bacter.* 191:4534–4545.

Shahid, M., C. Dumat, J. Silvestre, and E. Pinelli. 2012. Effect of fulvic acids on lead-induced oxidative stress to metal sensitive *Vicia faba* L. plant. *Biol. Fertil. Soils* 48:689–697.

Sharma, A.K. 2004. Bio-intensive nutrient management. In: Sharma (ed), *A Handbook of Organic Farming.* Agrios Publication, Pune, pp. 216–274.

Sharma, S.H.S., G. Lyons, C. McRoberts, et al. 2012. Biostimulant activity of brown seaweed species from Strangford Lough: Compositional analyses of polysaccharides and bioassay of extracts using mung bean (*Vigno mungo* L.) and pak choi (*Brassica rapa* chinensis L.). *J. Appl. Phycol.* 24:1081–1091.

Siddiqui, Z.A., G. Baghel, and M.S. Akhtar. 2007. Biocontrol of *Meloidogyne javanica* by rhizobium and plant growth-promoting rhizobacteria on lentil. *World J. Micro. Biotech.* 23:435–441.

Smith, S.E., and D.J. Read. 2008. *Mycorrhizal Symbiosis*, 3rd edn. Academic Press, Cambridge, MA, p. 816.

Smith, S.E., and F.A. Smith. 2012. Fresh perspectives on the roles of arbuscular mycorrhizal fungi in plant nutrition and growth. *Mycologia* 104:1–13. https://doi.org/10.3852/11-229.

Smith, S.E., I. Jakobsen, M. Grønlund, and F.A. Smith. 2011. Roles of arbuscular mycorrhizas in plant phosphorus nutrition: Interactions between pathways of phosphorus uptake in arbuscular mycorrhizal roots have important implications for understanding and manipulating plant phosphorus acquisition. *Plant Physiol.* 156:1050–1057. https://doi.org/10.1104/pp.111.174581.

Spaepen, S., and J. Vanderleyden. 2015. Auxin signaling in *Azospirillum brasilense*: A proteome analysis. In: F.J. de Bruijn (ed.), *Biological Nitrogen Fixation.* Wiley, Hoboken, NJ, pp. 937–940. https://doi.org/10.1002/9781119053095.ch91.

Spaepen, S., S. Bossuyt, K. Engelen, K. Marchal, and J. Vanderleyden. 2014. Phenotypical and molecular responses of *Arabidopsis thaliana* roots as a result of inoculation with the auxin-producing bacterium *Azospirillum brasilense. New Phytol.* 201:850–861 https://doi.org/10.1111/nph.12590.

Sprent, J.I., and E.K. James. 2007. Legume evolution: Where do nodules and mycorrhizas fit in? *Plant Physiol.* 144:575–581.

Steinberg, C.E., T. Meinelt, M.A. Timofeyev, M. Bittner, and R. Menzel. 2008. Humic substances. *Environ. Sci. Poll. Res.* 15(2):128–135.

Stevenson, F.J. 1994. *Humus Chemistry: Genesis, Composition, Reactions.* Wiley, New York.

Strzelczyk, E., M. Kampert, and C.Y. Li. 1994. Cytokinin-like substances and ethylene production by *Azospirillum* in media with different carbon sources. *Microbiol. Res.* 149:55–60. https://doi.org/10.1016/S0944-5013(11)80136-9.

Tahir, M.M., M. Khurshid, M.Z. Khan, M.K. Abbasi, and M.H. Hazmi.2011. Lignite-derived humic acid effect on growth of wheat plants in different soils. *Pedosphere* 2:124–131.

Tarafdar, J.C., M. Bareja, and J. Panwar. 2003. Efficiency of some phosphatase producing soil-fungi. *Ind. J. Micro.* 43:27–32.

Tarrand, J.J., N.R. Krieg, and J. Dobereiner. 1978. A taxonomic study of the *Spirillum lipoferum* group, with the descriptions of a new genus, *Azospirillum* gen. nov. and two species *Azospirillum lipoferum* (Beijerinck) comb. nov. and *Azospirillum brasilense* sp. nov. *Can. J. Microbiol.* 24:967–980.

Tate, R.L. 1995. *Soil Microbiology (Symbiotic Nitrogen Fixation)*, pp. 307–333. John Wiley & Sons, Inc., New York.

Thajuddin, N., and G. Subramanian. 2005. Cyanobacterial biodiversity and potential applications in biotechnology. *Curr. Sci.* 89:47–57.

Tien, T.M., M.H. Gaskins, and D. Hubbell. 1979. Plant growth substances produced by *Azospirillum brasilense* and their effect on the growth of pearl millet (*Pennisetum americanum* L.). *Appl. Environ. Microbiol.* 37:1016–1024. https://doi.org/10.1128/AEM.37.5.1016-1024.1979.

Torre, L.A., V. Battaglia, and F. Caradonia. 2016. An overview of the current plant biostimulant legislations in different European Member States. *J. Sci. Food. Agric.* 96:727–734. https://doi.org/10.1002/jsfa.7358.

Traon, D., L. Amat, F. Zotz, and P. Du Jardin. 2014. A legal framework for plant biostimulants and agronomic fertiliser additives in the EU-Report to the European Commission, DG Enterprise & Industry (No. Contract n° 255/PP/ENT/IMA/13/1112420).

Turan, M., M. Gulluce, N. von Wiren, and F. Sahin. 2012. Yield promotion and phosphorus solubilization by plant growth-promoting rhizobacteria in extensive wheat production in Turkey. *J. Plant Nutr. Soil Sci.* 175:818–826. https://doi.org/10.1002/jpln.201200054.

Vaishampayan, A., P.S. Rajeshwar, D. Donat Ha, et al. 2001. Cyanobacterial biofertilizers in rice agriculture. *Bot. Rev.* 67(4):453–516.

Vance, C.P. 1998. Legume symbiotic nitrogen fixation: Agronomic aspects. In: Spaink, H.P., et al. (eds.), *The Rhizobiaceae.* Kluwer Academic Publishers, Dordrecht, pp. 509–530.

Vejan, P., R. Abdullah, T. Khadiran, S. Ismail, and A. Nasrulhaq Boyce. 2016. Role of plant growth promoting rhizobacteria in agricultural sustainability–A review. *Molecules* 21:573. https://doi.org/10.3390/molecules21050573. PMID: 27136521

Wagner, S.C. 2012. Biological nitrogen fixation. *Nat. Educ. Knowl.* 3:15.

Wani, S.A., S. Chand, and T. Ali. 2013. Potential use of *Azotobacter chroococcum* in crop production: An overview. *Curr. Agric. Res. J.* 1(1):35–38.

Williams, P.J.B., and L.M.L. Laurens. 2010. Microalgae as biodiesel and biomass feedstock: Review and analysis of the biochemistry, energetic and economics. *Energy Environ. Sci.* 3(5):554–590. https://doi.org/10.1039/b924978h.

Wilson, L.T. 2006. Cyanobacteria: A potential nitrogen source in rice fields. *Texas Rice* 6:9–10.

Xie, C.H., A. Yokota. 2005. *Azospirillum oryzae* sp. nov. a nitrogen-fixing bacterium isolated from the roots of the rice plant *Oryza sativa. Int. J. Syst. Evol. Microbiol.* 55:1435–1438 https://doi.org/10.1099/ijs.0.63503-0.

Yakhin, O.I., A.A. Lubyanov, I.A. Yakhin, and P.H. Brown. 2017. Biostimulants in plant science: A global perspective. *Front. Plant Sci.* 7:2049. http://10.3389/fpls.2016.02049.

Yildirim, E. 2007. Foliar and soil fertilization of humic acid affect productivity and quality of tomato. *Acta Agric. Scand. Section B-Soil Plant Sci.* 57(2):182–186.

8 Microbial Biostimulants and their Role in Environmental Bioremediation

Faouzia Tanveer, Irum Iqrar, and Zabta Khan Shinwari
Quaid-i-Azam University

Iram Gul
Hazara University Mansehra

CONTENTS

8.1   INTRODUCTION

In the last decade, the field application of biostimulants has progressively been growing as a novel strategy for sustainable agricultural production (Yakhin et al. 2017). The emerging evidence suggests that biostimulants-based formulations may act as priming agents for plants, which exert positive effects on plant growth and productivity and also provide protective actions against various environmental stresses (Fleming et al. 2019). Biostimulants could be based on organic/non-nutrient constituents or microorganisms capable of improving plant health and productivity by regulating the plant's physiological features. They may provide benefits to the plant in terms of enhancing nutrient uptake, stimulating growth, and conferring stress tolerance, which in turn improve crop production and quality. Application of biostimulants in the environment may provide a more viable and resilient solution against a number of environmental stresses (Munjal 2020, Nephali et al. 2020). In the future, these could be adopted as an environmentally sustainable solution to improve farming practices and to restore contaminated environments. This chapter focuses on the role of microbial

DOI: 10.1201/9781003188032-8

biostimulants in environmental bioremediation, discussing different microorganisms used and their general mode of action for pollutant stress alleviation.

8.2 ENVIRONMENTAL POLLUTION: AN OVERVIEW

Pollution has now become a major concern in the world due to severe contamination of our environment with the release of different organic and inorganic pollutants, mainly from anthropogenic activities including industrial practices. At present, the physicochemical methods that are used for environmental remediation purposes are not environmentally friendly due to the excessive use of chemicals that are harmful to the environment or intensive physical techniques that are also not very cost-effective (Deb et al. 2020). In addition to that, growing abiotic stresses on crops due to rapidly changing climate greatly impede food security. Apart from salinity, drought, and extreme temperatures, these stresses also include the noxious pollutants that may result in decreased soil fertility and increased crop losses worldwide. Thus, novel strategies are needed to better deal with the undesirable effects of harsh environmental conditions on modern agriculture (Kerchev et al. 2020).

Among various pollutants, heavy metals are highly toxic and may have adverse health consequences for humans. These metals are released into the environment due to various human activities and predominantly include As, Cd, Co, Cr, Cu, Hg, Mn, Mo, Ni, Pb, and Zn (Siegel 2002). In developing countries, it is a common irrigation practice to use metal-contaminated water coming from the industrial effluents to serve the water requirements of the edible crops in the agricultural fields. This may be potentially harmful due to the movement of toxic pollutants up the food chain ultimately reaching humans and causing detrimental health effects (Saxena et al. 2019). Moreover, heavy metals accumulating in the agricultural soil may affect its physicochemical properties and beneficial microbial soil processes and reduce fertility and land use for agricultural production, eventually leading to food insecurity (Wuana and Okieimen 2011).

Organic pollutants also contribute to environmental pollution. These include several compounds such as pesticides (e.g., atrazine), polycyclic aromatic hydrocarbons (PAHs), and chlorinated solvents (e.g., trichloroethylene). Some of the organic pollutants have an adverse environmental impact as they persist in the environment for a longer period and are resistant to degradation. These are called persistent organic pollutants (POPs) and may bioaccumulate in biological systems causing detrimental effects on the environment including agricultural systems (Buccini 2003, Manz et al. 2001).

Bioremediation, i.e., use of biological organisms including microorganisms and plants for the removal of contaminants from the polluted environment, is considered to be the most ecologically sustainable approach (Deb et al. 2020). Literature suggests that microbes have a special contribution in the effective bioremediation of environmental pollutants as well as sustainable agriculture (Mishra et al. 2020) due to the involvement of several important mechanisms that are discussed in the next sections.

8.3 BIOSTIMULANTS

Biostimulants are natural formulations that have gained considerable interest in the last decade due to their beneficial effects on the plant (Mrid et al. 2021). Several authors have proposed different definitions of a biostimulant to differentiate it from the other available bioformulations intended for agricultural improvements. Conceptually, it is defined as a substance or microorganism other than nutrient or pesticide and without any soil improvement properties, but capable of promoting plant growth and health by regulating natural processes of the plant (Yakhin et al. 2017). La Torre et al. (2016) categorized biostimulants into six groups: humic substances, seaweed extracts, inorganic acids, amino acids, hydrolyzed proteins, and microbes. Biostimulants consisting of natural molecules or microbial inoculants are considered a novel class of agricultural inputs that can complement the existing agrochemicals such as synthetic fertilizers to improve the quality of crops and enhance their ability to tolerate biotic and abiotic stresses (Rouphael and Colla 2020, Du Jardin

2015). Increasing scientific publications, as well as a greatly expanding commercial market for agricultural biostimulants, suggests considerable potential (Povero et al. 2016), which could be further harnessed for environmental bioremediation. These biostimulants can enhance the plant defense system against environmental stresses by increased root and shoot development as well as improved yield and fruit quality (Shukla et al. 2019). The greatest advantage is that these are based on natural compounds or inoculants that may be non-toxic to the environment when applied at lower concentrations (Kerchev et al. 2020). The mechanism of action of a biostimulant is either directly by stimulating plant enzymes and hormones or indirectly by increasing nutrient availability in soil for plant uptake. However, detailed investigations at the molecular and cellular level under different environmental settings may provide more information for fully understanding these mechanisms (Yakhin et al. 2017).

8.4 MICROBE-BASED BIOREMEDIATION: AN OVERVIEW

In bioremediation, microorganisms are used to convert or transform the contaminants into less/non-toxic forms. Bacteria, fungi, and archaea are the most commonly used microorganisms in the bioremediation of contaminated sites (Figure 8.1; Tang et al. 2007). Bioremediation involves the degradation, removal, alteration, immobilization, or detoxification of different contaminants (organic and inorganic) through microorganisms and plants (Iqbal et al. 2019; Gul et al. 2020; Manzoor et al. 2020). Microorganisms are generally involved via their enzymatic pathways, acting as biocatalysts and facilitating the progression of biochemical reactions that degrade the desired pollutant (Abatenh et al. 2017).

8.4.1 HEAVY METALS

Microorganisms are diverse and found in all types of environmental conditions. The soil and water contaminated with heavy metals also contained a large diversity of microbes which play an important role in converting toxic heavy metals to less/non-toxic metals (Verma and Kuila 2019). Microorganisms (bacteria, algae, and fungi) are widely used for the removal of heavy metals from contaminated sites by the application of heavy metal-resistant bacteria as single or in the consortium. Table 8.1 presents different microbes that utilize heavy metals as a source of food in bioremediation.

Microorganisms in Bioremediation

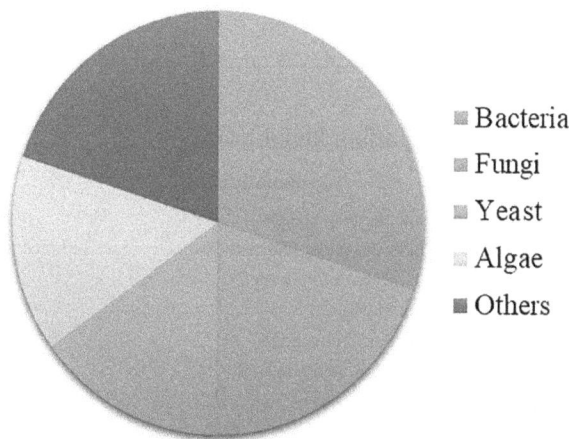

- Bacteria
- Fungi
- Yeast
- Algae
- Others

FIGURE 8.1 The most predominant microorganisms used in bioremediation.

TABLE 8.1

Microbes Utilizing Heavy Metals in Bioremediation

Microbes	Heavy Metals	References
Pseudomonas aeruginosa and *Bacillus* sp.	Copper and zinc	Philip et al. (2000)
Citrobacter sp. and *Zooglea* sp.	Cadmium, copper, cobalt, nickel, and zinc	Sar and D'Souza (2001)
Aspergillus niger	Cadmium, uranium, thorium, and zinc	Rajendran et al. (2003)
Rhizopus arrhizus	Cadmium, lead, mercury, and calcium	
Stereum hirsutum	Cadmium, cobalt, copper, and nickel	Gabriel et al. (1994, 1996)
Phormidium valderium	Cadmium and lead	
Ganoderma applanatum	Copper, mercury, and lead	

8.4.2 ORGANIC POLLUTANTS

Biostimulation is proven to be efficient in the degradation of organic pollutants including hydrocarbons, herbicides, fungicides, and polyester polyurethanes. Different studies have reported that a large diversity of microbes including bacteria and archaebacteria are capable of degrading hydrocarbons (Fowler et al. 2016). Some of the microbes able to degrade the xenobiotics are presented in Table 8.2. Different microbes are efficient, but the following are predominant (Adams et al. 2015):

a. Bacteria: *Acinetobacter*, *Bacillus*, *Vibrio*, *Arthrobacter*, *Achromobacter*, *Flavobacterium*, and *Pseudomonas*.

b. Fungi and yeast: *Trichoderma*, *Aspergillus*, *Cladosporium*, *Sporobolomyces*, *Penicillium*, and *Rhodotorula*.

8.5   ROLE OF MICROBIAL BIOSTIMULANTS IN BIOREMEDIATION

Microorganisms can be used as inoculants for agronomically important crops to reduce the effect of toxic pollutants on the agricultural soils and, at the same time, to improve plant health. Microbe-based biostimulants consist of beneficial fungi, bacteria, and yeast that may enhance plant nutrition and growth under stress conditions by improving nutrient uptake and use efficiency, better root system, and greater crop yield (Sangiorgio et al. 2020, Mrid et al. 2021). These microbial inoculants may be isolated from diverse environments including plants, water, soil, composted manure, and industrial effluents (Bulgari et al. 2019).

TABLE 8.2

Microorganisms Capable of Degrading Xenobiotics

Microbes	Xenobiotics	References
Bacillus sp.	Hydrocarbons and phenoxyacetate	Cybulski et al. (2003)
Pseudomonas	Hydrocarbons, benzene, polychlorinated biphenyl, and anthracene	
Azotobacter sp.	Benzene, cycloparaffins, and aromatics	Dean et al. (2002)
Rhodococcus sp.	Hydrocarbons and aromatics	
Nocardia sp.	Halogenated hydrocarbon diazinon	Janssen et al. (2005)
Flavobacterium	Aromatics, naphthalene, and biphenyls	
Arthrobacter sp.	Benzene, phenol, hydrocarbons, and cresol	
Methanogens	Polychlorinated biphenyls	
Corynebacterium	Aromatics	

There are several categories of microbes with proven potential in the bioremediation of polluted environments. In this regard, plant-associated bacteria including plant growth-promoting rhizobacteria (PGPR) (Basharat et al. 2018), endophytes (Sim et al. 2019) and arbuscular mycorrhizal fungi (AMF) (Fecih and Baoune 2019) have shown great potential for alleviating the toxicity of the pollutant in crops and could be developed as biostimulants to serve the purpose of bioremediation. Plants have evolved in a symbiotic relationship with microbes that play an important role for the plants by maintaining and regulating the ecosystem health. These symbiotic relations could be exploited for developing microbial biostimulants as an environmentally sustainable solution to counteract abiotic stresses imposed by climate change (Sangiorgio et al. 2020) as well as human activities including stress induced by heavy metals and/or organic pollutants. These microbes can contribute to plant nourishment through nitrogen fixation and nutrient solubilization, induction of plant hormones such as auxins and cytokinins, as well as amelioration of abiotic stress via production of volatile organic compounds (VOCs), which exert direct beneficial effects on the plant (Bulgari et al. 2019).

8.5.1 BIOSTIMULANT BACTERIA

Biostimulant bacteria may be classified as plant growth-promoting bacteria (PGPB) in general, plant growth-promoting rhizobacteria (PGPR) that specifically colonize the root rhizosphere of plants (Bhattacharyya and Jha 2012), and endophytic bacteria (EB) that colonize inside tissues of plants and live in a symbiotic relationship with their plant counterpart without causing infections in the host plant (Schulz and Boyle 2006). Several common genera have been studied in this regard, which most commonly include *Burkholderia, Serratia, Bacillus, Azotobacter, Azospirillum, Pseudomonas, Enterobacter, Streptomyces, Arthrobacter,* and *Paenibacillus* (Bonaldi et al. 2015, Van Oosten et al. 2017, Sharma et al. 2018).

In addition to direct plant growth regulation, several studies illustrate the role of plant-associated bacteria in remediating heavy metal-contaminated soils via specific mechanisms including siderophores production to sequester trace metal from the plant root surrounding soil and production of organic and inorganic molecules thereby increasing metal bioavailability and causing induced systemic tolerance (IST) in plants by limiting the heavy metal accumulation in plant roots and shoots (Guo et al. 2020, Rajkumar et al. 2012). A recent study reported heavy metal-resistant bacteria, i.e., *Bacillus subtilis* subsp. *spizizenii*, *Paenibacillus jamilae*, and *Pseudomonas aeruginosa*, for the potential bioremediation of cadmium (Cd) and lead (Pb) metals affecting the spinach plant. It was found that these bacteria induced heavy metal stress tolerance in spinach by improving several plant attributes, including relative water content, rates of transpiration and net photosynthesis, stomatal conductance, and membrane stability index. On the other hand, antioxidant activities were greatly repressed. The results indicated the positive impact of heavy metal-resistant bacteria isolated from effluent-contaminated soils on plant growth (Desoky et al. 2020). Another study reported the potential role of metal-immobilizing bacteria, *Serratia liquefaciens*, by reducing metal uptake (Cd and Pb) in potato tubers through decreased metal availability, increased urease activity, and surrounding pH (Cheng et al. 2020).

8.5.2 BIOSTIMULANT FUNGI

Different fungi have shown a considerable effect on plant yield and productivity with potential applications in agriculture (Kiruba and Thatheyus 2021). Among fungi, arbuscular mycorrhizal fungi (AMF) and *Trichoderma* sp. have been considered as the most efficient biostimulants with the potential to improve plant growth by efficient nutrient uptake and solubilization as well as the activity of plant growth hormones (Szczałba et al. 2019, Gupta 2020). AMF are the fungi that colonize plant roots and have been observed to guard their host plants under heavy metal stress using different strategies that include metal detoxification or transformation, mobilization or immobilization, and bioaccumulation or translocation (Dhalaria et al. 2020, Tiwari et al. 2020).

8.5.3 CONSORTIUM OF MICROBES

Microbes exist in the form of communities in their natural habitats. Thus, the use of two or more beneficial microbes together as a consortium offers a promising strategy to improve crop productivity. Moreover, mixed systems involving biostimulant microbes and organic compounds may help ameliorate pollutant stress more effectively and efficiently (Tabacchioni et al. 2021). In a study, a combination of compost and two commercially available microbial biostimulants was used to check the effect of consortium on the mobility of toxic elements, i.e., Cd and Pb within the two selected grass species grown in the contaminated industrial soil. The soil was attained from a lead–acid battery landfill site. The commercial mixed biostimulant system called Panoramix (Koppert Biological Systems, the Netherlands) comprising of AMF and PGPR, i.e., *Bacillus* sp. and *Trichoderma* sp., was observed to improve plant growth and nutrient uptake by lowering the bioavailability of selected heavy metals in soil, which in turn resulted in improved soil quality. The mixed consortium facilitated the effective restoration of the soil by stabilizing Cd, reducing the risk of downward leakage in the groundwater (Visconti et al. 2020).

8.6 MECHANISM OF ACTION

Bacteria can produce and release different types of metabolites such as organic acids, biosurfactants, and siderophores based on soil conditions. These metabolites can modify trace elements' behavior and bioavailability, providing two possible responses to increase the plant trace metal uptake and accumulation or to immobilize trace metals (Sessitsch et al. 2013), as illustrated in Figure 8.2. Metabolites may increase metal mobility, so the plant trace metal uptake is raised, and this process enhances the shoot biomass production and raises the extraction of metals from soil. These metal-tolerant microbes have extensively been studied for the bioremediation of metals. The major mechanisms include (i) stabilization; (ii) sorption; (iii) accumulation; (iv) transformation; (v) leaching; and (vi) microbially assisted chemisorption of metals (Patel and Kasture 2014). Many soil bacteria (*Pseudomonas aeruginosa*, *Pseudomonas fluorescens*, *Ralstonia metallidurans*, *Microbacterium paraoxydans* (Braud et al. 2009)) and some fungi (*Mucor* spp. CBRF59; *Glomus*

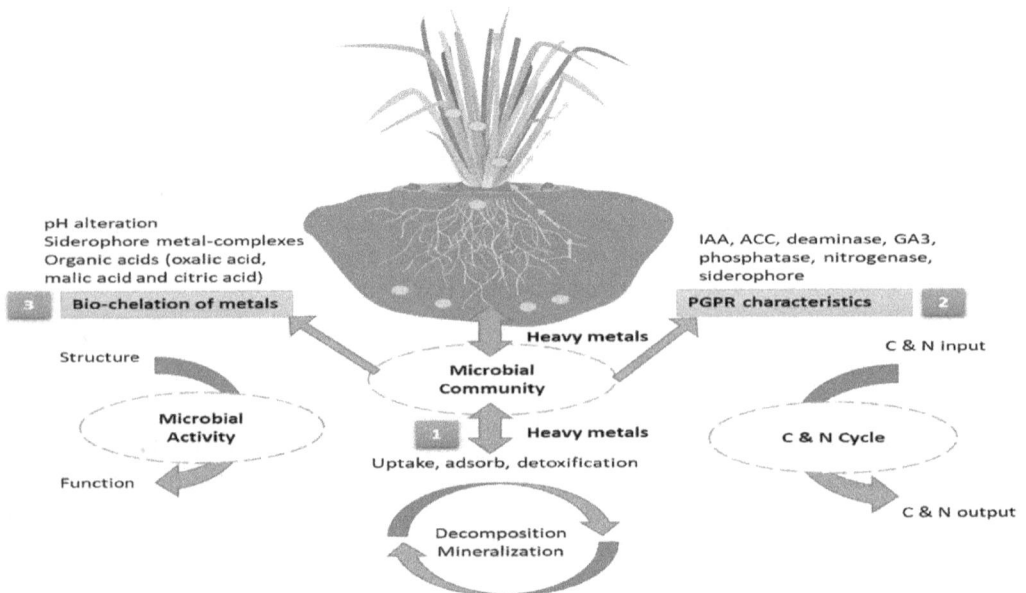

FIGURE 8.2   Microbial interaction in response to heavy metals exposure.

mosseae) have also been found to significantly increase the availability of heavy metals (Deng et al. 2011, Punamiya et al. 2010).

Heavy metal detoxification mechanisms in bacteria have widely been reported in the literature (Jarosławiecka and Piotrowska-Seget 2014). Lead efflux and precipitation are two components of the Pb detoxification mechanism in bacteria. The major Pb detoxification system in bacteria is led by *pbr* efflux system, which is surrounded by *mer* and *czc* families on pMOL30 (Monchy et al. 2007). *pbrA* transports Pb from inside cytoplasm to periplasm where the released inorganic phosphate groups by *pbrB* precipitate Pb and prevent its re-entry into cells. The export of Pb ions to the periplasm involves the joint function of *pbrA* and *PIB* ATPases, *cadA* and *zntA*, and their functions are not limited to Pb only. In *Pseudomonas putida* KT2440, the main efflux transporter identified for Cd and Pb is *cadA2* (Hynninen et al. 2009).

Microbes can produce organic acids such as oxalic acid, malic acid, citric acid, gluconic acid, and fumaric acid as the means of carboxylate acids that can change the pH and dissolved organic matter, which ultimately helps in enhancing the heavy metal availability and uptake. Organic acids as siderophores can form complexes with metals in soil solution (Rajkumar et al. 2012). The soluble organometal complexes increase the mobility and uptake of metals. Siderophores are the second most released metabolites from bacteria and can reach several molecular structures (Sessitsch et al. 2013). These metabolites are involved in metal chelation, pH alteration, and oxidation/reduction reactions. Moreover, siderophores are "iron carriers" and they are released under iron-limiting conditions such as soils. Furthermore, they are known as iron chelators due to a strong tendency to form a complex with iron. Hence, in the presence of trace elements, bacteria induce siderophore synthesis and guide the formation of siderophore–metal complexes with Al, Cd, Cu, Ga, In, Pb, and Zn (Rajkumar et al. 2012).

Different strains of *Microbacterium* (Sheng et al. 2008), *Pseudomonas* (Rajkumar and Freitas 2008), *Enterobacter* (Kumar et al. 2009), and *Klebsiella* (Farina et al. 2012) have been documented for PGPR characteristics in metal-contaminated soil. Pb-resistant bacteria with indole acetic acid (IAA), 1-aminocyclopropane-1-carboxylate (ACC) deaminase, siderophores (Sheng et al. 2008), hydrogen cyanide (HCN), nitrogenase activity, and phosphate solubilization have been shown to promote plant growth, tolerate Pb, and improve Pb uptake in plants and can be used as potential biostimulants. Furthermore, these compounds play a role in protecting plants from Ni, Pb, and Zn toxicity (Zhang et al. 2011).

AMF have shown the ability to cope with metal toxicity, thus contributing to the plant's heavy metal tolerance. Vesicles found in AMF are similar to vacuoles found in fungi and are capable of storing a large number of metals. Their ubiquitous hyphae form a network that helps in nutrient as well as water uptake, thus reducing the use of organic fertilizers on the plants. They also produce a glycoprotein called glomalin, which helps promote plant health in response to heavy metal by reducing the metal uptake as a result of a glycoprotein and metal complex formation. Moreover, they perform an important role in phytoremediation by increasing absorption area and antioxidant activity as well as heavy metal chelation (Dhalaria et al. 2020). Biodegradation of organic contaminants has also been observed in the presence of AMF, which provides a promising bioremediation technique for soils polluted with PAH, petroleum, and pesticides. The mechanism involved includes the metabolic degradation of organic pollutants by the enzymes derived from fungal exudation, by the improved root exudation as well as increased microbial activity in the rhizosphere, and the pollutant removal by plant uptake (Li et al. 2006).

8.7 COMMERCIALIZATION OF MICROBIAL BIOSTIMULANTS: POTENTIAL AND CHALLENGES

The idea that microbial biostimulants can increase crop yield as a green alternative to agrochemicals is generating a lot of interest. Microbial stimulants such as PGPR and AMF are widely considered not only as sustainable and efficient tools for ensuring yield stability under low-input conditions, but also as an innovative technology for improving crop tolerance to abiotic stressors (Bhattacharyya and Jha 2012, Berg et al. 2019). Microbial biostimulants are marketed as alternatives or substitutes to traditional fertilizers and pesticides, allowing farmers to extend their management toolbox. There's also the assumption that these goods have a lower environmental effect during processing and after they've been applied to agricultural systems (e.g., less nutrient runoff than chemical additives).

It is worth noting that useful microbes or microbial consortia-based commercial products express multiple functions including nutritional exchange, hormonal stimulation, induction of plant defenses, and competition with pathogens with additive and synergistic effects (Sangiorgio et al. 2020). Despite their enormous potential, microbial biostimulants have been slow to catch on, and their effects on crops are often unpredictable or unknown to farmers. One key to effective use of microbial products, as with any agricultural input, is applying them at the right time, in the right place, and at the right concentration. These products, however, contain live organisms that must survive and become involved in the environment, unlike most inputs. This complicates the production of microbial products in two major ways (Fox 2015, Waltz 2017).

8.7.1 Formulation and Shelf Life

Companies working on microbial products must figure out how to keep live microorganisms stable enough. Although some materials are sold in liquid form, dry powders or granules have a longer shelf life and are less likely to be contaminated. Dry formulations, on the other hand, are difficult in another way, since many microorganisms cannot withstand the stress of drying or persisting for an extended period without water. This is partially prevented in some plants by using dormant spores, which are more resistant to stress. Under the right conditions in the soil, certain spores can become active again. However, long-term storage is still problematic for most spores, and many beneficial microorganisms do not form this type of spore. Instead of active cells, several live products containing *Bacillus* and fungi incorporate spores. Other typical agricultural inputs can be incompatible with microbial products. When a broad-spectrum fungicide is used alongside a beneficial fungus, for example, it may reduce its activity and survival.

8.7.2 Survival and Establishment of Microbial Products

Microbial biostimulants must be able to thrive in the soil environment and, in most cases, grow around plant roots in the soil. To have measurable effects on crop production, they need to establish a large and active population. The pH, nutrient concentration, organic matter content, texture, temperature, and moisture of soils differ greatly. All of these factors have been shown to affect the survival of these various microorganisms. If the added microbes are not well suited to a specific soil type, their growth and activity can be slowed or stopped. Furthermore, soils are already home to a large number of resident microorganisms, and the introduced microorganisms may face significant competition.

8.8 CONCLUSIONS

The use of microbial biostimulants represents a long-term strategy for improving plant performance and productivity. Beneficial bacteria and fungi are examples of microbial biostimulants that can improve plant nutrition, growth, productivity, and stress tolerance. Furthermore, the use of microbial biostimulants may help to maintain agroecosystem ecological balance by reducing the use of pesticides and/or heavy metals in agriculture. Currently available products frequently improve cultural conditions in general, but their action mechanisms are largely unknown and their effects are frequently unreliable. In future, research based on the characterization of plant–microbe and microbial community interactions could result in more precisely targeted products.

REFERENCES

Abatenh, Endeshaw, Birhanu Gizaw, Zerihun Tsegaye, and Misganaw Wassie. 2017. "The role of microorganisms in bioremediation-A review." *Open Journal of Environmental Biology*, 2(1), 038–046.

Adams, Godleads Omokhagbor, Prekeyi Tawari Fufeyin, Samson Eruke Okoro, and Igelenyah Ehinomen. 2015. "Bioremediation, biostimulation and bioaugmention: A review." *International Journal of Environmental Bioremediation & Biodegradation* 3(1):28–39.

Basharat, Zarrin, Faouzia Tanveer, Azra Yasmin, Zabta Khan Shinwari, Tongtong He, and Yigang Tong. 2018. "Genome of Serratia nematodiphila MB307 offers unique insights into its diverse traits." *Genome* 61(7):469–476.

Berg, Shelby, Paul G. Dennis, Chanyarat Paungfoo-Lonhienne, Jay Anderson, Nicole Robinson, Richard Brackin, Adam Royle, Lawrence DiBella, and Susanne Schmidt. 2019. "Effects of commercial microbial biostimulants on soil and root microbial communities and sugarcane yield." *Biology and Fertility of Soils* 56:1–16.

Bhattacharyya, P.N., and Dhruva K. Jha. 2012. "Plant growth-promoting rhizobacteria (PGPR): Emergence in agriculture." *World Journal of Microbiology and Biotechnology* 28 (4):1327–1350.

Bonaldi, Maria, Xiaoyulong Chen, Andrea Kunova, Cristina Pizzatti, Marco Saracchi, and Paolo Cortesi. 2015. "Colonization of lettuce rhizosphere and roots by tagged Streptomyces." *Frontiers in Microbiology* 6:25.

Braud, Armelle, Karine Jézéquel, Stéphane Bazot, and Thierry Lebeau. 2009. "Enhanced phytoextraction of an agricultural Cr-and Pb-contaminated soil by bioaugmentation with siderophore-producing bacteria." *Chemosphere* 74 (2):280–286.

Buccini, John. 2003. "The development of a global treaty on persistent organic pollutants (POPs)." In: Fiedler, H. (ed.), *Persistent Organic Pollutants*, pp. 13–30. Springer, Heidelberg.

Bulgari, Roberta, Giulia Franzoni, and Antonio Ferrante. 2019. "Biostimulants application in horticultural crops under abiotic stress conditions." *Agronomy* 9 (6):306.

Cheng, Cheng, Hui Han, Yaping Wang, Linyan He, and Xiafang Sheng. 2020. "Metal-immobilizing and urease-producing bacteria increase the biomass and reduce metal accumulation in potato tubers under field conditions." *Ecotoxicology and Environmental Safety* 203: 111017.

Cybulski, Zefiryn, Ewa Dziurla, Ewa Kaczorek, and Andrzej Olszanowski. 2003. "The influence of emulsifiers on hydrocarbon biodegradation by Pseudomonadacea and Bacillacea strains." *Spill Science & Technology Bulletin* 8 (5–6):503–507.

Dean, Wendy L., Gavin Kelsey, and Wolf Reik. 2002. "Generation of monoparental embryos for investigation into genomic imprinting." In: Ward, A. (ed.), *Genomic Imprinting*, 1–19. Springer, Heidelberg.

Deb, Vishal Kumar, Ahmad Rabbani, Shashi Upadhyay, Priyam Bharti, Hitesh Sharma, Devendra Singh Rawat, and Gaurav Saxena. 2020. "Microbe-assisted phytoremediation in reinstating heavy metal-contaminated sites: Concepts, mechanisms, challenges, and future perspectives." In Arora, P. (ed.), *Microbial Technology for Health and Environment*, pp. 161–189. Springer, Singapore.

Deng, Zujun, Lixiang Cao, Haiwei Huang, Xinyu Jiang, Wenfeng Wang, Yang Shi, and Renduo Zhang. 2011. "Characterization of Cd-and Pb-resistant fungal endophyte Mucor sp. CBRF59 isolated from rapes (Brassica chinensis) in a metal-contaminated soil." *Journal of Hazardous Materials* 185 (2–3):717–724.

Desoky, El-Sayed M., Abdel-Rahman M. Merwad, Wael M. Semida, Seham A. Ibrahim, Mohamed T. El-Saadony, and Mostafa M. Rady. 2020. "Heavy metals-resistant bacteria (HM-RB): Potential bioremediators of heavy metals-stressed Spinacia oleracea plant." *Ecotoxicology and Environmental Safety* 198: 110685.

Dhalaria, Rajni, Dinesh Kumar, Harsh Kumar, Eugenie Nepovimova, Kamil Kuča, Muhammad Torequl Islam, and Rachna Verma. 2020. "Arbuscular mycorrhizal fungi as potential agents in ameliorating heavy metal stress in plants." *Agronomy* 10 (6):815.

Du Jardin, Patrick. 2015. "Plant biostimulants: Definition, concept, main categories and regulation." *Scientia Horticulturae* 196: 3–14.

Farina, Roberto, Anelise Beneduzi, Adriana Ambrosini, Samanta B. de Campos, Bruno Brito Lisboa, Volker Wendisch, Luciano K. Vargas, and Luciane M.P. Passaglia. 2012. "Diversity of plant growth-promoting rhizobacteria communities associated with the stages of canola growth." *Applied Soil Ecology* 55: 44–52.

Fecih, Thinhinane, and Hafida Baoune. 2019. "Arbuscular mycorrhizal fungi remediation potential of organic and inorganic compounds." In: Arora, P. (ed.), *Microbial Technology for the Welfare of Society*, pp. 247–257. Springer, Singapore.

Fleming, Thomas R., Colin C. Fleming, Camila C.B. Levy, Carlos Repiso, Franck Hennequart, José B. Nolasco, and Fuquan Liu. 2019. "Biostimulants enhance growth and drought tolerance in Arabidopsis thaliana and exhibit chemical priming action." *Annals of Applied Biology* 174 (2):153–165.

Fowler, S. Jane, Courtney R.A. Toth, and Lisa M. Gieg. 2016. "Community structure in methanogenic enrichments provides insight into syntrophic interactions in hydrocarbon-impacted environments." *Frontiers in Microbiology* 7: 562.

Fox, Jeffrey L. 2015. *Agricultural Probiotics Enter Spotlight*. Nature Publishing Group, Berlin.

Gabriel, Jiri, O. Kofroňová, P. Rychlovský, and M. Krenželok. 1996. "Accumulation and effect of cadmium in the wood-rotting basidiomycete Daedalea quercina." *Bulletin of Environmental Contamination and Toxicology* 57 (3):383–390.

Gabriel, J., M. Mokrejš, J. Bílý, and P. Rychlovský. 1994. "Accumulation of heavy metals by some wood-rotting fungi." *Folia Microbiologica* 39 (2):115–118.

Gul, Iram, Maria Manzoor, Jean Kallerhoff, and Muhammad Arshad. 2020. "Enhanced phytoremediation of lead by soil applied organic and inorganic amendments: Pb phytoavailability, accumulation and metal recovery." *Chemosphere*, 258, 127405

Guo, JunKang, Haris Muhammad, Xin Lv, Ting Wei, XinHao Ren, HongLei Jia, Saleem Atif, and Li Hua. 2020. "Prospects and applications of plant growth promoting rhizobacteria to mitigate soil metal contamination: A review." *Chemosphere* 246: 125823.

Gupta, Nibha. 2020. "Trichoderma as biostimulant: Factors responsible for plant growth promotion." In: Manoharachary, C., Singh, H.B., Varma, A. (eds.), *Trichoderma: Agricultural Applications and Beyond*, pp. 287–309. Springer, Cham.

Hynninen, Anu, Thierry Touzé, Leena Pitkänen, Dominique Mengin-Lecreulx, and Marko Virta. 2009. "An efflux transporter PbrA and a phosphatase PbrB cooperate in a lead-resistance mechanism in bacteria." *Molecular Microbiology* 74 (2):384–394.

Iqbal, Aneela, Muhammad Arshad, Raghupathy Karthikeyan, Terry J. Gentry, Jamshaid Rashid, Iftikhar Ahmed, and Arthur Paul Schwab. 2019. "Diesel degrading bacterial endophytes with plant growth promoting potential isolated from a petroleum storage facility." *Biotech*, 9(1): 35.

Janssen, Dick B., Inez J.T. Dinkla, Gerrit J. Poelarends, and Peter Terpstra. 2005. "Bacterial degradation of xenobiotic compounds: Evolution and distribution of novel enzyme activities." *Environmental Microbiology* 7 (12):1868–1882.

Jarosławiecka, Anna, and Zofia Piotrowska-Seget. 2014. "Lead resistance in micro-organisms." *Microbiology* 160 (1):12–25.

Jennifer Michellin Kiruba, N., and A. Joseph Thatheyus. 2021. "Fungi, fungal enzymes and their potential application as biostimulants." *Microbiome Stimulants for Crops* 2021:305–314. https://www.sciencedirect.com/science/article/pii/B9780128221228000248

Kerchev, Pavel, Tom van der Meer, Neerakkal Sujeeth, Arno Verlee, Christian V. Stevens, Frank Van Breusegem, and Tsanko Gechev. 2020. "Molecular priming as an approach to induce tolerance against abiotic and oxidative stresses in crop plants." *Biotechnology Advances* 40:107503.

Kumar, Kalpna V., Shubhi Srivastava, N. Singh, and H.M. Behl. 2009. "Role of metal resistant plant growth promoting bacteria in ameliorating fly ash to the growth of Brassica juncea." *Journal of Hazardous Materials* 170 (1):51–57.

La Torre, Anna, Valerio Battaglia, and Federica Caradonia. 2016. "An overview of the current plant biostimulant legislations in different European Member States." *Journal of the Science of Food and Agriculture* 96 (3):727–734.

Li, Qiuling, Wanting Ling, Yanzheng Gao, Fuchun Li, and Wei Xiong. 2006. "Arbuscular mycorrhizal bioremediation and its mechanisms of organic pollutants-contaminated soils." *Ying yong sheng tai xue bao= The Journal of Applied Ecology* 17 (11):2217–2221.

Manz, M., K.-D. Wenzel, U. Dietze, and G. Schüürmann. 2001. "Persistent organic pollutants in agricultural soils of central Germany." *Science of the Total Environment* 277 (1–3):187–198. https://www.sciencedirect.com/science/article/pii/S0048969700008779?casa_token=H763FDOmDKMAAAAA:cx-8CudO MO7VBjif2r_8rJceFN9yjFtHZmiIsYrfqEMaCnMrO_D7h3L_3D3SuQUld0zHyB_3Ig37

Manzoor, Maria, Iram Gul, Aamir Manzoor, Usman Rauf Kamboh, Kiran Hina, Jean Kallerhoff, and Muhammad Arshad. 2020. "Lead availability and phytoextraction in the rhizosphere of Pelargonium species." *Environmental Science and Pollution Research*, 27(32): 39753-39762.

Mishra, Akash, Shraddha Priyadarshini Mishra, Anfal Arshi, Ankur Agarwal, and Sanjai Kumar Dwivedi. 2020. "Plant-microbe interactions for bioremediation and phytoremediation of environmental pollutants and agro-ecosystem development." In: Bharagava, R., Saxena, G. (eds.), *Bioremediation of Industrial Waste for Environmental Safety*, pp. 415–436. Springer, Singapore.

Monchy, Sébastien, Mohammed A. Benotmane, Paul Janssen, Tatiana Vallaeys, Safiyh Taghavi, Daniel Van Der Lelie, and Max Mergeay. 2007. "Plasmids pMOL28 and pMOL30 of Cupriavidus metallidurans are specialized in the maximal viable response to heavy metals." *Journal of Bacteriology* 189 (20):7417–7425.

Mrid, Reda Ben, Bouchra Benmrid, Jawhar Hafsa, Hassan Boukcim, Mansour Sobeh, and Abdelaziz Yasri. 2021. "Secondary metabolites as biostimulant and bioprotectant agents: A review." *Science of The Total Environment* 777:146204.

Munjal, Renu. 2020. "Use of biostimulants in conferring tolerance to environmental stress." In: Hasanuzzaman, M. (ed.), *Plant Ecophysiology and Adaptation under Climate Change: Mechanisms and Perspectives II*, pp. 231–244. Springer, Singapore.

Nephali, Lerato, Lizelle A Piater, Ian A Dubery, Veronica Patterson, Johan Huyser, Karl Burgess, and Fidele Tugizimana. 2020. "Biostimulants for plant growth and mitigation of abiotic stresses: A metabolomics perspective." *Metabolites* 10 (12):505.

Patel, Shuchi, and Avani Kasture. 2014. "E (electronic) waste management using biological systems-overview." *International Journal of Current Microbiology and Applied Sciences* 3 (7):495–504.

Philip, Ligy, Leela Iyengar, and C. Venkobachar. 2000. "Site of interaction of copper on Bacillus polymyxa." *Water, Air, and Soil Pollution* 119 (1):11–21.

Povero, Giovanni, Juan F. Mejia, Donata Di Tommaso, Alberto Piaggesi, and Prem Warrior. 2016. "A systematic approach to discover and characterize natural plant biostimulants." *Frontiers in Plant Science* 7: 435.

Punamiya, Pravin, Rupali Datta, Dibyendu Sarkar, Summer Barber, Mandakini Patel, and Padmini Das. 2010. "Symbiotic role of Glomus mosseae in phytoextraction of lead in vetiver grass [Chrysopogon zizanioides (L.)]." *Journal of Hazardous Materials* 177 (1–3):465–474.

Rajendran, P., J. Muthukrishnan, and P. Gunasekaran. 2003. "Microbes in heavy metal remediation." *Indian Journal of Experimental Biology* 41:935–944. http://nopr.niscair.res.in/handle/123456789/17153

Rajkumar, M., S. Sandhya, M.N.V. Prasad, and H. Freitas. 2012. "Perspectives of plant-associated microbes in heavy metal phytoremediation." *Biotechnology Advances* 30 (6):1562–1574.

Rajkumar, Mani, and Helena Freitas. 2008. "Influence of metal resistant-plant growth-promoting bacteria on the growth of Ricinus communis in soil contaminated with heavy metals." *Chemosphere* 71 (5):834–842.

Rouphael, Youssef, and Giuseppe Colla. 2020. "Biostimulants in agriculture." *Frontiers in Plant Science* 6:11.

Sangiorgio, Daniela, Antonio Cellini, Irene Donati, Chiara Pastore, Claudia Onofrietti, and Francesco Spinelli. 2020. "Facing climate change: Application of microbial biostimulants to mitigate stress in horticultural crops." *Agronomy* 10 (6):794.

Sar, Pinaki, and Stanislaus F. D'Souza. 2001. "Biosorptive uranium uptake by a Pseudomonas strain: Characterization and equilibrium studies." *Journal of Chemical Technology & Biotechnology* 76 (12):1286–1294.

Saxena, Gaurav, Diane Purchase, Sikandar I Mulla, Ganesh Dattatraya Saratale, and Ram Naresh Bharagava. 2019. "Phytoremediation of heavy metal-contaminated sites: Eco-environmental concerns, field studies, sustainability issues, and future prospects." *Reviews of Environmental Contamination and Toxicology* 249:71–131.

Schulz, Barbara, and Christine Boyle. 2006. "What are endophytes?" In: Schulz, B.J.E., Boyle, C.J.C., and Sieber, T.N. (eds.), *Microbial Root Endophytes*, pp. 1–13. Springer, Heidelberg.

Sessitsch, Angela, Melanie Kuffner, Petra Kidd, Jaco Vangronsveld, Walter W. Wenzel, Katharina Fallmann, and Markus Puschenreiter. 2013. "The role of plant-associated bacteria in the mobilization and phytoextraction of trace elements in contaminated soils." *Soil Biology and Biochemistry* 60:182–194.

Sharma, Anket, Vinod Kumar, Neha Handa, Shagun Bali, Ravdeep Kaur, Kanika Khanna, Ashwani Kumar Thukral, and Renu Bhardwaj. 2018. "Potential of endophytic bacteria in heavy metal and pesticide detoxification." In: Egamberdieva, D., Ahmad, P. (eds.), *Plant Microbiome: Stress Response*, pp. 307–336. Springer, Singapore.

Sheng, Xia-Fang, Juan-Juan Xia, Chun-Yu Jiang, Lin-Yan He, and Meng Qian. 2008. "Characterization of heavy metal-resistant endophytic bacteria from rape (Brassica napus) roots and their potential in promoting the growth and lead accumulation of rape." *Environmental Pollution* 156 (3):1164–1170.

Shukla, Pushp Sheel, Emily Grace Mantin, Mohd Adil, Sruti Bajpai, Alan T Critchley, and Balakrishnan Prithiviraj. 2019. "Ascophyllum nodosum-based biostimulants: Sustainable applications in agriculture for the stimulation of plant growth, stress tolerance, and disease management." *Frontiers in Plant Science* 10:655.

Siegel, Frederic R. 2002. "Geochemistry in ecosystem analysis of heavy metal pollution." In *Environmental Geochemistry of Potentially Toxic Metals*, pp. 1–14. Springer, Heidelberg.

Sim, Carrie Siew Fang, Si Hui Chen, and Adeline Su Yien Ting. 2019. "Endophytes: Emerging tools for the bioremediation of pollutants." In: Bharagava, R., and Chowdhary, P. (eds.), *Emerging and Eco-Friendly Approaches for Waste Management*, pp. 189–217. Springer, Singapore.

Szczałba, Maciej, Tomas Kopta, Maciej Gąstoł, and Agnieszka Sękara. 2019. "Comprehensive insight into arbuscular mycorrhizal fungi, Trichoderma spp. and plant multilevel interactions with emphasis on biostimulation of horticultural crops." *Journal of Applied Microbiology* 127 (3):630–647.

Tabacchioni, Silvia, Stefania Passato, Patrizia Ambrosino, Liren Huang, Marina Caldara, Cristina Cantale, Jonas Hett, Antonella Del Fiore, Alessia Fiore, and Andreas Schlüter. 2021. "Identification of beneficial microbial consortia and bioactive compounds with potential as plant biostimulants for a sustainable agriculture." *Microorganisms* 9 (2):426.

Tang, Chuyang Y., Q. Shiang Fu, Craig S. Criddle, and James O. Leckie. 2007. "Effect of flux (transmembrane pressure) and membrane properties on fouling and rejection of reverse osmosis and nanofiltration membranes treating perfluorooctane sulfonate containing wastewater." *Environmental science & technology*, 41(6): 2008–2014.

Tiwari, Jaya, Ying Ma, and Kuldeep Bauddh. 2020. "Arbuscular mycorrhizal fungi: An ecological accelerator of phytoremediation of metal contaminated soils." *Archives of Agronomy and Soil Science*, 1–14. doi: 10.1080/03650340.2020.1829599.

Van Oosten, Michael James, Olimpia Pepe, Stefania De Pascale, Silvia Silletti, and Albino Maggio. 2017. "The role of biostimulants and bioeffectors as alleviators of abiotic stress in crop plants." *Chemical and Biological Technologies in Agriculture* 4 (1):1–12.

Verma, Samakshi, and Arindam Kuila. 2019. "Bioremediation of heavy metals by microbial process." *Environmental Technology & Innovation* 14:100369.

Visconti, Donato, Antonio Giandonato Caporale, Ludovico Pontoni, Valeria Ventorino, Massimo Fagnano, Paola Adamo, Olimpia Pepe, Sheridan Lois Woo, and Nunzio Fiorentino. 2020. "Securing of an industrial soil using turfgrass assisted by biostimulants and compost amendment." *Agronomy* 10 (9):1310.

Waltz, Emily. 2017. *A New Crop of Microbe Startups Raises Big Bucks, Takes on the Establishment.* Nature Publishing Group, Berlin.

Wuana, Raymond A., and Felix E. Okieimen. 2011. "Heavy metals in contaminated soils: A review of sources, chemistry, risks and best available strategies for remediation." *International Scholarly Research Notices.* doi: 10.5402/2011/402647.

Yakhin, Oleg I., Aleksandr A. Lubyanov, Ildus A. Yakhin, and Patrick H. Brown. 2017. "Biostimulants in plant science: A global perspective." *Frontiers in Plant Science* 7: 2049.

Zhang, Yan-feng, Lin-yan He, Zhao-jin Chen, Wen-hui Zhang, Qing-ya Wang, Meng Qian, and Xia-fang Sheng. 2011. "Characterization of lead-resistant and ACC deaminase-producing endophytic bacteria and their potential in promoting lead accumulation of rape." *Journal of hazardous materials*, 186 (2–3): 1720–1725.

9 Microbial Biostimulants in Bioremediation Process for Treatment of Municipal Solid Waste

El Asri Ouahid, Fatima Boubrik, and Jalal Mouadi
Ibn Zohr University

CONTENTS

9.1 INTRODUCTION

The socioeconomic development of each country is reflected in the quantity and quality of waste produced by daily activities. One of the significant problems facing municipalities worldwide is the increasing waste production and choosing the treatment process type. The municipalities are in a constant race with the exponential production of waste. Among waste types, municipal solid waste (MSW) includes all wastes produced, collected, transported, and stored within the jurisdiction of a municipal authority (Kumar 2020; Periathamby 2011). These wastes gravely threaten citizens' health and the different environmental ecosystems due to their physicochemical characteristics, which are qualified by toxic, mutagenic, and carcinogenic properties.

The MSW is qualified as biodegradable due to its high content of organic matter. This organic fraction accounts for between 40% and 70% of biodegradable materials (Wei et al. 2017). So, the MSW is a potential source of carbon, nitrogen, and energy for bacterial communities. Among the biotechnologies that use microbes in conversion is bioremediation (Chandra and Sobti 2020). This emerging and new approach is the latest trend in managing, treating, and valorizing MSW worldwide. This technique becomes more efficient by adding other substances or microbes that are named biostimulants. The latest additive in the bioremediation is allowing to facilitate the degradation of the substrates constituting waste. It can enhance the production of different enzymes and set up an excellent environment for the bioconversion process. In this work, we will focus only on the microbial biostimulants applied to bioremediation.

The microbial biostimulants can be isolated from various sources (soil, wastewater, leachate, solid waste, and manure). Several researchers have used these biostimulants during different

DOI: 10.1201/9781003188032-9

types of MSW bioremediation. The microbial mixture (*Bacillus casei*, *Candida rugopelliculosa*, *Lactobacillus buchneri*, *Trichoderma*, and white-rot fungi) in the bioremediation of MSW such as composting can improve the conversion process (Awasthi et al. 2014). Rastogi et al. (2020) demonstrated that inoculation in composting of MSW could produce in less time good quality and maturity of compost. Some microbes such as *Firmicutes*, *Bacteroidetes*, *Spirochaetes*, and *Fibrobacteres* are good microbial biostimulants added to the landfill technology to enhance conversion (Ransom-Jones et al. 2017). So, these microbial precursors make it possible to improve and optimize the bioremediation of MSW.

This chapter is an excellent tool for showing the role of microbial biostimulants in MSW bioremediation technologies. For this, we have divided this chapter into three parts: (i) In the first part, we provide an update to global MSW production and its capacity to suitable conversion. (ii) The second part contains two sub-parts: The first shows the principle, types, and benefits of the bioremediation process, and the second part presents the microbial biostimulant concept. (iii) In the third part, we will highlight the applications of microbial biostimulants in different types of bioremediation for the management, treatment, and recovery of MSW.

9.2 MUNICIPAL SOLID WASTE

The definition of MSW is different between researchers from developing countries and developed countries. In developing countries, the collection, adequate disposal, and municipal treatment are not yet developed (Stafford 2020). Periathamby (2011) defined MSW as all wastes produced, collected, transported, and stored within the jurisdiction of a municipal authority (Periathamby 2011). MSW is regrouped into industrial, residential, institutional, commercial, and municipal waste. So, the MSW is all wastes collected by the municipality or the waste that gets accumulated in the municipal waste disposal place.

MSW production rates vary from one country to another and from one city to another in the same nation. The USA has produced 262 million tons of MSW in 2018 (Baredar, Khare, and Nema 2020). China produces annually about 130 Mt million tons of MSW (Hui et al. 2006). Africa generated 169119 tons of MSW per day in 2012, and it will generate 441840 tons per day in 2025 (Stafford 2020). The global MSW production was 0.49 billion tons in 1997 (Suocheng, Tong, and Yuping 2001). After 9 years, this production was estimated at 2.02 billion tons with an 8% of increasing rate (Kolekar, Hazra, and Chakrabarty 2016). Mushtaq et al. (2020) estimated the future amount of about 3.4 billion tons in 2050. Therefore, despite the variations in production from one region to another, the global production is continually increasing. This increase in global production is due to several factors: (i) increase in industrialization, (ii) rapid urbanization, (iii) demographic rise, (iv) flourishing economy, (v) increase in the standard of living, (vi) increase in the consumption of products among citizens, and (vii) improvement in environmental legislation.

Generally, we can find in MSW five types of waste (paper and card, kitchen and garden waste, plastics, glass, and metals). The percentage of each class is different between countries. So, the physicochemical composition of MSW is also different from one country to another; it also depends on the factors mentioned above. Barlaz (1998) said that biodegradable fractions can reach 90% of MSW (Barlaz 1998). The amount of organic particles in MSW is about 82% (Ramachandra et al. 2018). The amount of biodegradables in MSW is, respectively, 72.5%, 68%, and 69% in Iran, Tunisia, and Turkey (Bölükbaş and Akıncı 2018). The MSW is characterized in developing countries by the dominance of biodegradable matter (> 70%), but in developed countries, we can find a slight distinction of organic waste of around 58% (Burnley 2007; Ramachandra et al. 2018). On the other hand, the MSW is characterized by its high moisture content (up to 75%) (Yuan et al. 2019). This amount of water can reduce the yield of its energy recovery by incineration and the feasibility of mechanical segregation for optimal valorization (El Asri et al. 2020). So, MSW is rich in water and biodegradable matter, which subtle to microbial conversion, and it has strong recovery potential by bioremediation processes.

9.3 BIOREMEDIATION AND MICROBIAL BIOSTIMULANTS

9.3.1 BIOREMEDIATION PROCESS: PRINCIPLE, TYPES, AND BENEFITS

The first appearance of the word bioremediation was in the early 1970s; it was used to refer to treatment of petroleum hydrocarbon contamination (Chapelle 1999). Bioremediation methods are currently applied to suppress a large spectrum of pollutants (Vishwakarma et al. 2020). Thassitou and Arvanitoyannis (2001) qualified this process by the emerging and new sustainable technology for management, treatment, and valorization of organic solid waste. Zouboulis et al. (2019) declared that bioremediation is appreciable primarily by the public because it is a natural way to treat wastes.

Bioremediation is described as treating toxic, hazardous, and otherwise organic wastes by biological processes. These wastes are susceptible to successful and natural conversion through the microbes community, particularly bacteria, yeasts, algae, and fungi. We can distinguish two types of bioremediation microbes: indigenous, i.e., present on the site, and allogeneic, i.e., isolated from somewhere else and added to the waste managed. In general, both types of microorganisms metabolize toxic chemicals and organic matter of waste to produce carbon dioxide or methane, water, biomass, and non-toxic form of elements. So, the bioremediation process uses eco-friendly microbial communities to degrade, eradicate, and manage organic pollutants-rich solid wastes.

Bioremediation technology is divided into two categories: in situ bioremediation, which includes six types of biotechnology: natural attenuation, bioventing, bioslurping, biosparging, phytoremediation, and bioaugmentation (Vishwakarma et al. 2020; Zouboulis, Moussas, and Psaltou 2019). These types of biotechnologies consist in treating the pollutant material in its place, so the treatment process is at the site of contamination. On the other hand, the ex situ bioremediation needs the ablation of the pollutant substrate, and it is moved to another place for treatment. This type has five biotechnologies: biopiling, windrow, bioreactors, landfarming, and composting (Vishwakarma et al. 2020; Zouboulis, Moussas, and Psaltou 2019). So, in the case of MSW, the bioremediation is of ex situ type because this waste is collected by the municipality of each town and then moved to a storage and treatment site. As soon as the waste arrives at the treatment place, it is generally subject to the addition of microbes. Thus, we will limit our discussion in this chapter to the technologies of ex situ bioremediation.

These ex situ bioremediation technologies have many benefits, which are as follows: (i) harmless detoxification and restoration of the contaminants, (ii) reduction in soil and groundwater contamination risk by the transportation of the solid waste to a particular site, (iii) reduction in waste volume and quantity, (iv) elimination of hazardous and toxic waste, (v) renovation of waste into a more serviceable shape such as fertilizer and green energy, (vi) being in harmony with environmental regulations, and (vii) economic feasibility and being less expensive. To get all these benefits, we need to ensure the effectiveness of the ex situ bioremediation of MSW. So, the efficiency of this process depends on numerous factors such as the typology of MSW, the physicochemical conditions of bioremediation technologies, and the performance of microorganisms of treatment. Several researchers have confirmed that the conversion of MSW by bioremediation is characterized by microbial dynamics and their interactions between different bacterial, archaeal, and fungal groups (Sekhohola-Dlamini and Tekere 2020). In the optimized environmental conditions and easily biodegradable MSW, the active microbial community is primordial. Rastogi et al. (2020) recommended that an efficacious microbial culture (mixture or waste specific) is significantly economical and eco-friendly in ex situ bioremediation. We are faced with the need to ensure an efficient and permanent microbial activity during the bioremediation of this waste; it is the microbial biostimulants.

9.3.2 MICROBIAL BIOSTIMULANT

Biostimulants are biological or mineral substances that can upgrade the natural process even when added in low concentrations. In the European market, the value of biostimulants is estimated at over 399 million euros, and the global market for biostimulants is 2,241 million dollars (Xu and Geelen

2018). The first application of biostimulants in the agriculture field was carried out by Schmidt and Chalmers in 1993 (Pereira et al. 2004; Schmidt, Ervin, and Zhang 2003). After 4 years, its first definition was described by Zhang and Schmidt in 1997 (Kauffman, Kneivel, and Watschke 2007). Currently, the last description of this word is developed by du Jardin (2015).

We can distinguish three types of biostimulants (chemical, botanical, and microbial biostimulants) according to their nature, physiological function, and economic and environmental benefits. The microbial biostimulant combines the use of microbes and their substances for improving the biological processes. The industry has included microorganisms in the definition of biostimulants (du Jardin 2015). In the past few decades, we used the word biostimulants to refer to substances and microbial inoculum use. So, the biostimulant concept is defined as organic or inorganic substances and microorganisms applied to upgrade the natural process when added in low concentrations.

Microbial biostimulant is one biostimulant type; it is assured by different communities of microorganisms, principally fungi and bacteria, and their enzymes and metabolites secretion. In 2017, the microbial biostimulants industry generated 2.9 billion dollars, which is presaged to double in 2022 (Berg et al. 2020). Sessitsch et al. (2018) declared that more than half of the commercial enzymes are of bacterial or fungal origin. Microbial biostimulants can increase the biological process both microbiologically and enzymatically, change the solubility of matter and exchange of micronutrients, and modify the primary substrate structure relating to the processing.

Microbial biostimulants are used in several fields: In agriculture, they stimulate growth, enhance stress tolerance, and improve nutrients uptake, enhancing crop quality (Calvo, Nelson, and Kloepper 2014; Van Oosten et al. 2017). The industrial field uses microbial biostimulants to produce commercially valuable products such as antibiotics, e.g., penicillin G acylase, enzymes, vaccines, drugs, toxins, and vitamins (Pinotti et al. 2002; Sood et al. 2011). In renewable energy production, mainly hydrogen, methane, and biofuel, microbial biostimulants are a necessary constituent (Elasri and Afilal 2016; Gottumukkala et al. 2019; Venkata Mohan and Pandey 2019). In the following part of this chapter, we will discuss the roles of biostimulants in improving the bioremediation technology of MSW.

9.4 MICROBIAL BIOSTIMULANTS IN THE BIOREACTOR LANDFILL

Landfilling is the most usual way for MSW disposal in the entire world. The amount of MSW disposed of in landfills can vary from 70% to 90% because it is the cheapest disposal management biotechnology (Abdel-Shafy and Mansour 2018; Greedy 2016). Sabour et al. (2020) confirmed that landfill research has increased threefold; the number of publications in the scientific literature has increased from 662 in 2000 to 2335 in 2017. Thus, landfilling has become a new trend for the bioremediation of MSW.

We can classify landfilling into two types: The old method is the open dumping of wastes and the modern one is sanitary landfilling. The first type is currently gradually being upgraded to sanitary landfills (Sabour, Alam, and Hatami 2020). The sanitary landfill is where the MSW is spread on the ground, then covered by thin ground layers. All these components are compacted to reduce their volume (Castrillón et al. 2010). This waste burial leads to anaerobic conditions, so anaerobic digestion is the dominant process in landfills (Prakash and Ranade 2021). Reinhart et al. (2002) declared that the bioreactor landfill provides an analogous approach and processing utilized in organic solid waste digestion. We know that anaerobic digestion depends on beading microbes present in a bioreactor landfill. Thus, it has become necessary to ensure the inoculation of these systems with a microbial biostimulant.

The addition of specialized microorganisms such as biostimulants in bioreactor landfill systems is a successful way. He et al. (2005) recommended using effective microorganisms (*Cellulomonas* sp., *Brevundimonas diminuta*, *Bacillus anthracis*, *Bacillus megaterium*, *Streptomyces* sp., and *Mucor* sp.) to obtain an enhanced biodegradability of MSW of 24%, shortened methane conversion time (91 days), and improved sustainable energy production (7 kW/m³) in the landfill system. Feng et al. (2019) cultured exogenous strains outside of landfills, which comprise *Methanobacterium*,

Methanoculleus, *Methanosaeta*, and *Methanosarcina*. These microbial stimulants are added artificially in the bioreactor to regulate the balance of community structure and improve methane production. In the MSW, we find debris plastics or microplastics, which is a biggest problem in sanitary landfills. Park and Kim (2019) isolated from a landfill site microbial biostimulants based mainly on *Bacillus* sp. and *Paenibacillus* sp. This inoculation can accelerate the microplastics degradation; the reduction is 14.7% and 22.8% of the dry weight and diameter of microplastics. Gajendiran et al. (2016) used the fungus *Aspergillus clavatus* to degrade plastics. So, we can use these microbes to increase the decomposition of plastic fraction in a bioreactor landfill. Ransom-Jones et al. (2017) stated that three phyla of microbes (*Firmicutes*, *Spirochaetes*, and *Fibrobacteres*) are key to landfill bioremediation and are the most cellulolytic community, and they produce more than 8,300 carbohydrate-active enzymes. Van der Gast et al. (2004) used a microbial consortium with four species (*Clavibacter michiganensis*, *Methylobacterium mesophilicum*, *Rhodococcus erythropolis*, and *Pseudomonas putida*). This inoculation can reduce the COD by 85% of the total organic load and 30%–40% more strongly than any indigenous microbial population alone.

We can see from the previous studies that microbial biostimulants, which are isolated from different sources (landfill system, contaminated area, or waste), can be a better choice for ex situ bioremediation of MSW. These microbial populations are more resistant to extreme ambient conditions and predation processes (van der Gast et al., 2004). Generally, the biostimulants used in landfill processes can increase the biodegradability of subsurface MSW, improve the degree of landfilled waste stabilization, reduce the life span of contaminants, permit landfill air space recovery, and enhance the rate and quality of methane production for energy recovery. So, improving conditions of the sanitary landfill by microbial biostimulants is an economical and environmentally friendly technique for the bioremediation of MSW.

9.5 MICROBIAL BIOSTIMULANTS FOR BIOREMEDIATION OF LANDFILL LEACHATE

Among the big problems of landfills of MSW is the production of a significant quantity of leachate. This last product constitutes a hazard to the environment and public health because it is immensely loaded with heavy metals, pathogenic bacteria, and a xenobiotic organic compound. The use of microbial biostimulants for leachate bioremediation is an excellent treatment technique for avoiding these problems. Zegzouti et al. (2020) used the spores of the fungus *Aspergillus flavus* as microbial additives to remove contaminants from the landfill leachate. *A. flavus* can reduce the removal rates of COD, BOD_5, and NH_4^+ by around 48%, 82%, and 99%, respectively (Zegzouti et al. 2020). Spina et al. (2018) confirmed that the two fungi *Pseudallescheria boydii* and *Phanerochaete sanguinea* are good and very fast microbial stimulants for the decolorization of leachate. Siracusa et al. (2020) confirmed that the use of *Lambertella* sp. can deplete the organic carbon in leachate. Microalgae can also be used as a means of bioremediation of leachate. Paskuliakova et al. (2018) used the *Chlamydomonas* sp. for removing the ammonia nitrogen completely by continuous growth on landfill leachate. The microalgae *Desmodesmus* sp. and *S. obliquus* also facilitate the bioremediation of leachate (Hernández-García et al. 2019). So, we confirm that the use of microbial biostimulants for leachate bioremediation is a good way because its handling is easy, simple, reliable, and of low cost, and it removes most of the biodegradable matter and nitrogen compounds in landfill leachate.

9.6 MICROBIAL BIOSTIMULANTS FOR COMPOSTING

Composting is one of the ex situ bioremediation types of MSW; it is based on the microbial degradation of organic waste matter under aerobic conditions to produce compost. This biological conversion represents 25% of the global waste product, i.e., 30 million tons per year (Razza et al. 2018). But some researchers declared that the part of MSW composting does not exceed 5.5% (Karimi et al. 2020).

The composting process is assured by two types of microbes: mesophilic and thermophilic bacteria and fungi (Partanen et al. 2010; Wei et al. 2017; Jurado et al. 2014). So, the microbes play a vital role in this process. The use of microbial biostimulants as an inoculum is useful in the composting of MSW because it provides several functions: (i) secretion of extra- or intracellular enzymes to facilitate the decomposition of these wastes, (ii) installation of good environmental conditions for composting, (iii) minimization of the initial lag time and reduced processing period, (iv) improvement of humification process of MSW, and (v) reduction in odorous emissions.

Several studies recommend using bacterial additives such as *Bacillus azotofixans*, *Bacillus megaterium*, and *Bacillus mucilaginosus* to reduce odorous gas emissions and stabilize composting products (Karnchanawong and Nissaikla 2014; Xi, Zhang, and Liu 2005). Xi et al. (2012) recommended bacterial inoculation by *Nitrobacter* and *Thiobacillus* to increase humic and fulvic acids to get a good humification degree of MSW (Xi et al. 2012). Song et al. (2018) demonstrated that inoculation with the microbial consortium composed of *Dysgonomonas* sp., *Pseudomonas caeni*, *Aeribacillus pallidus*, *Pseudomonas* sp., *Lactobacillus salivarius*, *Bacillus thuringiensis*, and *Bacillus cereus* allows avoiding the lag phase in the pile temperature and enormously shortens the composting period (Song et al. 2018). *Mycobacterium*, *Nocardia*, and *Rhodococcus* species have the capacity to convert recalcitrant pollutants of crude oil sludge, such as polycyclic aromatic hydrocarbons, by their multipurpose enzymatic activity (Cerniglia 2003). Obi et al. (2020) reported two microbial biostimulant types for the degradation of polycyclic aromatic hydrocarbons in MSW composting: The fungal type includes *Aspergillus*, *Bionectria*, *Doratomyces*, *Exophiala*, *Fusarium*, *Galactomyces*, *Geotrichum*, *Mucor*, *Penicillium*, *Trichoderma*, and *Trichurus*. The bacterial type includes *Stenotrophomonas*, *Pseudomonas*, *Bordetella*, *Brucella*, *Bacillus*, *Achromobacter*, *Advenella*, *Klebsiella*, *Mesorhizobium*, *Mycobacterium*, *Ochrobactrum*, *Pusillimonas*, and *Raoultella*. So, the use of these microorganisms for composting hazardous wastes is considered a sustainable waste management way.

9.7   CONCLUSIONS

We have shown in this work that the addition of an excellent microbial biostimulant makes the ex situ bioremediation type more attractive and efficient for the management and treatment of MSW. The microbial biostimulants are isolated from inexpensive sites to convert this hazardous waste into a less toxic form or to completely mineralize them. So, the microbial biostimulant can detoxify the hazards fraction of this waste, reduce the waste quantity, recover storage volume, improve the conversion of these free wastes into a more serviceable shape such as fertilizer and green energy, and reduce conversion time. All these roles provided by these microorganisms are in harmony with environmental regulations and their economic feasibility. So, the bioremediation of MSW by microbial biostimulants is one way of circular economy; it is the most cost-effective and eco-friendly approach to the management and treatment of MSW.

REFERENCES

Abdel-Shafy, Hussein I., et Mona S.M. Mansour. 2018. Solid waste issue: Sources, composition, disposal, recycling, and valorization. *Egyptian Journal of Petroleum* 27 (4): 1275–90. https://doi.org/10.1016/j.ejpe.2018.07.003.

Awasthi, Mukesh Kumar, Akhilesh Kumar Pandey, Jamaluddin Khan, Pushpendra Singh Bundela, Jonathan W.C. Wong, et Ammaiyappan Selvam. 2014. Evaluation of thermophilic fungal consortium for organic municipal solid waste composting. *Bioresource Technology* 168: 214–21. https://doi.org/10.1016/j.biortech.2014.01.048.

Baredar, Prashant, Vikas Khare, et Savita Nema. 2020. Optimum sizing and modeling of biogas energy system. In *Design and Optimization of Biogas Energy Systems*, pp. 33–78. Elsevier. https://doi.org/10.1016/B978-0-12-822718-3.00002-2.

Barlaz, Morton A. 1998. Carbon storage during biodegradation of municipal solid waste components in laboratory-scale landfills. *Global Biogeochemical Cycles* 12 (2): 373–80. https://doi.org/10.1029/98GB00350.

Berg, Shelby, Paul G. Dennis, Chanyarat Paungfoo-Lonhienne, Jay Anderson, Nicole Robinson, Richard Brackin, Adam Royle, Lawrence DiBella, et Susanne Schmidt. 2020. Effects of commercial microbial biostimulants on soil and root microbial communities and sugarcane yield. *Biology and Fertility of Soils* 56 (4): 565–80. https://doi.org/10.1007/s00374-019-01412-4.

Bölükbaş, Ayşenur, et Görkem Akıncı. 2018. Solid waste composition and the properties of biodegradable fractions in Izmir City, Turkey: An investigation on the influencing factors. *Journal of Environmental Health Science and Engineering* 16(2): 299–311. https://doi.org/10.1007/s40201-018-0318-2.

Burnley, Stephen J. 2007. A review of municipal solid waste composition in the United Kingdom. *Waste Management* 27 (10): 1274–85. https://doi.org/10.1016/j.wasman.2006.06.018.

Calvo, Pamela, Louise Nelson, et Joseph W. Kloepper. 2014. Agricultural uses of plant biostimulants. *Plant and Soil* 383 (1–2): 3–41. https://doi.org/10.1007/s11104-014-2131-8.

Castrillón, L., Y. Fernández-Nava, M. Ulmanu, I. Anger, et E. Marañón. 2010. Physico-chemical and biological treatment of MSW landfill leachate. *Waste Management* 30 (2): 228–35. https://doi.org/10.1016/j.wasman.2009.09.013.

Cerniglia, C.E. 2003. Recent advances in the biodegradation of polycyclic aromatic hydrocarbons by mycobacterium species. In *The Utilization of Bioremediation to Reduce Soil Contamination: Problems and Solutions*, édité par Václav Šašek, John A. Glaser, et Philippe Baveye, pp. 51–73. Dordrecht: Springer. https://doi.org/10.1007/978-94-010-0131-1_4.

Chandra, Ram, et R.C. Sobti. 2020. *Microbes for Sustainable Development and Bioremediation*. Boca Raton, FL: CRC Press.

Chapelle, Francis H. 1999. Bioremediation of petroleum hydrocarbon-contaminated ground water: The perspectives of history and hydrology. *Ground Water* 37(1): 122–32. https://doi.org/10.1111/j.1745-6584.1999.tb00965.x.

du Jardin, Patrick. 2015. Plant biostimulants: Definition, concept, main categories and regulation. *Scientia Horticulturae* 196: 3–14. https://doi.org/10.1016/j.scienta.2015.09.021.

El Asri, Ouahid, Mohamed Elamin Afilal, Hayate Laiche, et Arbi Elfarh. 2020. Evaluation of physicochemical, microbiological, and energetic characteristics of four agricultural wastes for use in the production of green energy in Moroccan farms. *Chemical and Biological Technologies in Agriculture* 7 (1): 21. https://doi.org/10.1186/s40538-020-00187-3.

Elasri, Ouahid, et Mohamed Elamin Afilal. 2016. Potential for biogas production from the anaerobic digestion of chicken droppings in Morocco. *International Journal of Recycling of Organic Waste in Agriculture*. https://doi.org/10.1007/s40093-016-0128-4.

Feng, Shoushuai, Shaoxiang Hou, Xing Huang, Zheng Fang, Yanjun Tong, et Hailin Yang. 2019. Insights into the microbial community structure of anaerobic digestion of municipal solid waste landfill leachate for methane production by adaptive thermophilic granular sludge. *Electronic Journal of Biotechnology* 39: 98–106. https://doi.org/10.1016/j.ejbt.2019.04.001.

Gajendiran, Anudurga, Sharmila Krishnamoorthy, et Jayanthi Abraham. 2016. Microbial degradation of low-density polyethylene (LDPE) by aspergillus clavatus strain JASK1 isolated from landfill soil. *3 Biotech* 6 (1): 52. https://doi.org/10.1007/s13205-016-0394-x.

Gottumukkala, Lalitha Devi, Anil K. Mathew, Amith Abraham, et Rajeev Kumar Sukumaran. 2019. Biobutanol production: Microbes, feedstock, and strategies. In *Biofuels: Alternative Feedstocks and Conversion Processes for the Production of Liquid and Gaseous Biofuels*, pp. 355–77. Elsevier. https://doi.org/10.1016/B978-0-12-816856-1.00015-4.

Greedy, Derek. 2016. Landfilling and landfill mining. *Waste Management & Research* 34 (1): 1–2. https://doi.org/10.1177/0734242X15617878.

He, Ruo, Dong-sheng Shen, Jun-qin Wang, Yong-hua He, et Yin-mei Zhu. 2005. Biological degradation of MSW in a methanogenic reactor using treated leachate recirculation. *Process Biochemistry* 40 (12): 3660–66. https://doi.org/10.1016/j.procbio.2005.02.022.

Hernández-García, Andrea, Sharon B. Velásquez-Orta, Eberto Novelo, Isaura Yáñez-Noguez, Ignacio Monje-Ramírez, et María T. Orta Ledesma. 2019. Wastewater-leachate treatment by microalgae: Biomass, carbohydrate and lipid production. *Ecotoxicology and Environmental Safety* 174: 435–44. https://doi.org/10.1016/j.ecoenv.2019.02.052.

Hui, Yuan, Wang Li'ao, Su Fenwei, et Hu Gang. 2006. Urban solid waste management in Chongqing: Challenges and opportunities. *Waste Management* 26 (9): 1052–62. https://doi.org/10.1016/j.wasman.2005.09.005.

Jurado, Macarena, María J. López, Francisca Suárez-Estrella, María C. Vargas-García, Juan A. López-González, et Joaquín Moreno. 2014. Exploiting composting biodiversity: Study of the persistent and biotechnologically relevant microorganisms from lignocellulose-based composting. *Bioresource Technology* 162: 283–93. https://doi.org/10.1016/j.biortech.2014.03.145.

Karimi, Mohsen, Jose L. Diaz de Tuesta, Carmem Natália de Pina Gonçalves, Helder T. Gomes, Alírio E. Rodrigues, et José A. C. Silva. 2020. Compost from municipal solid wastes as a source of biochar for CO_2 capture. *Chemical Engineering & Technology* 43 (7): 1336–49. https://doi.org/10.1002/ceat.201900108.

Karnchanawong, Somjai, et Siriwan Nissaikla. 2014. Effects of microbial inoculation on composting of household organic waste using passive aeration bin. *International Journal of Recycling of Organic Waste in Agriculture* 3 (4): 113–19. https://doi.org/10.1007/s40093-014-0072-0.

Kauffman, Gordon L., Daniel P. Kneivel, et Thomas L. Watschke. 2007. Effects of a biostimulant on the heat tolerance associated with photosynthetic capacity, membrane thermostability, and polyphenol production of perennial ryegrass. *Crop Science* 47 (1): 261–67. https://doi.org/10.2135/cropsci2006.03.0171.

Kolekar, K.A., T. Hazra, et S.N. Chakrabarty. 2016. A review on prediction of municipal solid waste generation models. *Procedia Environmental Sciences* 35: 238–44. https://doi.org/10.1016/j.proenv.2016.07.087.

Kumar, Sunil. 2020. *Municipal Solid Waste Management in Developing Countries*. Boca Raton, FL: CRC Press.

Mushtaq, Jasir, Abdul Qayoom Dar, et Naved Ahsan. 2020. Spatial–temporal variations and forecasting analysis of municipal solid waste in the mountainous City of North-Western Himalayas. *SN Applied Sciences* 2 (7): 1161. https://doi.org/10.1007/s42452-020-2975-x.

Obi, Linda, Harrison Atagana, Rasheed Adeleke, Mphekgo Maila, et Emomotimi Bamuza-Pemu. 2020. Potential microbial drivers of biodegradation of polycyclic aromatic hydrocarbons in crude oil sludge using a composting technique. *Journal of Chemical Technology & Biotechnology* 95 (5): 1569–79. https://doi.org/10.1002/jctb.6352.

Park, Seon Yeong, et Chang Gyun Kim. 2019. Biodegradation of micro-polyethylene particles by bacterial colonization of a mixed microbial consortium isolated from a landfill site. *Chemosphere* 222: 527–33. https://doi.org/10.1016/j.chemosphere.2019.01.159.

Partanen, Pasi, Jenni Hultman, Lars Paulin, Petri Auvinen, et Martin Romantschuk. 2010. Bacterial diversity at different stages of the composting process. *BMC Microbiology* 10 (1): 94. https://doi.org/10.1186/1471-2180-10-94.

Paskuliakova, Andrea, Ted McGowan, Steve Tonry, et Nicolas Touzet. 2018. Microalgal bioremediation of nitrogenous compounds in landfill leachate – The importance of micronutrient balance in the treatment of leachates of variable composition. *Algal Research* 32: 162–71. https://doi.org/10.1016/j.algal.2018.03.010.

Pereira, M.A., D.Z. Sousa, M. Mota, et M.M. Alves. 2004. Mineralization of LCFA associated with anaerobic sludge: Kinetics, enhancement of methanogenic activity, and effect of VFA. *Biotechnology and Bioengineering* 88 (4): 502–11. https://doi.org/10.1002/bit.20278.

Periathamby, Agamuthu. 2011. Municipal waste management. In *Waste*, pp. 109–25. Elsevier. https://doi.org/10.1016/B978-0-12-381475-3.10008-7.

Pinotti, Laura M., Rosineide G. Silva, Roberto C. Giordano, et Raquel L.C. Giordano. 2002. Inoculum studies in production of penicillin G acylase by bacillus megaterium ATCC 14945. *Applied Biochemistry and Biotechnology* 98–100 (1–9): 679–86. https://doi.org/10.1385/ABAB:98-100:1-9:679.

Prakash, Om, et Dilip R. Ranade. 2021. *Anaerobes and Anaerobic Processes*. CRC Press, Boca Raton, FL.

Ramachandra, T.V., H.A. Bharath, Gouri Kulkarni, et Sun Sheng Han. 2018. Municipal solid waste: Generation, composition and GHG emissions in Bangalore, India. *Renewable and Sustainable Energy Reviews* 82: 1122–36. https://doi.org/10.1016/j.rser.2017.09.085.

Ransom-Jones, Emma, Alan J. McCarthy, Sam Haldenby, James Doonan, et James E. McDonald. 2017. Lignocellulose-degrading microbial communities in landfill sites represent a repository of unexplored biomass-degrading diversity. *MSphere* 2 (4): e00300-17. https://doi.org/10.1128/mSphere.00300-17.

Rastogi, Mansi, Meenakshi Nandal, et Babita Khosla. 2020. Microbes as vital additives for solid waste composting. *Heliyon* 6 (2): e03343. https://doi.org/10.1016/j.heliyon.2020.e03343.

Razza, Francesco, Lorenzo D'Avino, Giovanni L'Abate, et Luca Lazzeri. 2018. The role of compost in bio-waste management and circular economy. In *Designing Sustainable Technologies, Products and Policies*, édité par Enrico Benetto, Kilian Gericke, et Mélanie Guiton, pp. 133–43. Cham: Springer International Publishing. https://doi.org/10.1007/978-3-319-66981-6_16.

Reinhart, Debra R., Philip T. McCreanor, et Timothy Townsend. 2002. The bioreactor landfill: Its status and future. *Waste Management & Research: The Journal for a Sustainable Circular Economy* 20 (2): 172–86. https://doi.org/10.1177/0734242X0202000209.

Sabour, Mohammad Reza, Ehsan Alam, et Amir Mostafa Hatami. 2020. Global trends and status in landfilling research: A systematic analysis. *Journal of Material Cycles and Waste Management* 22 (3): 711–23. https://doi.org/10.1007/s10163-019-00968-5.

Schmidt, R.E., E.H. Ervin, et Xunzhong Zhang. 2003. Questions and answers about biostimulants. *Golf Course Manage* 71(6): 91–94.

Sekhohola-Dlamini, Lerato, et Memory Tekere. 2020. Microbiology of municipal solid waste landfills: A review of microbial dynamics and ecological influences in waste bioprocessing. *Biodegradation* 31 (1–2): 1–21. https://doi.org/10.1007/s10532-019-09890-x.

Sessitsch, Angela, Günter Brader, Nikolaus Pfaffenbichler, Doris Gusenbauer, et Birgit Mitter. 2018. The contribution of plant microbiota to economy growth. *Microbial Biotechnology* 11 (5): 801–5. https://doi.org/10.1111/1751-7915.13290.

Siracusa, Giovanna, Qiuyan Yuan, Ilaria Chicca, Alessandra Bardi, Francesco Spennati, Simone Becarelli, David Bernard Levin, Giulio Munz, Giulio Petroni, et Simona Di Gregorio. 2020. Mycoremediation of old and intermediate landfill leachates with an ascomycete fungal isolate, *Lambertella* Sp. *Water* 12 (3): 800. https://doi.org/10.3390/w12030800.

Song, Caihong, Yali Zhang, Xunfeng Xia, Hui Qi, Mingxiao Li, Hongwei Pan, et Beidou Xi. 2018. Effect of inoculation with a microbial consortium that degrades organic acids on the composting efficiency of food waste. *Microbial Biotechnology* 11 (6): 1124–36. https://doi.org/10.1111/1751-7915.13294.

Sood, S., R. Singhal, S. Bhat, et A. Kumar. 2011. Inoculum preparation. In *Comprehensive Biotechnology*, pp. 151–64. Elsevier. https://doi.org/10.1016/B978-0-08-088504-9.00090-8.

Spina, Federica, Valeria Tigini, Alice Romagnolo, et Giovanna Varese. 2018. Bioremediation of landfill leachate with fungi: Autochthonous vs. allochthonous strains. *Life* 8 (3): 27. https://doi.org/10.3390/life8030027.

Stafford, William H.L. 2020. WtE best practices and perspectives in Africa. In *Municipal Solid Waste Energy Conversion in Developing Countries*, 185–217. Elsevier. https://doi.org/10.1016/B978-0-12-813419-1.00006-1.

Suocheng, Dong, Kurt W. Tong, et Wu Yuping. 2001. Municipal solid waste management in china: Using commercial management to solve a growing problem. *Utilities Policy* 10 (1): 7–11. https://doi.org/10.1016/S0957-1787(02)00011-5.

Thassitou, P.K., et I.S. Arvanitoyannis. 2001. Bioremediation: A novel approach to food waste management. *Trends in Food Science & Technology* 12 (5–6): 185–96. https://doi.org/10.1016/S0924-2244(01)00081-4.

van der Gast, Christopher J., Andrew S. Whiteley, et Ian P. Thompson. 2004. Temporal dynamics and degradation activity of an bacterial inoculum for treating waste metal-working fluid. *Environmental Microbiology* 6 (3): 254–63. https://doi.org/10.1111/j.1462-2920.2004.00566.x.

Van Oosten, Michael James, Olimpia Pepe, Stefania De Pascale, Silvia Silletti, et Albino Maggio. 2017. The role of biostimulants and bioeffectors as alleviators of abiotic stress in crop plants. *Chemical and Biological Technologies in Agriculture* 4 (1). https://doi.org/10.1186/s40538-017-0089-5.

Venkata Mohan, S., et Ashok Pandey. 2019. Sustainable hydrogen production. In *Biohydrogen*, pp. 1–23. Elsevier. https://doi.org/10.1016/B978-0-444-64203-5.00001-0.

Vishwakarma, Gajendra Singh, Gargi Bhattacharjee, Nisarg Gohil, et Vijai Singh. 2020. Current status, challenges and future of bioremediation. In *Bioremediation of Pollutants*, 403–15. Elsevier. https://doi.org/10.1016/B978-0-12-819025-8.00020-X.

Wei, Yunmei, Jingyuan Li, Dezhi Shi, Guotao Liu, Youcai Zhao, et Takayuki Shimaoka. 2017. Environmental challenges impeding the composting of biodegradable municipal solid waste: A critical review. *Resources, Conservation and Recycling* 122: 51–65. https://doi.org/10.1016/j.resconrec.2017.01.024.

Xi, B., G. Zhang, et H. Liu. 2005. Process kinetics of inoculation composting of municipal solid waste. *Journal of Hazardous Materials* 124 (1–3): 165–72. https://doi.org/10.1016/j.jhazmat.2005.04.026.

Xi, Bei-Dou, Xiao-Song He, Zi-Min Wei, Yong-Hai Jiang, Ming-Xiao Li, Dan Li, Ye Li, et Qiu-Ling Dang. 2012. Effect of inoculation methods on the composting efficiency of municipal solid wastes. *Chemosphere* 88 (6): 744–50. https://doi.org/10.1016/j.chemosphere.2012.04.032.

Xu, Lin, et Danny Geelen. 2018. Developing biostimulants from agro-food and industrial by-products. *Frontiers in Plant Science* 9: 1567. https://doi.org/10.3389/fpls.2018.01567.

Yuan, Jing, Difang Zhang, Ruonan Ma, Guoying Wang, Yun Li, Shuyan Li, Huan Tang, Bangxi Zhang, Danyang Li, et Guoxue Li. 2019. Effects of inoculation amount and application method on the biodrying performance of municipal solid waste and the odor emissions produced. *Waste Management* 93: 91–99. https://doi.org/10.1016/j.wasman.2019.05.029.

Zegzouti, Yassine, Aziz Boutafda, Amine Ezzariai, Loubna El Fels, Miloud El Hadek, Lalla Amina Idrissi Hassani, et Mohamed Hafidi. 2020. Bioremediation of landfill leachate by Aspergillus flavus in submerged culture: Evaluation of the process efficiency by physicochemical methods and 3D fluorescence spectroscopy. *Journal of Environmental Management* 255: 109821. https://doi.org/10.1016/j.jenvman.2019.109821.

Zouboulis, Anastasios I., Panagiotis A. Moussas, et Savvina G. Psaltou. 2019. Groundwater and soil pollution: Bioremediation. In *Encyclopedia of Environmental Health*, 369–81. Elsevier. https://doi.org/10.1016/B978-0-12-409548-9.11246-1.

10 Microbial Biostimulants for Bioremediation of Organic and Inorganic Compounds

Samantha Pardo, Mauricio Pérez-Albornoz,
and David Morales-Pérez
Salesian Polytechnic University

Moises Bustamante-Torres
Yachay Tech University and National Autonomous University of Mexico

Emilio Bucio
National Autonomous University of Mexico

CONTENTS

DOI: 10.1201/9781003188032-10

10.1 INTRODUCTION

The release of organic and inorganic compounds into the environment started years ago when the industrial age was at its peak. The activities that were done by humans changed, and the machines took their place. Nowadays, technologies have revolutionized machines with time. Due to this, industries require new chemical compounds to automate a large number of activities carried out by people. Additionally, these new compounds have brought considerable damage to the environment in aspects such as water, soil, and air. So, the use of microorganisms for bioremediation has been studied in more cases considering wastewater and soil than air. On the other hand, scientific research using microbes as bioindicators has contributed to improving researches concerning air pollution. Finally, activities that produce and use more quantity of organic and inorganic compounds are involved in food industry, oil industry and textile industry and are examined in this chapter.

10.2 ENVIRONMENTAL POLLUTION

This problem is defined as the presence of various physical, biological, and chemical compounds in the environment that represent a danger to human health as well as to the life that develops in nature, negatively altering the characteristics of hydrological, forest, soil, and wind resources. The environment is in constant change because the human being advances rapidly in its technological and territorial expansion where additional needs arise in the social environment causing the place where they are developed to deteriorate as a result of these activities; it is of vital importance to protect both renewable and non-renewable resources, including the remediation of them, to prevent the pollution generated from causing an imbalance by the action of agents that attenuate the nature [1].

There are different ways in which environmental pollution is evidenced, and these are the products of anthropogenic activities and the nature itself. The sources that produce environmental alteration on the part of human beings are industrial activities (among the representatives are the mining sector, food, pharmaceuticals, oil extraction), commerce, the agricultural sector, and mobile sources; on the other hand, the pollution that comes from nature is caused by processes that are generated over time, such as uncontrolled forest fires in which much of the vegetation and wildlife are lost, volcanic eruptions that produce gas emissions into the atmosphere, acid rain and other natural phenomena that can cause significant changes in the environment, and other natural phenomena that can cause significant changes in the environment [2].

The deterioration of environmental conditions in different parts of the planet is reflected as severe consequences of human actions, being even more noticeable in global warming. That is why there is currently a crisis in environmental issues, mainly caused by humanity; therefore, the problems unleashed must be taken from a broad perspective to establish the interconnections among them, integrally attaching remediation. The emissions of polluting agents can also cause problems globally, as is the clear example of greenhouse gases on the destruction of the ozone layer, mainly in those countries considered as world powers. Figure 10.1 represents the environmental problems and the diffusion of a pollutant in different media concerning its nature and the characteristics of the substance, which leads to the establishment of a cycle of affectation [2].

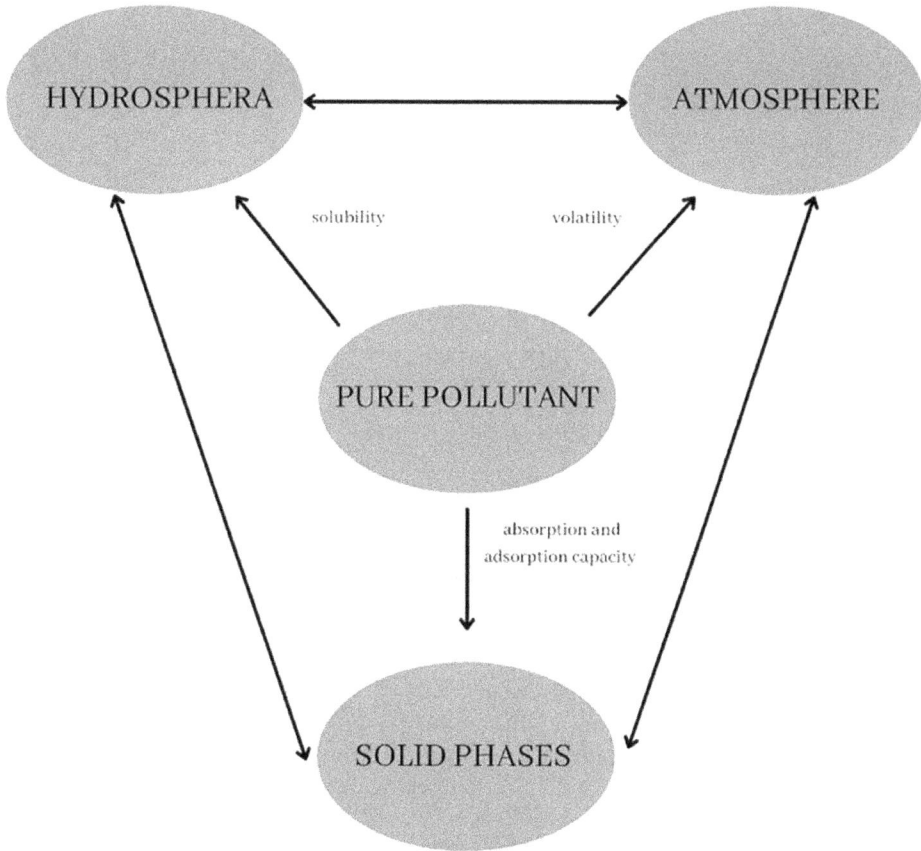

FIGURE 10.1 Interaction of pollutants with natural resources.

10.2.1 INORGANIC COMPOUNDS

These compounds are established by various combinations of chemical elements; such combination is usually toxic to the health of living beings and therefore leads to the contamination of the environment from the use of radionuclides to the use of metals for various anthropogenic activities. One of the characteristic problems of these compounds is the difficulty in elimination. In the case of agriculture, it is common to use products consisting mainly of inorganic compounds to counteract insects and fungi that affect crops and thus prevent diseases caused by pests, but if used indiscriminately, they can become dangerous [3].

The inorganic contaminants that are of most significant concern are metalloids and heavy metals because when exposed, they present a risk due to their high toxic level in small quantities that can easily be absorbed by the microorganisms found in the soil, and these contaminants can also present changes when the pH is altered as shown in Figure 10.2. In addition to the finding of these contaminants in the soil, they can also be found in bodies of water. They are dissolved in water and come from household discharges, crop activities, erosion, and business sectors. They mainly include sulphides, nitrates, and chlorides. On the other hand, these components also come from acid waste and gases such as ammonia (present in garbage dumps), hydrogen sulphide, and sulphur oxides [4].

Other problems arising from these compounds when released by mining industries and transport increase the risks to the environment because in the soil, there is an affectation in the quality and nutrients. However, how harmful the pollution is depends not only on the concentration that is found, but also on the characteristics of the environment in which it is located. Unlike the agricultural soil

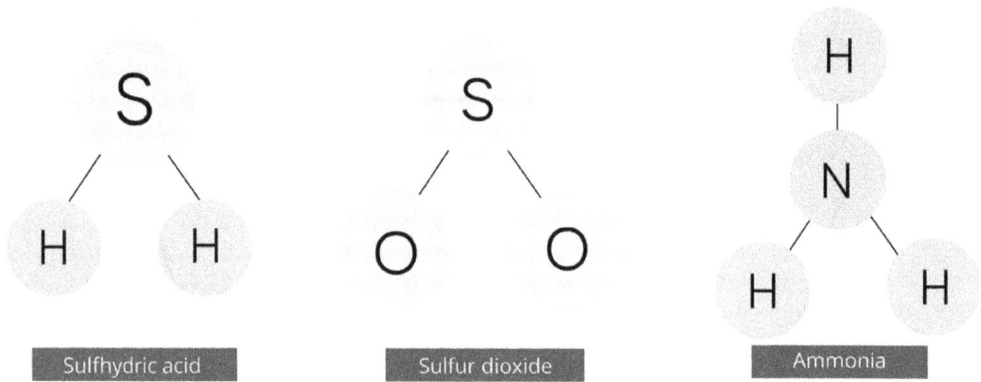

FIGURE 10.2 Most common inorganic compounds.

in the field, the soil in the city or in more urban areas will have different adverse effects because the compounds will move differently for each case [5].

10.2.2 ORGANIC COMPOUNDS

These compounds are characterized by covalent carbon–hydrogen bonds combined with other elements such as oxygen or nitrogen [6]. Most of these compounds are found in nature, although humans can synthesize them artificially as shown in Figure 10.3. Among the pollutants from organic compounds are pesticides and herbicides that are used in agriculture, generating negative impacts on ecosystems; when used for agricultural production, traces of these chemicals are often left in resources, mainly in the soil and water, where their adverse consequences reach the food consumed by people, affecting the health of consumers; on the other hand, the combustion of certain compounds originating in the hydrocarbon industry has collateral effects on health and nature [7].

Organic pollutants are classified into two groups: chlorinated pollutants, which have chlorine atoms in their structure and are divided into alkenes and aliphatic compounds, represented by dioxins (produced during the incineration of plastics and in the production of PVC pipes), polycyclic aromatic hydrocarbons (PAHs) (produced by incomplete combustion in the oil sector, coal, and particular wastes), dichlorodiphenyltrichloroethane (used in pest control in the agricultural sector), and polychlorinated biphenyls (used as an insulating material in the electrical industry, and in refrigerants); on the other hand, non-chlorinated compounds are divided into alkenes and aliphatic hydrocarbon compounds, represented by ethane (part of natural gas), benzenes (used in the manufacture of plastics), xylenes (used in the rubber industry, made from petroleum), toluene (from the manufacture of fuels from petroleum), and ethylbenzene (produced in the manufacture of pest control products and paints) as we can see the detailed classification in Figure 10.4 [8].

There are both positive and negative aspects of the use of organic compounds. In the first case, due to industrial progress we can have products for the use of humanity as is the case of cosmetics, the area of medicine, in the movement of vehicles, in the manufacture of paper, which have helped humans in their daily lives over the years. On the other hand, the excessive use of compounds can have adverse effects, as they can persist in the environment for many years without degrading. This is where another problem may arise: bioaccumulation in organisms that are in soil and water which were exposed to compounds in high or low amounts, come to have highly toxic levels with the passage of time producing alterations inside, as is the case of malformations present in communities of the Amazon by the use of pesticides [9,10].

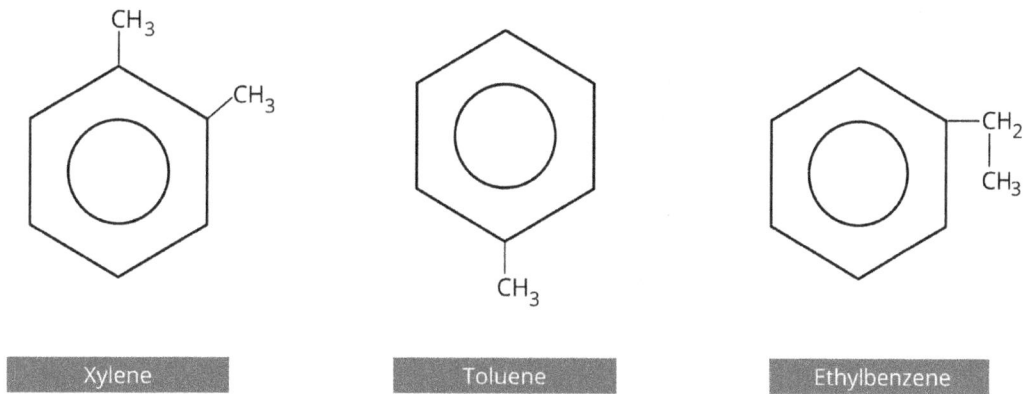

FIGURE 10.3   Organic compounds.

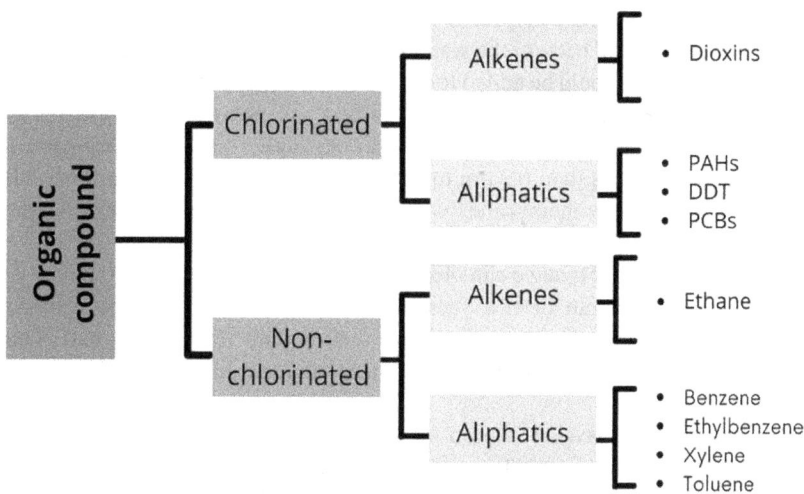

FIGURE 10.4   Classification of organic compounds.

10.3   ENVIRONMENTAL REMEDIATION

It is the action by which we try to provide a solution to the problems of contamination in natural resources (water, air, and soil) as shown in Figure 10.5.

Some of the processes used for the remediation of inorganic pollutants are bioaccumulation and bioadsorption. By using several microorganisms, they retain heavy metals in their interior, and there are also a series of extracellular materials that absorb these pollutants. On the other hand, phytoelimination (also known as phytoremediation) uses plant species to remove inorganic compounds, efficiently filtering out the contaminants that are in contact with the soil. Thus, this process reduces erosion caused not only by these agents, but also by runoff; in another technique in this procedure, contaminants are absorbed by plants and thus proceed to their separation, ending with the remediation mechanism [4].

Methylation is the addition of a methyl group to detoxify metals with a high degree of toxicity and thus save the organisms exposed to these agents. In contrast, demethylation consists of breaking the methyl bonds, which leads to elimination of the contaminants. Complexation is used to decrease the degree of solubility of certain metals and thus their concentration.

Like the previous process, ligand degradation results in the disabling of the metal contaminant from spreading in the medium where it is found, leading to a decrease in its concentration.

The oxidation–reduction method is used to modify the valences of the elements, both increasing and decreasing them, respectively, which makes them less toxic and thus facilitates their treatment [3].

The remediation of organic and inorganic pollution involves a series of processes that, in some cases, are similar; the use of biochar from biowaste works by adsorbing pollutants into the interior, thus facilitating decontamination [7]. Another process used is the chemical oxidation of the existing contamination by a hydrocarbon material; this is where specific chemical agents are added, which have the function of decontaminating the resources that have been affected, something straightforward but of high cost. Phytoremediation uses plants for the decontamination of soils and water by organic agents. In contrast to the previous method, this technique is of low cost because the plants are responsible for absorbing the pollutants into their tissues while reducing the high levels of substances in the environment [11].

10.3.1   Conventional Methods

Different methods are commonly used for the removal of both organic and inorganic compounds. The primary treatment is not very helpful in this particular case since the wastewater would be with high contamination if sent back to the waterways, which is why both secondary treatment and advanced methods (tertiary) should be added to the first, obtaining better results during remediation.

Secondary treatment encompasses several processes; reverse osmosis works by the principle of pressure on the contaminated water, which passes through a series of semi-permeable membranes with thicknesses of less than 0.2 μm to remove heavy metals [12]; similarly, electrodialysis works with semi-permeable membranes with the difference from the previous method being the use of electrodes.

The activated sludge process that we can observe in Figure 10.6 uses a rectangular tank in which the flocculus has been placed with air or oxygen. It is here where the microorganisms make contact with the water to be treated, which results in sludge that remains in the lower part. Therefore, one part is removed and the other is used again so that these microorganisms remain active, feeding on the organic compounds.

Oxidation lagoons, which have been used for a long time, are divided into aerobic, anaerobic, facultative, and aerated lagoons. Facultative ponds are retained for approximately 6 months, where it depends on the climatic conditions since the algae are used to release oxygen through the photosynthesis

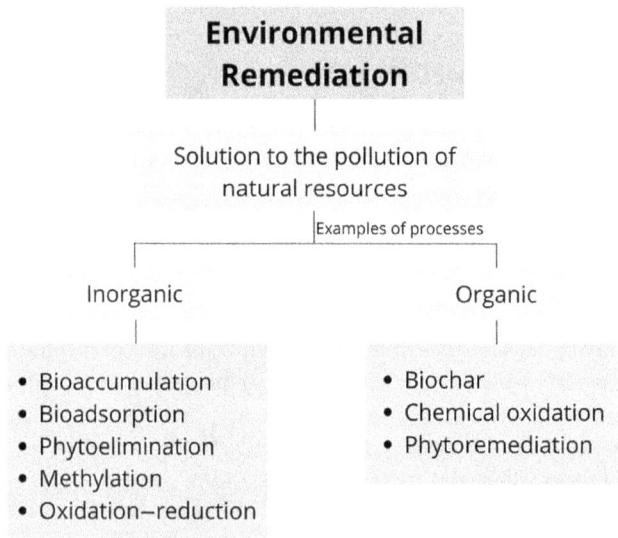

Environmental Remediation

Solution to the pollution of natural resources

Examples of processes

Inorganic

- Bioaccumulation
- Bioadsorption
- Phytoelimination
- Methylation
- Oxidation–reduction

Organic

- Biochar
- Chemical oxidation
- Phytoremediation

FIGURE 10.5   Environmental remediation techniques for organic and inorganic compounds.

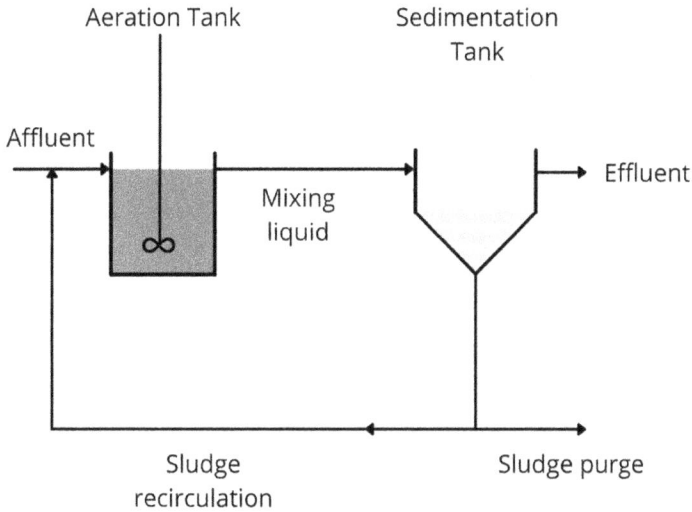

FIGURE 10.6 Sludge treatment as a conventional method for wastewater treatment.

process, giving rise to the development of microorganism populations. The aerated ponds are usually retained for up to a week, allowing the solids to be suspended, and they are then moved to a second pond in which they are placed on the floor, allowing organic degradation in the absence of oxygen, and the last pond is used to establish clarity of the effluent. Aerobic ponds have a retention time of up to 5 days. The process delivers the oxygen for photosynthesis, producing waters with many microbial populations benefiting from hot, dry climates. Anaerobic ponds are usually retained from 20 days to 2,5 months and are used in cases of strong pollutants causing foul odours [12].

Trickling filtration is characterized by the microorganisms that are not suspended as in previous methods. This layer is sent to another tank to end up as a sediment in the primary tank using trickling filters to facilitate the process as shown in Figure 10.7.

Membrane bioreactors are also used, with the difference that this process is of high cost and excellent results. The use of membranes allows the pollutants and solids present in the water to be filtered, thus eliminating the use of tanks where the sludge settles, giving rise to the development of populations of microorganisms responsible for the degradation of organic compounds [13].

Tertiary treatment is used for the elimination of organic pollutants that are considered to have a high environmental impact, such as volatile organic compounds, persistent organic compounds, and heavy metals [12].

In this treatment, coagulation is used, which consists of adding a coagulant, in this case aluminium or iron oxide, to have a neutralizing effect on the load, followed by flocculation, which neutralizes and adsorbs the floccules, which leads to their sedimentation, and therefore, the solids obtained will be discarded in the filtration process with the use of sand beds [13].

10.3.2 Bioremediation

The bioremediation process is carried out by microorganisms and plant species responsible for the degradation, transformation, and metabolization of pollutants found on land surfaces and in aquatic environments.

This process is characterized by the fact that bacteria can convert substances that can alter ecosystems into harmless products or with a minimum toxicity. They also use the substrates to generate energy, which leads to their growth and thus gives way to their incorporation into the biogeological cycles [14]. Microorganisms are found in different parts of the environment, but those needed for the degradation of pollutants are found in places such as landfills, and soils that have

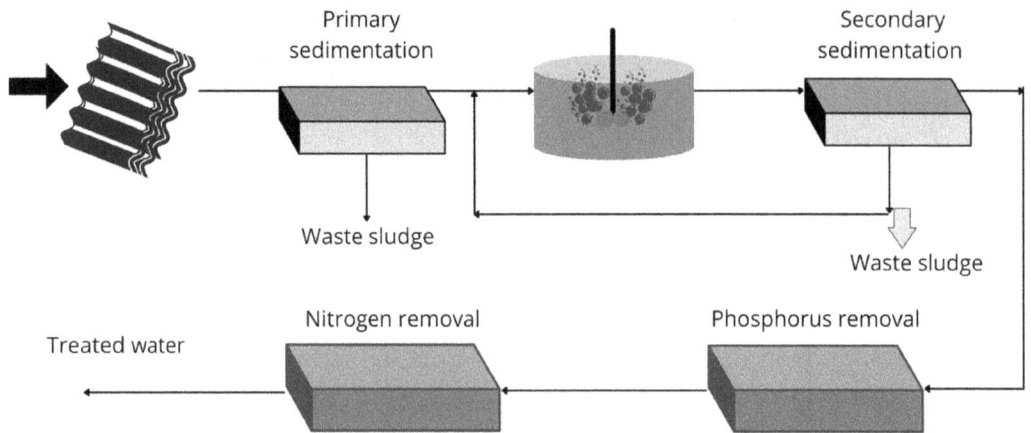

FIGURE 10.7 Conventional methods for the treatment of some organic and inorganic compounds in wastewater.

been affected by the use of pesticides, especially those of the agricultural sector, mines, and oil spill sites, since in the presence or absence of oxygen, they transform contamination into products with a low level of toxicity.

Bioremediation is classified into two categories and is detailed in Figure 10.8: *in situ* and *ex situ* bioremediation.

The *in situ* treatment refers to the fact that the problem caused by contamination has not had any change or excavation in the environment, and this type of bioremediation is divided into intrinsic, enhanced, biosparging, bioaugmentation, bioventing, and biosuction [14].

Intrinsic bioremediation refers to treatment without human interventions, using microbiological techniques with and without oxygenation.

Enhanced bioremediation, as its name explains, establishes an improvement in the process since nutrients and oxygen are applied to the microorganisms used for a good development [15].

Bioaugmentation uses microorganisms from both inside and outside the area where the environment was affected and modified for treatment.

In biosparging, oxygen is injected to bring the pollutants to the surface so that they can be degraded due to the enhanced reactions of the microorganisms [16,17].

The bioventing process injects oxygen in small quantities through a control system allowing the degradation of pollutants, especially light hydrocarbons [18].

Biosuction is a technique that uses indirect air injection to enhance and stimulate microorganism populations to eliminate VOCs.

The *ex situ* bioremediation consists in collecting part of the area affected by the pollutant by excavation for future treatment and disposal elsewhere; this process is also classified into several processes such as composting, biofiltration, use of bioreactors and bio-stacking, and landfarming [14].

Biofilters using microorganisms within these tools are intended to treat gaseous contamination [19].

Microorganisms have grown adequately within parameters controlled in the bioreactors. These microorganisms can treat and degrade pollutants such as benzene, xylene, and volatile organic pollutants, which have affected water bodies and soil; this remediation technique is the most effective [20].

Bio-stacking is a technique used for treating soils contaminated by hydrocarbons. It involves making a homogeneous mixture of the affected area and placing the contaminated part in piles, where nutrients are injected with aeration to improve the elimination of the microorganisms together with controlled irrigation and leachate collection [14].

Landfarming is used for the decontamination of soils affected by PAHs. The affected soil is excavated and then spread over a wide area where the pollutant will be degraded by the action of microorganisms that will have controlled conditions of temperature, pH, and humidity. Therefore, it is necessary to move the soil to allow effective aeration, so it is considered to be of low cost and simple to perform [21,22].

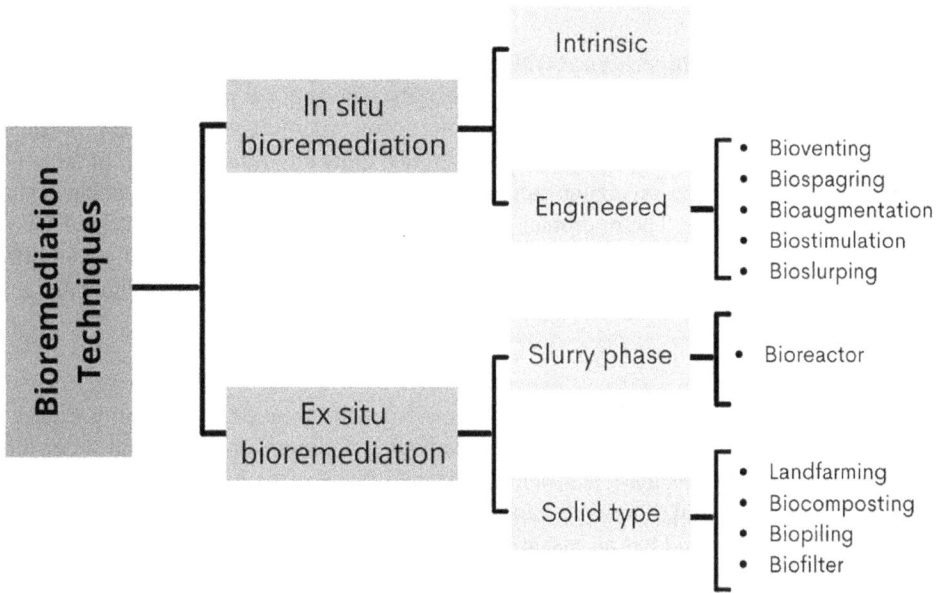

FIGURE 10.8 Classification of bioremediation techniques.

10.4 MICROBES-DEPENDENT CLEANING SYSTEM

Organic and inorganic wastes include metals and metalloids, xenobiotic contaminants, salts, leachate, sewage, sludge, and other conventional wastes. Some redundant or backup treatment may be necessary depending on the acuteness of toxicity to offset the variability of biological systems [23]. Conventional methods are usually not enough to treat these contaminants. Therefore, microbial cleaning takes advantage of naturally occurring microbes to remove a wide variety of contaminants from various surfaces [24]. Table 10.1 describes the seven main microorganisms used for environmental remediation.

Bacteria and fungi are the most common microorganisms used for bioremediation. However, bacteria are the most employed compared to fungi. Fungi strains have been reported to be highly used in dye discolouration or removal. In general, microorganisms have successfully been used for bioremediation in petrochemical plants, chemical plants, refineries, food processing plants, marine barges, machine shop parts washers, truck washers, wood-treating plants and groundwater remediation applications [24]. Such studies tactfully attack the core problem of pollution, and the benefits of decontamination add a healthy atmosphere to humanity [30].

TABLE 10.1
Microbial Groups Commonly Used to Remove Diverse Pollutants from the Ecosystem

Microbial Groups	Description	Contaminant Degraded	Reference
Archaea	Unicellular prokaryotic cells	Produce methane during their metabolism	[24]
Bacteria	Unicellular prokaryotic cells	Hydrocarbon Mercury (II)	[24,25]
Fungi	Heterotrophs	Dyes Hydrocarbons	[24,26,27]
Protista	Non-photosynthetic eukaryotes	Hydrocarbons Wastewaters from pig-slaughtering plants	[24,28]
Viruses	Made up of nucleic acids (DNA or RNA) and proteins	Fe(II) bioreduction Organic/inorganic contaminants	[29]

10.4.1 BIOREMEDIATION

Environmental remediation is concerned with treating pollutants in water, air, or soil [31]. Bioremediation is an effective option to enhance the results of the conventional process to treat the ecosystem through eco-friendly alternatives. This process involves microorganisms, plants, or microbial or plant enzymes to detoxify contaminants in the soil and other environments [32]. In other words, it is a collective phenomenon involving processes that use biological systems to either restore or clean up contaminated sites. The microbial community has consistently been reported for bioremediation [33]. The reported efforts to enhance bioremediation have been sorted out in different ways, such as adsorption, biodegradation, biotransformation, biosorption, bioaugmentation, and phytoremediation.

10.4.1.1 By Adsorption

Adsorption involves the movement of contaminants from the solution to the soil particles [34]. This bioremediation technique is highly used for removing pollutants from soil and water through an effective driving force for mass transfer, which is high since the beginning. However, as the adsorption bed gets saturated, it loses its capacity of separation, and eventually, no further separation is done [35]. This method has an essential advantage because it is quick, simple, eco-friendly, and cost-effective. Figure 10.9 illustrates the adsorption process between the adsorbate and the adsorbent and their interactions.

Physical adsorption attaches the biomolecules to the surface through weak bonds [36]. The adsorption mechanism is often physical, but may also be chemical [37]. Chemisorption involves the strong interaction of H electrons and metal electrons in such a way that it leads to splitting of the molecule, formation of covalent bonds with metal electrons, and perturbations in the metal electron system [38].

10.4.1.2 By Biodegradation

Biodegradation refers to a single biological reaction or sequence of reactions that result in converting an organic substrate to a simpler molecule [39]. In principle, this process is carried out by many microorganisms. It is strongly dependent on the conditions for the microorganisms in the medium. The biodegradation is furthermore dependent on aerobic or anaerobic conditions [40].

Biodegradation can be carried out based on different steps, as shown in Figure 10.10. During the biodegradation of organic contaminants, the by-products can be considered as part of either the metabolism (energy available produced) or the co-metabolism (no energy available) of the microorganisms [41].

Primary biodegradation is determined based on the analytical signal, which corresponds to primary compounds. Moreover, the amount of CO_2 increases as the primary compounds and their metabolites are degraded by microorganisms [42].

10.4.1.2.1 By Mineralization

Biodegradation processes usually are catalysed by enzymes that convert chemicals via a series of intermediates into end products [43]. Mineralization term is related to the conversion of biomass into a gaseous state, water, salt and minerals, and residual biomass. The mineralization is complete

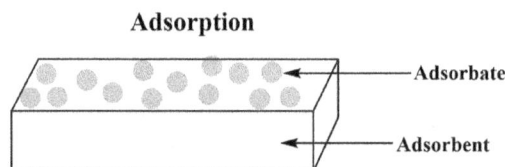

Adsorption

FIGURE 10.9   Molecules adhered to the surface of the material.

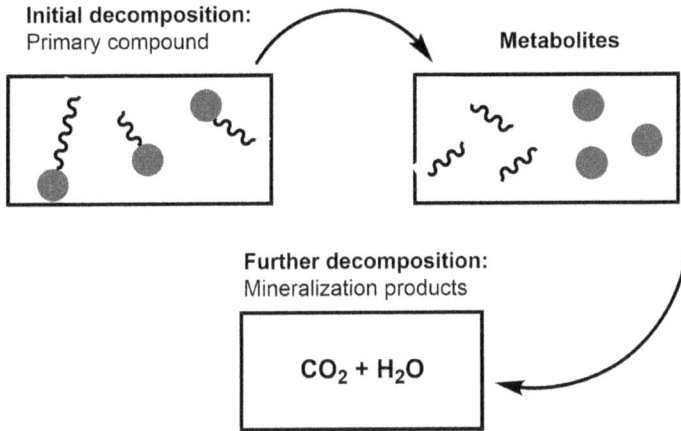

FIGURE 10.10 The general biodegradation process until mineralization.

when all solid carbon is converted into carbon dioxide and water, respectively, and methane is produced during aerobic or anaerobic degradation [44].

10.4.1.3 By Biotransformation

Biotransformation involves converting one chemical into another without complete mineralization and is seen frequently during microbial metabolism of unnatural synthetic chemicals (xenobiotics) [43]. In other words, biotransformation refers to the process by which xenobiotic (foreign) or endogenous chemicals are enzymatically modified (metabolized) to chemicals that differ in their excretability, biological activity, and toxicity [45]. Biotransformation usually involves the cytochrome P450-dependent monooxygenases enzymes to detoxify the metabolism in the presence of chemical compounds [46]. Moreover, biotransformation can be applied either *in situ* or *ex situ*.

The *in situ biotransformation* is used to treat the contaminated soil or groundwater in the location in which it is found. Meanwhile, the *ex situ biotransformation* requires excavation of contaminated soil or pumping of groundwater before they can be treated [47].

10.4.1.4 By Biosorption

Biosorption is a property of specific, inactive, non-living microbial biomass to bind and concentrate heavy metals from even very dilute aqueous solution [28]. It is considered a simple metabolic passive physicochemical process in which the pollutants (sorbate) bind to the surface of the biosorbent, which is of biological origin [49]. Besides, this technique provides some advantages and a low-cost process to remove toxic heavy metal ions from polluted solutions [50].

The materials employed in this process usually are obtained through electrostatic interactions between the pollutant and the functional groups presented in the outer cell wall. The most common functional groups are carboxylate, hydroxyl, amine, and phosphoryl groups [43] on the biosorbent surface.

Biosorption is simply the exchange of heavy metal ions attached to biomass. Besides, compared to removing metal ions through membrane technologies, reverse osmosis, and chemical precipitation, it presents significant advantages such as minimized biological sludge and chemical sludge, recycling of sorbent, and high efficiency [43]. Figure 10.11 shows some examples of the chemical method used during biosorption (physisorption, precipitation, chelation, complexation, and ion exchange).

Besides, biosorption is an effective alternative to remove dyes from aqueous solution. One study previously performed based on biosorption deals with a renewable, low-cost biosorbent derived from the waste biomass of *Corynebacterium glutamicum* [54].

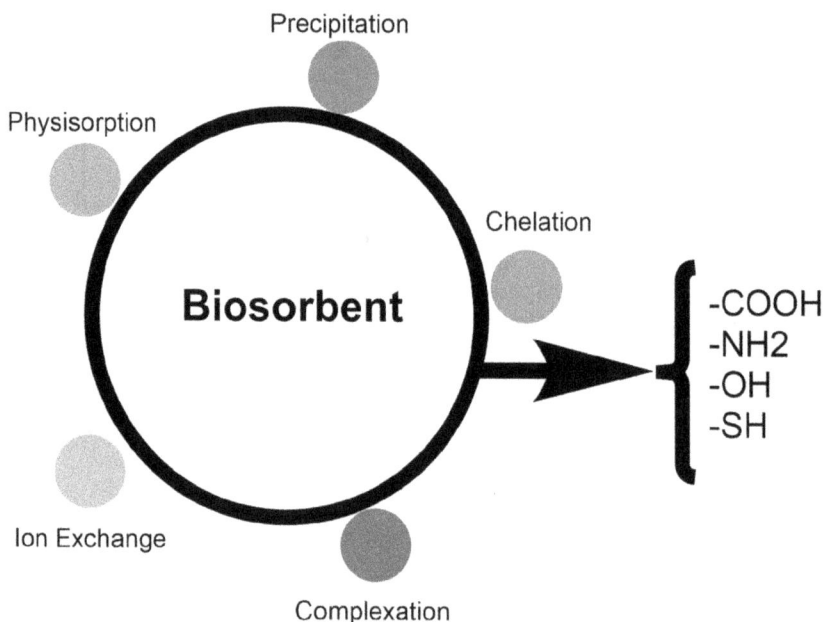

FIGURE 10.11 Interaction between the biosorbent and the different chemical techniques employed to interact with heavy metals.

10.4.1.5 By Bioaugmentation

Bioaugmentation is a controversial and important technique of bioremediation. The rationale for adding oil-degrading microorganisms is that indigenous microbial communities may not be capable of degrading complex mixtures such as crude oil [55]. Therefore, the addition of more indigenous microbial groups is necessary.

Bioaugmentation consists in the addition of pregrown microbial cultures to enhance microbial populations at a site to improve contaminant clean-up and reduce the time of the process and the costs [56]. Figure 10.12 represents the bioaugmentation technique by adding indigenous microorganisms to a polluted system, resulting in a contaminant-free environment.

Bioaugmentation has been a success with chlorinated ethenes such as trichloroethene (TCE) and tetrachloroethene (PCE) [58]. Many microbial species have been reported to dechlorinate PCE and TCE, while *Dehalococcoides* is the only identified genus reported to dechlorinate PCE/TCE to the benign product ethane [57].

However, complete microbial cell bioaugmentation has several limitations such as a rapid decrease in bacterial viability and abundance after inoculation. Hence, this method is not a successful practice for bioremediation [33]. There are several points of view to consider to get success during the bioaugmentation processes described in Table 10.2.

10.4.1.6 By Phytoremediation

Phytoremediation technique is based on a plant's ability to degrade and/or remove toxic chemicals from polluted areas. It is a "green" approach to environmental contamination problems. For assessing the effectiveness of plants as agents of bioremediation, the status of edaphic factors and soil macro- and microorganism communities cannot be neglected. Besides, a sufficient supply of essential plant nutrients is vital to get a successful process [61].

Phytoremediation could be a link among researchers from different institutes and farmers. The connection between farmers and researchers will increase the applications of polluted soils for agricultural purposes [62]. This bioremediation process has significant advantages because

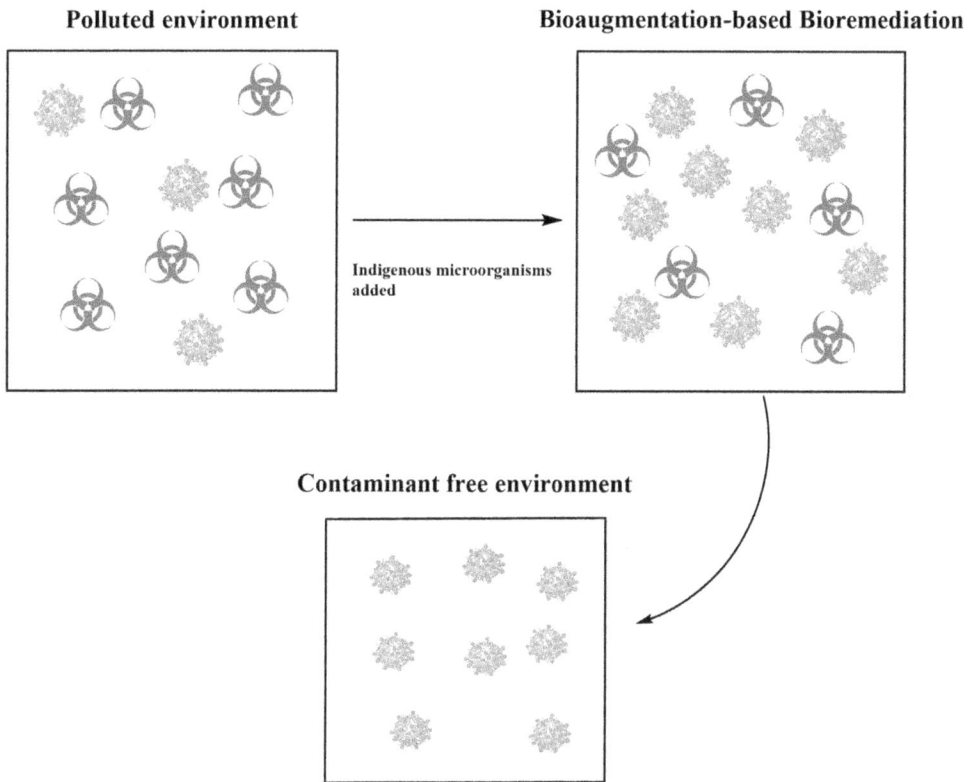

FIGURE 10.12 Schematic representation of the addition of indigenous microorganisms to a polluted environment for reducing the contamination in the medium.

TABLE 10.2
Criteria Required to Get a Successful Bioaugmentation Process

Selection Criteria	Description	References
Ecological basis	Basic knowledge about the microbial metabolic processes	[59,60]
Monitoring techniques	Polymerase chain reaction (PCR) to determine the activity and/or survival of the added microorganisms	
Plant management	Anticipating the seasonal effects on microbial and/or consortium communities	
Selection criteria	Storage and preservation of cultures under the best conditions, ensuring higher survival and activity	
Single inoculation versus continuous application in the environment	*Supplement required*	
Immobilization techniques	To increase the metabolic activity of microorganisms	
Gene transfer	*Gene bioaugmentation*	
Membrane reactors	The use of start-up degraders to accelerate the reaction	

it is tenfold cheaper than conventional technologies for the remediation of contaminated soils and wastes [63].

Nanotechnology offers remarkable solutions to some huge problems. This scientific field seeks to solve problems that affect plants, soil, or microorganisms. Therefore, the term "nanophytoremediation" integrates the scientific disciplines nanotechnology and phytoremediation [64].

10.5   MICROBIAL BIOREMEDIATION

Bioremediation is a microorganism-mediated transformation or degradation of contaminants into non-hazardous or less hazardous substances [65]. The mechanisms by which microorganisms present changes in metal speciation and mobility are fundamental components of biogeochemical cycles for metals and other elements, including C, N, S, and P, with additional implications for plant productivity and human health [66]. These various mechanisms that microbes use to biodegrade different metals are detailed in Figure 10.13.

Karigar and Rao [65] further pointed out that using some types of technologies such as using microorganisms to remove pollutants is a safe and efficient alternative to commonly used treatments. Microorganisms have the ability to affect metals through mobilization or immobilization processes; these microbes influence the soluble or insoluble phases of metals [66]. Microorganisms can efficiently sequester heavy metals because they don't tolerate high heavy metal concentrations, so the solution to this is their capacity to develop high heavy metal binding [67].

Bioremediation processes depend on microbes that enzymatically attack the pollutants and transform them into products that won't harm the environment [65]. Bioremediation of organic pollutants is a cost-effective alternative to conventional techniques, because the pollutants are mineralized or transformed into CO_2 and H_2O [68]. For example, benzene and related compounds can be completely mineralized to CO_2 [69].

10.5.1   EXTERNAL FACTORS

Wherever external conditions allow microorganisms' growth, bioremediation techniques can be practical; sometimes, it's necessary to manipulate the external conditions to permit microbial growth and a faster degradation [65].

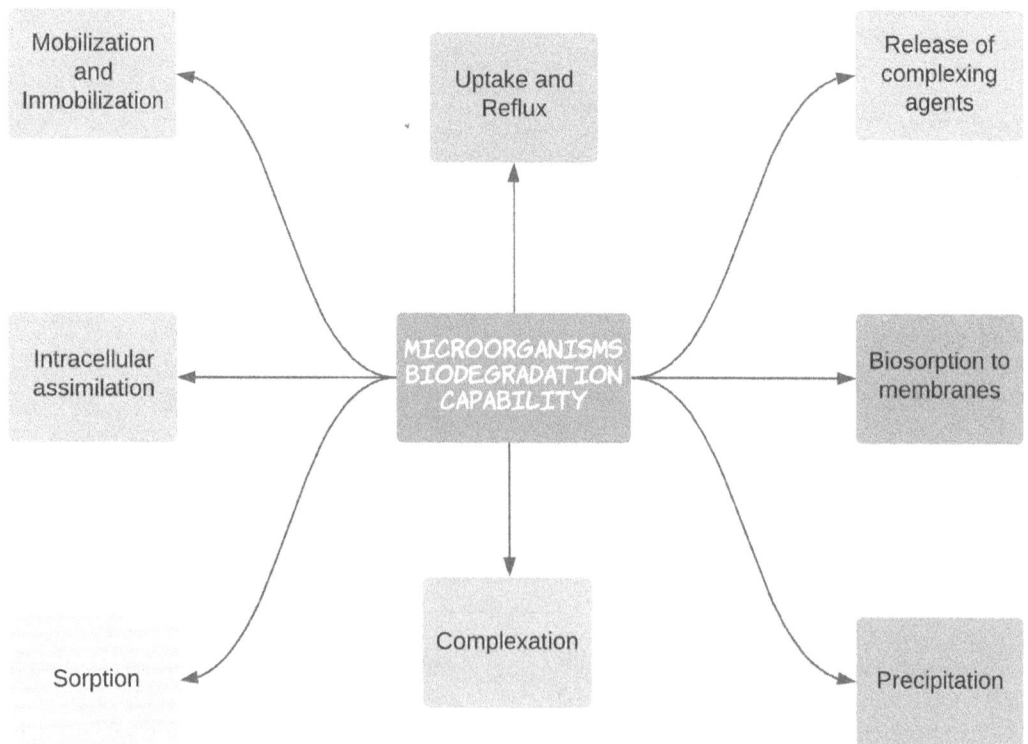

FIGURE 10.13   Bacteria bioremediation mechanisms.

TABLE 10.3
Environmental Conditions That Bacteria Require

Factors	Condition for Microbial Activity
Heavy metals	Max. 2000 ppm
pH	5.5–8.5
Nutrients	Nitrogenous and phosphorous
Temperature	15°C–45°C
Oxygen	Aerobic, pore space of 10%
Soil moisture	25%–28% of water-holding capacity

Bacterial growth is limited by many factors such as pH, O_2, soil, moisture, temperature, right amount of nutrients, bioavailability of contaminants, and existence of toxic chemicals [65]. More information about the optimal conditions that bacteria require to biodegrade pollutants is detailed in Table 10.3.

10.5.1.1 pH

The pH is one of the critical aspects that influence the bioavailability of contaminants, microbial activity, and characteristics of the pollutants [68]. The environment pH is able to drastically affect the microorganisms and their bioremediation rate; the vast majority of microbes survive and reproduce in a neutral pH [68]. Lananan et al. [70] reported in their study of wastewater bioremediation that the regulation of pH was vital since it affects the equilibrium between ammonia and ammonium concentrations in the wastewater. The bioadsorption of metal cations increases with an increase in the pH; however, in high alkali medium, sedimentation of metal complexes occurs [71].

Some bacteria need a specific pH to reproduce efficiently as reported in the study by Robinson et al. [72], where they concluded that the bicarbonate requirement depends on the minimum design pH with the requirement increasing significantly with an increase in design pH towards a more neutral value more favoured by dehalogenating bacteria.

The pH classification of microorganisms (neutrophiles, acidophiles, or alkaliphiles) is detailed below in Table 10.4.

10.5.1.2 Temperature

Although microorganisms can exist in extreme environments, most of them prefer optimal conditions: a situation that is difficult to achieve outside the laboratory [65]. This is evidenced in the work of Kermani et al. [73]; in his research, there were isolated several mesophilic gram-negative and cadmium-resistant bacteria.

Atlas [74] indicated that temperature affects the rates of microbial hydrocarbon-degrading activities. At low temperatures, oil becomes dense and reduces the volatilization of toxic alkanes, interrupting the biodegradation. And at higher temperatures, it increases the rates of hydrocarbon metabolism (30–40°C), where it becomes extremely toxic to bacteria [74].

TABLE 10.4
Bacteria pH Classification

Bacteria Type	pH Value
Neutrophiles	5–8
Optimal growth	7
Acidophiles	<5.5
Alkaliphiles	>8.5

10.5.1.3   Bioavailability

Bioavailability is the amount of a pollutant that can be degraded by microorganisms [75]. So, bioavailability is affected by many physical or chemical factors of both the pollutant and the substrate [76].

Furthermore, Maier [75] proposed three cases that would result from the bioavailability of contaminants:

- Biodegradation won't happen. The bioavailability of the pollutant is not enough to waste energy in the induction of biodegradation.
- At low bioavailable concentrations, some microorganisms may degrade the pollutant, but it will be limited because the strain is not reproducing; therefore, there won't be next generations that will continue the remediation.
- Optimal remediation rates will happen when there is enough bioavailable contaminant to induce biodegradation [75].

10.5.2   MICROBIAL DEGRADATION OF ORGANIC AND INORGANIC COMPOUNDS

Up ahead are detailed some organic compounds which can lead to contamination and their chemical representation that we can observe in Figure 10.14.

One of the groups of organic compounds in which microbial degradation can exist is the PAHs, which are extensively dispersed in the environment and may persist for a long time. PAHs are the best representation for which specific limitations in bioremediation exist due to low bioavailability [77]. This has raised a considerable concern that many of the PAHs are carcinogenic [78]. PAHs are organic pollutants composed of two or more benzenes and are found in petroleum [79].

Furthermore, other compounds which are hazardous by nature are established by Boyle et al. [80], who pointed out that several recalcitrant organic halogen compounds in the environment that may pose significant threats to the global ecosystem are the polychlorinated biphenyls (PCBs). The two fundamental processes where PCBs can be reduced and transformed with the usage of bacteria are aerobic oxidative processes and anaerobic reductive processes, and both of them happen naturally in the environment [81].

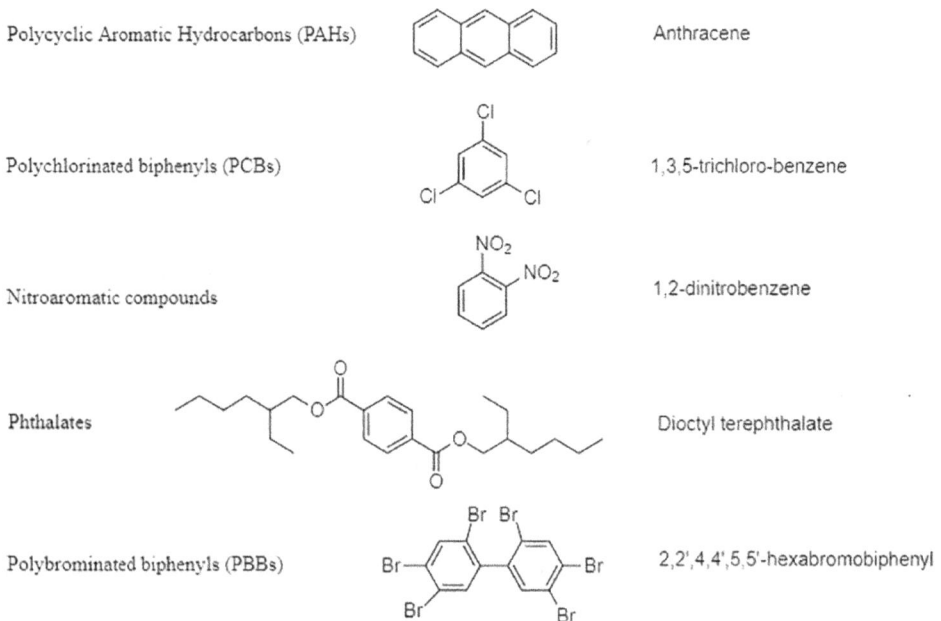

FIGURE 10.14   Organic compounds.

Another category of compound that is dangerous for the environment is nitroaromatic compounds, which are organic molecules that consist of one or more nitro groups ($-NO_2$) attached to a benzene [82]. Recent research revealed that many anaerobic bacteria can reduce the nitro group via nitroso and hydroxylamino intermediates to the corresponding amines [83].

Liang et al. [84] pointed out that phthalates are an environmental concern due to the massive amounts required to produce various plastics; these compounds are hepatotoxic, teratogenic, and carcinogenic in nature. Biodegradation is the process through which the phthalate could efficiently be removed from the environment [85]. Several studies indicated that bacteria fungi, yeast, and algae could transform phthalates under aerobic, anaerobic, and anoxic conditions [84].

The last groups of compounds that are going to be mentioned here is the polybrominated biphenyls (PBBs) and polybrominated diphenyl ethers (PBDEs), which are used as flame retardants [86]. Wiegel and Wu [87] further asserted that different microorganisms appear to be responsible for different dechlorination activities and the occurrence of various dehalogenation routes.

10.5.2.1 Water

To design an efficient bioremediation treatment when working with either effective microorganisms or microalgae, it is required to supply extra O_2 and CO_2, to sustain their growth [71]. Low water solubility is one factor that can limit the availability of the substrate to bacterial cells and hence constrain biodegradation [75].

In recent years, the number of studies on effective processes to clean up water and minimize pollution has continued to increase. Even though microbes can't destroy metals, they can change their chemical properties, making them less hazardous [88].

10.5.2.2 Soil

Gadd [66] remarked that different microbial processes can lead to the dissolution of insoluble metal compounds and minerals, including oxides, phosphates, sulphides, and desorption of metal species from exchange sites on, for example, clay minerals or organic matter in the soil. Bioremediation has been recommended as a cost-effective method when remediating soils [89].

Many contaminants are sources of soil contamination, such as pesticides and fertilizers, industrial practices, waste and wastewater sludge disposal, and accidental release of the pollutant into the environment [90].

In the research confirmed by Haghollahi et al. [89], it is mentioned that the properties of the soil are affected when it is in contact with petroleum hydrocarbons; the nature of these compounds can immobilize the nutrients present in the soil.

Since microorganisms help eliminate problems associated with chemical fertilizers and pesticides, they are now widely applied in natural farming and organic agriculture [71]. This is validated by Hrynkiewicz and Baum [67], who mentioned that microorganisms could be used as biosurfactants in metal-contaminated soils polluted with Cd and Pb. To understand this in a better way, biosurfactants are surfactants that are produced in the external part of the cell by microorganisms; some biosurfactants are applied nowadays because of their biodegradability, low toxicity, and efficiency in enhancing biodegradation and solubilization of substances [91].

Haritash and Kaushik [92] confirmed that some bacterial species can degrade PAHs and most of them can be found in the same spot of pollution (*Pseudomonas aeruginosa, Pseudomonas fluorescens, Mycobacterium spp., Haemophilus spp., Rhodococcus spp.,* and *Paenibacillus spp.*). Haghollahi et al. [89] mentioned some of the factors that can affect the soil bioremediation, such as the soil type, moisture, bacterial growth, nutrient availability, and oxygen transport.

10.5.2.3 Food Industry

Several chemicals are now entering in the food chain through industrial and agricultural treatments that the food is receiving [90]. The enzyme catalase has been employed in bioremediation, where it is an indicator of the hydrocarbon degradation of the crude oil-polluted soil [93]. This enzyme has been used to control the process, for example, of milk and cheese production, determining the quality of those products. In addition, Massoud et al. [71] proposed that *Saccharomyces cerevisiae* is a great choice for bioremediation because it's completely safe for humans.

10.5.2.4   Oil Industry

Soil and water pollution by petroleum hydrocarbons is a big issue in our society nowadays [65]. In petroleum-contaminated environments, gram-negative bacterial species tend to grow and adapt [94].

Atlas [74] distinguished that the persistence of petroleum pollutants depends on the quantity and quality of the hydrocarbon mixture and the properties of the affected ecosystem.

Microbial degradation of oil occurs through the attack on aliphatic or light aromatic fractions of the oil, with high molecular weight aromatics, resins, and asphaltenes considered to be recalcitrant [74].

10.5.2.5   Others

One of the major issues affecting the mining industry is the treatment of acid mine drainage (AMD). It presents high acidity and high concentration of metals (Cu, Fe, Zn, Al, Pb, and As) [95]. Sulphate-reducing microorganisms play a vital role in controlling and treating mine waste, generating alkalinity and neutralizing the acidic waste [96].

An example of sulphate-reducing microorganism is given in the study by Natarajan [97], where *Acidithiobacillus ferrooxidans*, *Acidithiobacillus thiooxidans*, and *Leptospirillum ferrooxidans* were isolated from the above mine sites and could be used to precipitate dissolved metals such as Cu, Zn, Fe, and As [97].

Each activity performs to generate a quantity of pollution that ends up in the environment. However, bioremediation is a promising way to be eco-friendly. For example, using *Paenibacillus azoreducers* has been shown to be adequate to blue and green dyes treatment, where COD and BOD levels were brought down below 30mg/L, and the researchers concluded that both dyes were utterly degraded.

10.6   PROMISING TECHNIQUES IN BIOREMEDIATION BASED ON MICROORGANISMS

10.6.1   Genetic Engineering

Several techniques have been studied for the degradation of plastic waste. The concern increases every moment due to the production of this type of waste in great amounts worldwide and its persistence in the environment. Genetic engineering tools have been used to manipulate the genes of microorganisms and optimize their potential for biodegrading plastics in conjunction with recombinant DNA technology [98,99]. Studies carried out have discovered plastic-degrading enzymes such as peroxidase and laccase using recombinant rDNA technology. They have been incorporated into the host microorganism *E. coli* to improve its bioremediation [100]. On the other hand, research has indicated contributions to the degradation of PET plastics by the enzyme cutinase, which has led to the use of genetic engineering to eradicate lumps, one of the main limitations in the degradation by the enzyme, and improve the percentage of bioremediation. The results showed a better bioremediation capacity at high temperatures by the recombined enzyme [99,101].

10.6.2   Biosensors

The term biosensor was first used in the year 1977 when the first device was developed using live microorganisms immobilized on the surface of an ammonium-sensitive electrode. This device was used to detect the amino acid arginine, and its creators called it a "bio-selective sensor". Later, to shorten it, it was called a "biosensor", and this term has remained ever since to designate the junction between a biological material and a physical transducer.

Concerning environmental applications, biosensors can be used to discover specific types of harmful extensions of materials that exist or dominate in the environment, thus determining various pollutants, toxic intermediates, and heavy metals in wastes or sources of water [102].

10.6.3 Advanced Oxidation Processes Combined with Microorganisms

Wastewater containing dyes is one of the main threats to the environment. Even at deficient concentrations, these dyes affect the aquatic life and the food chain. Conventional methods are insufficient to remove these persistent organic pollutants, such as adsorption, coagulation, filtration, electrocoagulation, photolysis, ozonation, photocatalysis, and membrane filtration [103]. Recently, removing various organic pollutants by the oxidation of electrochemically generated hydroxyl groups has attracted widespread attention.

The effectiveness of the advanced oxidation process (AOPs) is associated with the quantity of hydroxyl group (OH). The research on textile industry wastewater usually focuses on the development of new treatment strategies using synthetic wastewater. The process mainly contains azo dyes and surfactants, which are responsible for forming foam on the surface of rivers, abnormal growth of algae, and toxicity to certain aquatic organisms. In addition, although complete mineralization using chemical oxidation is usually expensive, it is widely reported that combining it with biological treatments can reduce operating costs and is a technically feasible option [104].

10.7 CONCLUSIONS

Industrialization and urban growth have increased pollution. The constant efforts by governments and researchers have increased as well. Currently, there is a great need to design eco-friendlier and low-cost techniques to reduce the pollution due to organic and inorganic compounds. Bioremediation is a powerful technique that uses natural resources, microorganisms, or even plants, considering their remarkable properties such as adsorption, biodegradation, biosorption, bioaugmentation, and phytoremediation for pollution reduction. Besides, the synergetic combinations with other techniques (nanotechnology) have brought promising results.

Summing up everything that has been stated so far, bioremediation has become a remarkable strategy employed in restoring polluted environments. The particular reason for this circumstance is the use of biodegradative microorganisms. Bioremediation is effective due to its low cost, high efficiency, and eco-friendly process. The action of known microbes can biodegrade different types of pollutants. These microorganisms possess different mechanisms to transform inorganic and organic compounds and diminish the concentration of the pollutant. Several studies have been conducted in the field of bioremediation, and the present review actively demonstrates that bacteria have the potential to degrade pollutants of diverse concentrations. However, not always bacteria have the biodegradability needed for a certain pollutant, because of the conditions and the environment where the bacteria are tested; for example, there is a limit in the bioavailability, the pH or temperature is not optimal for that strain of bacteria.

Organic compounds such as benzene and its derivatives have carcinogenic properties. Also, these compounds are persistent and bioaccumulate in plants, animals, and humans. That is why decontamination processes are vital to guarantee the health of ecosystems and to avoid illnesses in people. Many bacteria can manage to restore the areas polluted with organic compounds such as PAHs, nitroaromatic compounds, PCBs, and other contaminants that contain benzene. These resistant bacteria are commonly found in the exact spot of pollution even though degradation is slow sometimes; with controlled conditions in the laboratory, it reaches higher values. Some of the factors that need to be considered because they affect soil bioremediation are the microbial population, nutrient availability, soil type, salinity, and oxygen transport.

REFERENCES

1. C. Orozco, A. Pérez, M. N. González, F. Rodríguez, and J. Alfayate, Contaminación Ambiental. Una visión desde la Química. 2003.
2. O. Aluko, "Environmental pollution and waste management," *Introd. Course Environ. Sci.*, Vol 6, Issue 1. pp. 43–59, 2001.
3. J. Alguacir, "Biotratamiento de contaminantes de origen inorgánico," vol. 34. doi: 10.3989/revmetalm. 1998.v34.i5.810.
4. A. Kumar, I. J. Schreiter, A. Wefer-Roehl, L. Tsechansky, C. Schüth, and E. R. Graber, "Production and utilization of biochar from organic wastes for pollutant control on contaminated sites," *Environ. Mater. Waste Resour. Recover. Pollut. Prev.*, pp. 91–116, 2016. doi: 10.1016/B978-0-12-803837-6.00005-6.
5. X. Tang, et al., "Inorganic and organic pollution in agricultural soil from an emerging e-waste recycling town in Taizhou area, China," *J. Soils Sediments*, vol. 10, no. 5, pp. 895–906, 2010. doi: 10.1007/s11368-010-0252-0.
6. K. Ge. *Organic Contaminants in the Environment.* Springer, Dordrecht, 1967.
7. S. Mandal, A. Kunhikrishnan, N. S. Bolan, H. Wijesekara, and R. Naidu, *Application of Biochar Produced From Biowaste Materials for Environmental Protection and Sustainable Agriculture Production.* Elsevier Inc., Amsterdam, 2016.
8. S. Chowdhury, N. Khan, G.H. Kim, J. Harris, P. Longhurst, and N. S. Bolan, *Zeolite for Nutrient Stripping From Farm Effluents.* Elsevier Inc., Amsterdam, 2016.
9. M. G. Snousy, "An overview on the organic contaminants," *SDRP J. Earth Sci. Environ. Stud.*, vol. 1, no. 2, 2017. doi: 10.25177/jeses.1.2.1.
10. P. J. van den Brink and R. M. Mann, "Impacts of agricultural pesticides on terrestrial ecosystems," *Ecol. Impacts Toxic Chem. (Open Access)*, pp. 63–87, 2011. doi: 10.2174/978160805121210063.
11. E. C. Rada, G. Andreottola, I. A. Istrate, P. Viotti, F. Conti, and E. R. Magaril, "Remediation of soil polluted by organic compounds through chemical oxidation and phytoremediation combined with DCT," *Int. J. Environ. Res. Public Health*, vol. 16, no. 17, 2019. doi: 10.3390/ijerph16173179.
12. A. J. Englande, P. Krenkel, and J. Shamas, *Wastewater Treatment &Water Reclamation.* Elsevier Inc., Amsterdam, 2015.
13. M. Pell and A. Wörman, *Biological Wastewater Treatment Systems*, Second Edn, vol. 6. Elsevier B.V., Amsterdam, 2011.
14. B. Tyagi and N. Kumar, Bioremediation: Principles and applications in environmental management, 2021. doi: 10.1016/B978-0-12-820524-2.00001-8.
15. A. Stein and N. Kerle, "Environmental remediation," *Encycl. Quant. Risk Anal. Assess.*, 2008. doi: 10.1002/9780470061596.risk0317.
16. H. Ali, E. Khan, and M. A. Sajad, "Phytoremediation of heavy metals-Concepts and applications," *Chemosphere*, vol. 91, no. 7, pp. 869–881, 2013. doi: 10.1016/j.chemosphere.2013.01.075.
17. C. M. Kao, C. Y. Chen, S. C. Chen, H. Y. Chien, and Y. L. Chen, "Application of in situ biosparging to remediate a petroleum-hydrocarbon spill site: Field and microbial evaluation," *Chemosphere*, vol. 70, no. 8, pp. 1492–1499, 2008. doi: 10.1016/j.chemosphere.2007.08.029.
18. P. Höhener and V. Ponsin, "In situ vadose zone bioremediation," *Curr. Opin. Biotechnol.*, vol. 27, pp. 1–7, 2014. doi: 10.1016/j.copbio.2013.08.018.
19. R. Boopathy, "Factors limiting bioremediation technologies," *Bioresour. Technol.*, vol. 74, no. 1, pp. 63–67, 2000. doi: 10.1016/S0960–8524(99)00144-3.
20. S. V. Mohan, K. Sirisha, N. C. Rao, P. N. Sarma, and S. J. Reddy, "Degradation of chlorpyrifos contaminated soil by bioslurry reactor operated in sequencing batch mode: Bioprocess monitoring," *J. Hazard. Mater.*, vol. 116, no. 1–2, pp. 39–48, 2004. doi: 10.1016/j.jhazmat.2004.05.037.
21. J. Singh, D. Sharma, G. Kumar, and N. R. Sharma, "Microbial bioprospecting for sustainable development," *Microb. Bioprospecting Sustain. Dev.*, pp. 1–397, 2018. doi: 10.1007/978-981-13-0053-0.
22. G. A. Silva-Castro, I. Uad, J. Gónzalez-López, C. G. Fandiño, F. L. Toledo, and C. Calvo, "Application of selected microbial consortia combined with inorganic and oleophilic fertilizers to recuperate oil-polluted soil using land farming technology," *Clean Technol. Environ. Policy*, vol. 14, no. 4, pp. 719–726, 2012. doi: 10.1007/s10098-011-0439-0.
23. S. McCutcheon and S. Jørgensen, "Phytoremediation", In *Encyclopedia of Ecology*, pp. 568–582, 2008. doi: 10.1016/b978-0-444-63768-0.00069-x.
24. R. Kohli, "Microbial cleaning for removal of surface contamination", In *Developments in Surface Contamination and Cleaning*, pp. 139–161, 2013. doi: 10.1016/b978-1-4377-7879-3.00004–2.

25. Prokaryotes and Environmental Bioremediation", *Biology LibreTexts*, 2020. [Online]. Available: https://bio.libretexts.org/Bookshelves/Introductory_and_General_Biology/Book%3A_General_Biology_(Boundless)/22%3A_Prokaryotes%3A_Bacteria_and_Archaea/22.5%3A_Beneficial_Prokaryotes/22.5C%3A_Prokaryotes_and_Environmental_Bioremediation.

26. K. Ritz, "FUNGI", In *Encyclopedia of Soils in the Environment*, pp. 110–119, 2005. doi: 10.1016/b0-12-348530-4/00147-8.

27. Y. Zhang, et al., "Binding mechanisms and QSAR modeling of aromatic pollutant biosorption on Penicillium oxalicum biomass", *Chem. Eng. J.*, vol. 166, no. 2, pp. 624–630, 2011. doi: 10.1016/j.cej.2010.11.034.

28. M. Villarroel Hipp and D. Silva Rodríguez, "Bioremediation of piggery slaughterhouse wastewater using the marine protist, Thraustochytrium kinney VAL-B1", *J. Adv. Res.*, vol. 12, pp. 21–26, 2018. doi: 10.1016/j.jare.2018.01.010.

29. X. Liang, et al., "Viral and bacterial community responses to stimulated Fe(III)-bioreduction during simulated subsurface bioremediation", *Environ. Microbiol.*, vol. 21, no. 6, pp. 2043–2055, 2019. doi: 10.1111/1462-2920.14566.

30. C. Gupta, D. Prakash, and S. Gupta, "Microbes: "A Tribute" to clean environment", In *Springer Briefs in Environmental Science*, pp. 17–34, 2017. doi: 10.1007/978-3-319-58415-7_2.

31. M. Bustamante-Torres, D. Romero-Fierro, S. Hidalgo-Bonilla, and E. Bucio, "Basics and green solvent parameter for environmental remediation", In *Green Sustainable Process for Chemical and Environmental Engineering and Science*, pp. 219–237, 2021. doi:10.1016/b978-0-12-821884-6.00007-3.

32. S. Gouma, S. Fragoeiro, A. Bastos, and N. Magan, "Bacterial and fungal bioremediation strategies", In *Microbial Biodegradation and Bioremediation*, pp. 301–323, 2014. doi: 10.1016/b978-0-12-800021-2.00013-3.

33. G. Vishwakarma, G. Bhattacharjee, N. Gohil, and V. Singh, "Current status, challenges and future of bioremediation", In *Bioremediation of Pollutants*, pp. 403–415, 2020. doi: 10.1016/b978-0-12-819025-8.00020-x.

34. D. Liyanage and J. Walpita, "Organic pollutants from E-waste and their electrokinetic remediation", In *Handbook of Electronic Waste Management*, pp. 171–189, 2020. doi: 10.1016/b978-0-12-817030-4.00006-1.

35. P. Pal, "Arsenic removal technologies on comparison scale and sustainability issues", In *Groundwater Arsenic Remediation*, pp. 291–301, 2015. doi: 10.1016/b978-0-12-801281-9.00007-2.

36. N. Sandhyarani, "Surface modification methods for electrochemical biosensors", In *Electrochemical Biosensors*, pp. 45–75, 2019. doi: 10.1016/b978-0-12-816491-4.00003-6.

37. L. Xia, et al., "Antifouling membranes for bioelectrochemistry applications", In *Microbial Electrochemical Technology*, pp. 195–224, 2019. doi: 10.1016/b978-0-444-64052-9.00008-x.

38. A. Pisarev, "Hydrogen adsorption on the surface of metals", In *Gaseous Hydrogen Embrittlement of Materials in Energy Technologies*, pp. 3–26, 2012. doi: 10.1533/9780857095374.1.3.

39. C. Shackelford, "Geoenvironmental engineering", *Ref. Module Earth Syst. Environ. Sci.*, 2013. doi: 10.1016/b978-0-12-409548-9.05424-5.

40. S. Jørgensen, "Biodegradation", In *Encyclopedia of Ecology*, pp. 366–367, 2008. doi: 10.1016/b978-008045405-4.00260-3.

41. B. Van Aken and R. Bhalla, "Microbial degradation of polychlorinated biphenyls", In *Comprehensive Biotechnology*, pp. 151–166, 2011. doi: 10.1016/b978-0-08-088504-9.00378-0.

42. D. Cierniak, et al., "How to accurately assess surfactant biodegradation-impact of sorption on the validity of results", *Appl. Microbiol. Biotechnol.*, vol. 104, no. 1, pp. 1–12, 2019. doi: 10.1007/s00253-019-10202-9.

43. R. Crawford, "Biodegradation", In *Comprehensive Biotechnology*, pp. 3–13, 2011. doi: 10.1016/b978-0-08-088504-9.00368-8.

44. A. Kumari and D. Chaudhary, "Engineered microbes and evolving plastic bioremediation technology", In *Bioremediation of Pollutants*, pp. 417–443, 2020. doi: 10.1016/b978-0-12-819025-8.00021-1.

45. J. Rourke and C. Sinal, "Biotransformation/metabolism", In *Encyclopedia of Toxicology*, pp. 490–502, 2014. doi: 10.1016/b978-0-12-386454-3.00007-5.

46. R. Beiras, "Biotransformation", In *Marine Pollution*, pp. 205–214, 2018. doi: 10.1016/b978-0-12-813736-9.00012-x.

47. J. Speight, "Biological Transformations", In *Reaction Mechanisms in Environmental Engineering*, pp. 269–306, 2018. doi: 10.1016/b978-0-12-804422-3.00008-0.

48. S. Zainith, G. Saxena, R. Kishor, and R. Bharagava, "Application of microalgae in industrial effluent treatment, contaminants removal, and biodiesel production: Opportunities, challenges, and future prospects", In *Bioremediation for Environmental Sustainability*, pp. 481–517, 2021. doi: 10.1016/b978-0-12-820524-2.00020-1.

49. G. Naja and B. Volesky, "Biosorption for industrial applications", In *Comprehensive Biotechnology*, pp. 685–700, 2011. doi: 10.1016/b978-0-08-088504-9.00399-8.

50. M. Han, S. Won, and Y. Yun, "Broadening of the optimal pH range for reactive dye biosorption by chemical modification of surface functional groups of Corynebacterium glutamicum biomass", In *New Developments and Application in Chemical Reaction Engineering*, pp. 161–164, 2006. doi: 10.1016/s0167-2991(06)81558-2.

51. M. Nikolopoulou and N. Kalogerakis, "Petroleum spill control with biological means", In *Comprehensive Biotechnology*, pp. 197–210, 2019. doi: 10.1016/b978-0-444-64046-8.00356-6.

52. J. Speight, "Removal of organic compounds from the environment", In *Environmental Organic Chemistry for Engineers*, pp. 387–432, 2017. doi: 10.1016/b978-0-12-804492-6.00009-5.

53. X. Mao, "Bioaugmentation of chlorinated ethene-contaminated groundwater", In *Comprehensive Biotechnology*, pp. 158–180, 2019. doi: 10.1016/b978-0-444-64046-8.00473-0.

54. R. Maier and T. Gentry, "Microorganisms and organic pollutants", In *Environmental Microbiology*, pp. 377–413, 2015. doi: 10.1016/b978-0-12-394626-3.00017-x.

55. A. Marican and E. Durán-Lara, "A review on pesticide removal through different processes", *Environ. Sci. Pollut. Res.*, vol. 25, no. 3, pp. 2051–2064, 2017. doi: 10.1007/s11356-017-0796-2.

56. M. Herrero and D. Stuckey, "Bioaugmentation and its application in wastewater treatment: A review", *Chemosphere*, vol. 140, pp. 119–128, 2015. doi: 10.1016/j.chemosphere.2014.10.033.

57. V. Gunarathne, et al., "Phytoremediation for E-waste contaminated sites", In *Handbook of Electronic Waste Management*, pp. 141–170, 2020. doi: 10.1016/b978-0-12-817030-4.00005-x.

58. S. Imadi, S. Shah, A. Kazi, M. Azooz, and P. Ahmad, "Phytoremediation of saline soils for sustainable agricultural productivity", In *Plant Metal Interaction*, pp. 455–468, 2016. doi: 10.1016/b978-0-12-803158-2.00018-7.

59. V. Pandey, D. Patel, D. Maiti, and D. Singh, "Case studies of perennial grasses—phytoremediation (holistic approach)", In *Phytoremediation Potential of Perennial Grasses*, pp. 337–347, 2020. doi: 10.1016/b978-0-12-817732-7.00016-x.

60. I. Borišev, et al., "Nanotechnology and remediation of agrochemicals", In *Agrochemicals Detection, Treatment and Remediation*, pp. 487–533, 2020. doi: 10.1016/b978-0-08-103017-2.00019-2.

61. C. S. Karigar and S. S. Rao, "Role of microbial enzymes in the bioremediation of pollutants: A review," vol. 2011, 2011. doi: 10.4061/2011/805187.

62. G. M. Gadd, "Microbial influence on metal mobility and application for bioremediation," *Geoderma*, vol. 122, no. 2–4, pp. 109–119, 2004. doi: 10.1016/j.geoderma.2004.01.002.

63. K. Hrynkiewicz and C. Baum, "Application of microorganisms in bioremediation of environment from heavy metals," pp. 215–227, 2014. doi: 10.1007/978-94-007-7890-0.

64. G. Ajoku and M. Oduola, "Kinetic Model of pH Effect on Bioremediation of Crude Petroleum Kinetic model of pH effect on bioremediation of crude petroleum contaminated soil. 1. Model development," *Am. J. Chem. Eng.*, 2013. doi: 10.11648/j.ajche.20130101.12.

65. J. Bollag, T. Mertz, and L. Otjen, "Role of microorganisms in soil bioremediation", Publisher: American Chemical Society. Pennsylvania State University, University Park. pp. 2–10, 1994.

66. F. Lananan, et al., "International Biodeterioration & Biodegradation Symbiotic bioremediation of aquaculture wastewater in reducing ammonia and phosphorus utilizing Effective Microorganism (EM-1) and microalgae (Chlorella sp.)," *Int. Biodeterior. Biodegradation*, pp. 1–8, 2014. doi: 10.1016/j.ibiod.2014.06.013.

67. R. Massoud, M. R. Hadiani, K. K. Darani, and P. Hamzehlou, "Bioremediation of heavy metals in food industry: Application of Saccharomyces cerevisiae," *Electron. J. Biotechnol.*, pp. 1–5, 2018. doi: 10.1016/j.ejbt.2018.11.003.

68. C. Robinson, D. A. Barry, P. L. Mccarty, J. I. Gerhard, and I. Kouznetsova, "Science of the total environment pH control for enhanced reductive bioremediation of chlorinated solvent source zones," *Sci. Total Environ.*, vol. 407, no. 16, pp. 4560–4573, 2009. doi: 10.1016/j.scitotenv.2009.03.029.

69. A. J. N. Kermani, M. F. Ghasemi, A. Khosravan, A. Farahmand, and M. R. Shakibaie, "Cadmium bioremediation by metal-resistant mutated bacteria isolated from active sludge of industrial effluent," *Iran. J. Environ. Health. Sci. Eng.*, vol. 7, no. 4, pp. 279–286, 2010.

70. R. M. Atlas, "Microbial hydrocarbon degradation-bioremediation oil spills," *J. Chem. Technol. Biotechnol.*, vol. 52, pp. 149–156, 1991.

71. R. M. Maier, "Bioavailability and its importance to bioremediation," In *Bioremediation*, pp. 59–78, 2000. doi: 10.1007/978-94-015-9425-7_4.

72. J. W. Blackburn and W. R. Hafker, "The impact of biochemistry, bioavailability and bioactivity on the selection of bioremediation techniques," *Trends Biotechnol.*, vol. 11, no. 8, pp. 328–333, 1993. doi: 10.1016/0167-7799(93)90155-3.

73. J. J. Ortega-Calvo, M. C. Tejeda-Agredano, C. Jimenez-Sanchez, and E. Congiu, "Is it possible to increase bioavailability but not environmental risk of PAHs in bioremediation?," *J. Hazard. Mater.*, vol. 261, pp. 733–745, 2013. doi: 10.1016/j.jhazmat.2013.03.042.

74. K. A. Y. L. Shuttleworth and C. E. Cerniglia, "Environmental aspects of PAH biodegradation," *Appl. Biochem. Biotechnol.*, vol. 54, pp. 291–302, 1995.

75. H. Z. Hamdan, D. A. Salam, A. Rao, L. Semerjian, and P. Saikaly, "Science of the Total Environment Assessment of the performance of SMFCs in the bioremediation of PAHs in contaminated marine sediments under different redox conditions and analysis of the associated microbial communities," *Sci. Total Environ.*, 2020, doi: 10.1016/j.scitotenv.2016.09.232.

76. A. W. Boyle, C. J. Silvin, J. P. Hassett, J. P. Nakas, and S. W. Tanenbaum, "Bacterial PCB biodegradation," *Biodegradation*, vol. 3, pp. 285–298, 1992.

77. D. A. Abramowicz, "Aerobic and anaerobic PCB biodegradation in the environment," *Environ Health Perspect.*, vol. 103, no. 22, pp. 97–99, 1995.

78. K. Ju and R. E. Parales, "Nitroaromatic compounds, from synthesis to biodegradation," *Microbiol. Mol. Biol. Rev.* vol. 74, no. 2, pp. 250–272, 2010. doi: 10.1128/MMBR.00006-10.

79. J. Spain, "Biodegradation of Nitroaromatic Compounds," 1995. Annual Reviews of Microbiology, Armstrong Laboratory, Florida, United States of America

80. D. Liang, T. Zhang, and H. H. P. Fang, "Phthalates biodegradation in the environment," *Appl. Microbiol. Biotechnol.*, vol. 80, pp. 183–198, 2008. doi: 10.1007/s00253-008-1548-5.

81. S. Benjamin, S. Pradeep, M. Sarath, and S. Kumar, "A monograph on the remediation of hazardous phthalates," *J. Hazard. Mater.*, vol. 298, pp. 58–72, 2015. doi: 10.1016/j.jhazmat.2015.05.004.

82. J. De Boer, K. De Boer, and J. P. Boon, "Polybrominated biphenyls and diphenylethers," vol. 3, 2000. doi: 10.1007/3-540-48915-0_4.

83. J. Wiegel and Q. Wu, "Microbial reductive dehalogenation of polychlorinated biphenyls," vol. 32, 2000. doi: 10.1111/j.1574-6941.2000.tb00693.x.

84 L. Coelho, N. Coelho, D. Melo, H. Rezende, and P. Sousa, "Bioremediation of polluted waters using microorganisms," In S. Naofumi (ed.), *Advances in Bioremediation of Wastewater and Polluted Soil*, Rijeka: InTech, 2015, pp. 1–22.

85. A. Haghollahi, M. Hassan, and M. Schaf, "The effect of soil type on the bioremediation of petroleum contaminated soils," *J. Environ. Manage.*, vol. 180, 2016. doi: 10.1016/j.jenvman.2016.05.038.

86. P. K. Thassitou and I. S. Arvanitoyannis, "Bioremediation: A novel approach to food waste management," *Trends Food Sci. Technol.*, vol. 12, no. 5–6, pp. 185–196, May 2001. doi: 10.1016/S0924-2244(01)00081-4.

87. M. Ishiguro and L. K. Koopal, "Surfactant adsorption to soil components and soil," *Adv. Colloid Interface Sci.*, 2016. doi: 10.1016/j.cis.2016.01.006.

88. A. K. Haritash, and C. P. Kaushik. 2009. "Biodegradation aspects of polycyclic aromatic hydrocarbons (PAHs): A review." *J. Hazard. Mater.*, vol. 169, pp. 1–15. doi: 10.1016/j.jhazmat.2009.03.137.

89. J. Kaushal, S. G. Singh, and S. K. Arya, "Catalase enzyme: Application in bioremediation and food industry," *Biocatal. Agric. Biotechnol.*, 2018.doi: 10.1016/j.bcab.2018.07.035.

90. S. J. Macnaughton, J. R. Stephen, A. D. Venosa, G. A. Davis, Y. Chang, and D. C. White, "Microbial population changes during bioremediation of an experimental oil spill," *Appl. Environ. Microbiol.* vol. 65, no. 8, pp. 3566–3574, 1999.

91. C. García, D. A. Moreno, A. Ballester, M. L. Blázquez, and F. González, "Bioremediation of an industrial acid mine water by metal-tolerant sulphate-reducing bacteria," *Miner. Eng.*, vol. 14, no. 9, pp. 997–1008, 2001. doi: 10.1016/S0892-6875(01)00107-8.

92. A. S. Ayangbenro, O. S. Olanrewaju, and O. O. Babalola, "Sulfate-reducing bacteria as an effective tool for sustainable acid mine bioremediation," *Front. Microbiol.*, vol. 9, pp. 1–10, 2018. doi: 10.3389/fmicb.2018.01986.

93. K. A. Natarajan, "Microbial aspects of acid mine drainage and its bioremediation," *Trans. Nonferrous Met. Soc. China (English Ed.)*, vol. 18, no. 6, pp. 1352–1360, 2008. doi: 10.1016/S1003-6326(09)60008-X.

94. R. A. Wilkes and L. Aristilde, "Degradation and metabolism of synthetic plastics and associated products by Pseudomonas sp.: Capabilities and challenges," *J. Appl. Microbiol.*, vol. 123, no. 3, pp. 582–593, Sep. 2017. doi: 10.1111/jam.13472.

95. S. Jaiswal, B. Sharma, and P. Shukla, "Integrated approaches in microbial degradation of plastics," *Environ. Technol. Innov.*, vol. 17, p. 100567, Feb. 2020. doi: 10.1016/j.eti.2019.100567.

96. B. Sharma, A. K. Dangi, and P. Shukla, "Contemporary enzyme-based technologies for bioremediation: A review," *J. Environ. Manage.*, vol. 210, pp. 10–22, 2018. doi: 10.1016/j.jenvman.2017.12.075.

97. A. N. Shirke, et al., "Stabilizing leaf and branch compost cutinase (LCC) with glycosylation: Mechanism and effect on PET hydrolysis," *Biochemistry*, vol. 57, no. 7, pp. 1190–1200, Feb. 2018. doi: 10.1021/acs.biochem.7b01189.

98. P. Malik, V. Katyal, V. Malik, A. Asatkar, G. Inwati, and T. K. Mukherjee, "Nanobiosensors: Concepts and variations," *ISRN Nanomater.* vol. 2013, 2013. doi: 10.1155/2013/327435.

99. P. V. Nidheesh, M. Zhou, and M. A. Oturan, "An overview on the removal of synthetic dyes from water by electrochemical advanced oxidation processes," *Chemosphere*, vol. 197, pp. 210–227, Apr. 2018. doi: 10.1016/j.chemosphere.2017.12.195.

100. I. Oller, S. Malato, and J. A. Sánchez-Pérez, "Combination of Advanced Oxidation Processes and biological treatments for wastewater decontamination-A review," *Sci. Total Environ.*, vol. 409, no. 20, pp. 4141–4166, 2011. doi: 10.1016/j.scitotenv.2010.08.061.

11 Microbial Biostimulants for the Bioremediation of Petroleum-Contaminated Environments

Brian A. Wartell
Community College of Baltimore County

CONTENTS

11.1 INTRODUCTION

Petroleum spills are frequent across the globe because petroleum is heavily used as a fuel (Ahmed and Fakhruddin 2018). As a consequence, shorelines, waterways, and even soils and sediments can be greatly impacted by the hazardous chemicals contained within (Sharma et al. 2020, Atoufi and Lampert 2020, Emoyan et al. 2021). Areas affected by petroleum contamination can limit plant growth, has damaging and lasting effects on marine or other aquatic life, and creates many acute and chronic impacts on human health (Ahmed and Fakhruddin 2018). Spillage and leakage of crude oil or other petroleum products such as heating oil or gasoline can lead to a variety of environmental, ecological, and health issues (Ossai et al. 2020). Crude oil is especially problematic as it contains thousands of different compounds, if not more, many of which are quite toxic and impactful on the environment (Cheng et al. 2014, Overton et al. 2016). It is also highly variable and contains many different types of compounds, including a diversity of both alkanes and aromatic hydrocarbons (Overton et al. 2016, Brzeszcz and Kaszycki 2018).

Many of these components are highly toxic and difficult to remediate or are resistant to degradation. Many complex hydrocarbons, especially polycyclic aromatic hydrocarbons (PAHs), are particularly resistant to degradation under anaerobic or anoxic conditions (Wartell, Boufadel, and Rodriguez-Freire 2021, Nzila 2018). Yet, often, the removal of petroleum hydrocarbons via other means such as excavation, extraction, or flushing is not feasible, is very costly, or can lead to other environmental or health dangers (Chen and Zhong 2019). Bioremediation, or its mechanism, biodegradation, is therefore deemed to be a more eco-friendly and cost-effective alternative (Yuniati 2018). Additionally, biodegradation is often an ideal solution as it is a process occurring naturally in many environments (Thapa, Kc, and Ghimire 2012, Ron and Rosenberg 2014). Biodegradation, however, has its limitations, especially under anaerobic conditions (Agarry and Latinwo 2015). Reaction rates, kinetic unfavorability, population counts, and environmental conditions are just some examples of obstacles toward successful bioremediation (Talley 2016, Ramadass et al. 2018).

As a means of enhancing the effectiveness of bioremediation, biostimulation has been proposed to bolster degradation rates and enable metabolization of otherwise recalcitrant compounds (Ramírez-García, Gohil, and Singh 2019). This technique can be used as an effective tool to stimulate a microorganism's metabolic activity or population counts and enhance their ability to degrade hydrocarbons (Wu et al. 2016). Under anaerobic conditions, biostimulation incorporates supplemental electron acceptors (e.g., sulfate or nitrate) and, under both aerobic and anaerobic conditions, can utilize specific nutrients (e.g., nitrogen or phosphorus), or utilize co-substrates such as fatty acids (Bianco et al. 2019, Kronenberg et al. 2017).

11.2 HOW BIOSTIMULATION CAN ADDRESS OBSTACLES TO ANAEROBIC DEGRADATION

Anaerobic conditions can occur in a wide variety of environments (Al-Hawash et al. 2018). In aquatic environments, it can occur via stratification or via the depletion of oxygen by aerobic organisms (Ghattas et al. 2017). In terrestrial environments, for example, when an oil spill occurs, the oil, or at least its heavier components, can penetrate into the subsurface creating pockets with limited or virtually no oxygen (Xia and Boufadel 2011). Under aerobic conditions, many petroleum hydrocarbons can be rapidly degraded or transformed, but under anaerobic conditions, these potentially toxic compounds could persist undegraded for decades (Boufadel et al. 2019, Prince and Walters 2016). In contrast to aerobic environments, where bioremediation technology can be relatively quick and quite feasible (Alegbeleye, Opeolu, and Jackson 2017), anaerobic degradation is frequently slow, is markedly more susceptible to nutrient limitation, and has reduced enzymatic activity and capabilities (Dhar et al. 2020). This is primarily attributed to the lack of oxygen available to use for biochemical reactions. In place of oxygen, anaerobic organisms utilize other electron acceptors. These electron acceptors (often referred to specifically as terminal electron acceptors (TEAs)), are transformed to a reduced state, and this reduction feeds the metabolic and respiratory processes of the respective organism (Liu et al. 2020).

Biostimulation is a proposed means of enhancing the biodegradation process. Under anaerobic conditions, it can involve the supplementation of specific electron donors or co-substrates and can help to ensure that degradation takes place over a realistic time frame (Zhang and Lo 2015).

Increasing levels of necessary electron acceptors, which may be scarce in certain environments, can play a strong role in effective degradation capability (Alegbeleye, Opeolu, and Jackson 2017). Under nitrate-reducing or other conditions, utilizing sufficient nitrogenous compounds, a key factor to consider is the molar carbon-to-nitrogen (C/N) ratio. Yang et al. (2013) reported that high C/N values can stimulate degradation and found that its efficiency can often correlate linearly with increasing C/N values. Additionally, under nitrate-reducing conditions, the accumulation of nitrite (NO_2^-) can often be either inhibitory or create a lag phase (Zedelius et al. 2011). Yang et al. (2013), however, noted that in some circumstances, nitrite was shown to stimulate or enhance the denitrification process (by functioning as a key electron donor).

Under other anaerobic conditions, such as sulfate-reducing conditions, sulfate-reducing bacteria (SRB) also require nitrogen as it is vital for any biodegradation to occur. It is primarily used to form proteins and nucleic acids and thus essential for the growth of these microorganisms (Schaechter 2009). However, the lack of nitrogen as a nutrient can lead to ineffective SRB activity (Robinson-Lora and Brennan 2009), and as nitrate addition can often be highly detrimental under sulfate-reducing conditions, other nitrogenous compounds are often used. These include ammonium-based compounds, nitric acid, aniline, urea, peptone, chitin, and certain plant extracts. SRB, in cooperation with fermentative bacteria, can metabolize these nitrogenous compounds to ammonia, amines, and amino acids, ultimately yielding nucleic acids and also enzyme co-factors, which help to facilitate growth and allow for degradation pathways (Dev et al. 2015). Although nitrate is often inhibitory to sulfate-reducers, sulfate is often not inhibitory toward nitrate-reducers (Zedelius et al. 2011). In fact, it has been reported that sulfate addition under nitrate-reducing conditions can contribute positively to overall methane production (Kraft 2014). However, it is reported that certain compounds such as sulfide (the primary sulfate reduction product) can, in fact, inhibit denitrification or disrupt nitrate reduction efficiency (Zedelius et al. 2011).

Metals such as manganese and iron can be used by microorganisms as TEAs (via dissimilatory metal reduction) (Tremblay and Zhang 2020). The reduction of these metals creates electrochemical gradients, providing the biochemical energy needed for the organism's growth (Singh 2017). Co-substrates added, such as glucose, acetate, or citric acid, can function as supplemental electron donors and are well established in facilitating the iron reduction process (Lovley et al. 1993, Yan et al. 2012). Additionally, they can be also very useful in facilitating the role of iron in certain microbial processes (Zhang and Lo 2015).

Even under other anaerobic conditions, the lack of sufficient metals such as iron or manganese in some environments (e.g., those rich in sandstone) can limit the degradation potential even if suitable organisms are present (Jiménez et al. 2012). Therefore, to address this, if feasible, supplemental metal oxides can be added to the environment (Ni et al. 2021).

11.3 EXAMPLES OF BIOSTIMULATING AGENTS AND THEIR EFFECTS ON ANAEROBIC DEGRADATION

In a creative approach, Zhang et al. (2015) created utilized acetate and iron reduction, while also using nitrate and sulfate as electron acceptors. One process fueled another, with one reaction utilizing the product of another reaction, thereby leading to high TPH degradation rates. Zhang and Lo (2015) also researched the effectiveness of acetate (10 mM) and methanol (20 mM) as co-substrates for alkane degradation in marine sediment. They discovered both substances to increase total degradation rates, but that acetate was more favorable than methanol. (Methanol was much more effective in facilitating the degradation of alkanes $\leq C_{30}$.) Li et al. (2016) added glucose as a co-substrate to soil contaminated with petroleum hydrocarbons, which increased the hydrocarbon degradation rate by 200%. This was performed in a microbial fuel cell, but could hopefully be adapted to *in situ* environments. Paulo et al. (2017) found that the addition of lactate and yeast extract could increase the methanogenic degradation of 1-hexadecene by as much as sevenfold.

Müller et al (2017) successfully investigated the effect of ammonium acetate and a derivate of acid mine drainage to stimulate BTEX and PAH degradation under sulfate-reducing and iron-reducing conditions. Li et al. (2016) added glucose (which can be converted to acetate) to petroleum-contaminated soil and observed degradation rates increased by 200%. Bianco et al. (2019) observed the effect of biostimulation on four PAHs, namely phenanthrene, anthracene, fluoranthene, and pyrene. Specifically, they added anaerobic digestate (manure and whey) or OFMSW (organic fraction of municipal solid waste). In all cases, they observed higher degradation rates with either of these additives compared to nutrient addition alone. Of the four compounds, phenanthrene had the highest rate of removal (69%), which was achieved with the addition of OFMSW and nutrients (Bianco et al. 2019). Agarry and Owabor (2011) also investigated the effect of manure as a biostimulating agent. Specifically, they tested the following compounds for their effect on naphthalene and anthracene degradation: Tween 80 (a surfactant), NPK fertilizer, silicone oil, and pig dung. All four compounds were shown to increase degradation alone, with pig dung having the greatest effect when utilized independently, and pig dung + Tween 80, when used in combination (Agarry and Owabor 2011).

Ambrosoli et al. (2005) investigated the benefit of adding supplemental acetate and glucose, as electron donors, to soil contaminated with biphenyl, fluorene, phenanthrene, and pyrene. They found that the addition of either acetate or glucose could lead to improved degradation rates, but only when nitrate $\left(NO_3^-\right)$ was not supplied, as nitrate caused degradation rates to be slightly to moderately reduced (Ambrosoli et al. 2005).

Yuan and Chang (2007) noticed, in a similar manner, that the addition of electron donors (lactate, acetate, and pyruvate) negatively affected degradation rates under nitrate-reducing conditions, despite being beneficial under methanogenic and sulfate-reducing conditions. They explained that the presence of nitrate promoted the methanogen growth, which outcompeted the nitrogen-reducing bacteria, and in general, did not grow well in the presence of nitrate (Yuan and Chang 2007). In contrast, compounds such as benzoate have been shown to boost naphthalene degradation rates when using nitrate as the primary electron acceptor, although to a lesser extent than under sulfate-reducing conditions (Langenhoff, Zehnder, and Schraa 1996).

Numerous studies have demonstrated that the supplementation of electron donors, such as acetate or glucose, can yield a positive result on PAH degradation rates (Yuan and Chang 2007, Bianco et al. 2019). In one of the earlier studies, Yuan et al. (2000) discovered that the addition of acetate, glucose, or yeast extract could improve phenanthrene degradation, with complete mineralization occurring within only 28 hours. Yuan and Chang (2007) found that an augmentation of lactate, acetate, and pyruvate would enhance PAH degradation rates under sulfate-reducing and methanogenic conditions, but was inhibitory under nitrate-reducing conditions. Su et al. (2012) demonstrated much improved degradation of phenanthrene and anthracene via the supplementation of acetate, lactate, and spent mushroom compost (SMC).

Lei et al. (2005) noted enhanced phenanthrene degradation via the addition of ethanol or acetic acid, albeit specifically under sulfate-reducing conditions. They also noticed that degradation could occur without the addition of either of these co-substrates, yet only after a long lag phase (14 weeks) (Lei et al. 2005). In contrast to Lei et al.'s (2005) and other studies, there, in fact, are numerous studies that demonstrate enhanced PAH degradation under nitrate-reducing conditions utilizing these compounds as biostimulants (Ma et al. 2011, Abercron et al. 2016).

Bach et al. (2005) investigated how different additives could stimulate microbial growth and improve PAH degradation rates. Phosphate and nitrogen (as slow-release fertilizers), yeast extract, lactate, and dextrin were all added in different combinations as supplemental electron donors. Tween 80 (a surfactant) was also added to enhance bioavailability and extractability. All additives, with the exception of dextrin, were found to improve PAH degradation rates over unamended systems (Bach et al. 2005). More recently, many researchers have been exploring the potential of surfactants to aid hydrocarbon degradation, under both aerobic and anaerobic conditions (Ajona and Vasanthi 2021, Cui et al. 2021). Most studies reported have utilized surfactants under aerobic conditions, but they demonstrate the feasibility of using them under anaerobic conditions as well (Varjani 2017).

Although not as commonly reported, anaerobic degradation of petroleum hydrocarbons can also be enhanced using reduced humic substances (HS). HS can function as both electron donors (Coates et al. 2002) and electron acceptors (Lovley et al. 1996). HS can be found in many environments (Karr 2011) and may play a key role in certain hydrocarbon-degrading processes (Jednak et al. 2017). Biochar has also effectively been used to increase biodegradation rates, including under many anaerobic conditions, most notably nitrate-reducing conditions (Yang et al. 2018). Biochar contains charged surface groups that can function as either electron acceptors or electron donors (Kappler et al. 2014), and these create an affinity for aromatic hydrocarbons (Wu, Li, and Liu 2018).

Cellulose has also been tested for its effects of anaerobic biodegradation. Li et al. (2020) investigated the impact of cellulose on PAH degradation during the anaerobic digestion of sewage sludge. The addition of cellulose improved the degradation of 2-ringed, 3-ringed, and 4-ringed PAHs, yielding PAH removals of 14.82%, 20.75%, and 19.35%, respectively (Li et al. 2020). Adelaja et al. (2015) used microcrystalline cellulose (in microbial fuel cells) to successfully improve phenanthrene biodegradation, roughly doubling the degradation efficiency.

Peptone, another potential biostimulant, is frequently used to stimulate bacterial growth (Zhong et al. 2007, Li et al. 2019), but is rarely used as a stimulant of hydrocarbon degradation. In fact, of the few studies utilizing peptone for hydrocarbon degradation, nearly all are performed under aerobic conditions. However, some of the bacterial strains studied are facultative anaerobes and can therefore function under anaerobic conditions (Basha, Rajendran, and Thangavelu 2010).

11.4 HOW BIOSTIMULATION CAN ADDRESS OBSTACLES TO AEROBIC DEGRADATION

Bioremediation can be a slow and/or difficult process, even under aerobic conditions (Sharma 2019). In part, this can be due to environmental conditions such as pH or temperature (Varjani 2017, Ghosal et al. 2016), or even soil conditions (Bamforth and Singleton 2005). A lack of sufficient oxygen at a particular site can also hamper aerobic degradation rates (Prince and Walters 2016). However, a very frequent obstacle is the specificity of an organism's enzymes to utilize a particular substrate (e.g., hydrocarbon). For example, some microorganisms are more adept at utilizing alkanes vs. ringed structures, while some have oxygenases that are catered more toward aromatics (Xu et al. 2018). To address this, it is essential to have a diverse microbial population at a contaminant site, and bio-augmentation, or the addition of additional specific microorganisms, may be needed (Zawierucha and Malina 2011). Additionally, bacterial growth is highly dependent upon adequate nitrogen for metabolic processes (Dev et al. 2015) and limited nitrogen sources can reduce the success of biodegradation (Varjani 2017).

Utilizing biostimulation to enhance degradation rates has been proven to be particularly useful to combat marine oil spills (Mahjoubi et al. 2018, Nikolopoulou and Kalogerakis 2009). One of the greatest obstacles to aerobic degradation of marine oil spills is nutrient limitation and lack of a proper C:N:P ratio (Ławniczak et al. 2020). Nutrient delivery, including micro-encapsulation via slow-release particles, can be particularly effective, especially when combined with other techniques such as surfactant addition and bioaugmentation (Mapelli et al. 2017). Adjacent areas affected by marine oil spills, such as marshes, can utilize specific vegetation and/or fertilizers to stimulate remediation techniques (Horel, Mortazavi, and Sobecky 2012), and shorelines respond well to enhanced nutrient delivery, particularly as pore water within the beach areas contains ample hydrocarbon-degraders (Iskander, Khalil, and Boufadel 2021).

11.5 EXAMPLES OF BIOSTIMULATING AGENTS AND THEIR EFFECTS ON AEROBIC DEGRADATION

Biostimulation often targets microorganisms already capable of degrading hydrocarbons, but aims to increase the rate or range of degradation. Many studies have compared different substrates and their abilities to stimulate biodegradation rates. For example, Horel et al. (2014) looked at three different compounds to enhance the removal of crude oil in sandy sediments by the Gulf of Mexico. They investigated the potential of fish-based amendments (derived from *Chloroscombrus chrysurus*), plant-based amendments (derived from *Spartina alterniflora*), or various inorganic nutrients (INs). The fish-based amendments proved to be the most beneficial, increasing degradation rates by 123%. INs were the second most beneficial, increasing degradation rates by 52%. In contrast, the plant-based amendments increased degradation rates by only 25% (Horel, Mortazavi, and Sobecky 2014). Haller et al. (2020) studied the effects of three different potential biostimulants for their effect to enhance diesel degradation in tropical soil. Of three materials studied (whey, compost tea, and pyroligneous acid), only whey was shown to significantly improve biodegradation rates (Haller et al. 2020). Other materials that have been tested for their biostimulating effects include manures (Obiakalaije, Makinde, and Amakoromo 2015, Adams, Awode, and Agboola 2018), microalgal biomass (Decesaro et al. 2017), sorghum husk (Adams, Awode, and Agboola 2018), and agricultural wastes (Dadrasnia and Agamuthu 2014).

Some biostimulation techniques are more traditional, particularly when dealing with aquatic environments. Aleruchi and Gideon (2015) investigated crude oil-spiked groundwater in microcosms, utilizing an NPK fertilizer (15:15:15), yielding a 20% increase in oil degradation. Moliterni et al. (2012) investigated two different soil types (clay and silty loam) spiked with diesel fuel. They used a basal enrichment medium broth and balanced it to have a proper C:N:P ratio. The nature of the soils affected the sorption of the petroleum compounds, making it difficult to compare the results, although both conditions had significant TPH removal (Moliterni et al. 2012). At times, simply providing sufficient aeration to stimulate the aerobic organisms is enough to shift microbial activity to enhance degradation rates (Fitch 2016).

Biostimulation can be combined with bioaugmentation, but many environments contain sufficient hydrocarbon-degraders already present within an environment, especially under aerobic conditions (Adams et al. 2015). There is also a process known as autochthonous bioaugmentation (ABA), which exclusively uses microorganisms indigenous to a particular site. Nikolopoulou et al. (2013) studied the addition of uric acid and lecithin, combined with biosurfactants (rhamnolipids), to greatly enhance the biodegradation rates of crude oil. Utilizing these additives and indigenous organisms (i.e., ABA), they mimicked an oil-contaminated beach and their experiments showed that within 15 days, up to 99% of the alkanes were degraded, compared to a removal of only 20% of alkanes in 60 days under non-stimulated conditions. They also noted a reduction in the lag phase, which is often present during the degradation of hydrocarbons (Nikolopoulou, Pasadakis, and Kalogerakis 2013). The presence, absence, or an acclimation period can be highly dependent on a particular environment, its conditions, and its respective organisms (Brown et al. 2020).

Most often, combination treatments prove to be the most useful for biostimulation to be effective. For many studies, one of the additives is a biosurfactant used to enhance bioavailability (Bezza and Chirwa 2017, Adams et al. 2015). Another very promising additive is biochar, which can be made from a wide variety of carbon-rich sources, including wood. Kong et al. (2018) successfully utilized biochar made from sawdust and wheat straw to accelerate PAH degradation rates. In a similar study, Mukome et al. (2020) used biochar derived from both ponderosa pines and walnut shells to remediate oil-contaminated soils. The biochar made from pine wood was shown to effectively stimulate TPH degradation, whereas the biochar made form walnut shells proved to be inhibitory (Mukome et al. 2020). Brown et al. (2017) studied multiple amendment techniques to remediate soils contaminated with petroleum hydrocarbons and found that the most successful treatment incorporated nutrient supplementation, biosurfactants, and the addition of biochar.

As noted previously, PAHs are among the more toxic and recalcitrant hydrocarbons and many researchers have devoted studies toward enhancing PAH degradation. Lang et al. (2016) studied multiple techniques to degrade four PAH compounds: phenanthrene, anthracene, fluoranthene, and pyrene. The most successful technique employed both bioaugmentation and biostimulation. The microorganisms used for bioaugmentation were *Rhodococcus erythropolis* T902.1, part of a very diverse group of bacteria shown to be able to degrade many toxic and/or recalcitrant compounds. Mineral salt medium was used as the primary biostimulating agent (Lang et al. 2016). Bioaugmentation grants additional organisms capable of potentially degrading the pollutants, but thereby also provides additional pathways by which a particular compound, such as PAHs, can be degraded (Roy et al. 2018). These pathways can generate metabolites which can sometimes also stimulate further degradation (Yao et al. 2015).

Bacteria and related microorganisms are not the only organisms utilized for bioremediation. Another set of organisms where biostimulation is relevant is fungi. For brevity purposes, only a couple of examples will be cited, but the use of fungi for bioremediation purposes, dubbed mycoremediation, is a fascinating topic and is too lengthy to be discussed here. Sayara et al. (2011) studied PAH degradation in simulated contaminated creosote. They investigated the use of a white-rot fungus added to the naturally occurring microbes in the soil to degrade the PAHs over time with the addition of one of two stimulants: municipal solid waste (OFMSW) and rabbit food. The OFMSW proved to be a far better biostimulant and resulted in the degradation of 89% of the total PAHs in 30 days when performed under composting conditions. Without any biostimulant, only 29.5% of the total PAHs were degraded (Sayara et al. 2011). Another example utilizing fungi was a study performed by Fan et al. (2014), where they investigated the potential of the yeast strain *Candida tropicalis* SK21 as a bioaugmenting organism. The yeast strain was isolated from petroleum-contaminated soil and applied to a clay loam soil heavily contaminated with saturated hydrocarbons and aromatics. The addition of the yeast boosted TPH degradation over the course of 180 days from 61% to 83% (Fan, Xie, and Qin 2014).

11.6 SUMMARY

Biostimulation can be a very valuable tool to enhance hydrocarbon degradation, under both aerobic and anaerobic conditions. Its effectiveness varies greatly dependent upon the nature of the stimulant, the environmental conditions, and the particular strains of microorganisms present. Additionally, some compounds are more recalcitrant under certain conditions and may require a combination of techniques, including the employment of surfactants. Each case should be assessed individually, but biostimulation, even if not necessary, is still often beneficial as it can reduce lag time and decrease the time for degradation to occur.

REFERENCES

Abercron, Martirani-Von, Daniel Pacheco, Patricia Benito-Santano, Patricia Marín, and Silvia Marqués. 2016. "Polycyclic aromatic hydrocarbon-induced changes in bacterial community structure under anoxic nitrate reducing conditions." *Frontiers in Microbiology* 7:1775.

Adams, Feyisayo V., Maryam F. Awode, and Bolade O. Agboola. 2018. "Effectiveness of sorghum husk and chicken manure in bioremediation of crude oil contaminated soil." In Naofumi Shiomi (ed.), *Advances in Bioremediation and Phytoremediation*: 99. doi: 10.5772/intechopen.71832

Adams, Godleads Omokhagbor, Prekeyi Tawari Fufeyin, Samson Eruke Okoro, and Igelenyah Ehinomen. 2015. "Bioremediation, biostimulation and bioaugmention: A review." *International Journal of Environmental Bioremediation & Biodegradation* 3 (1):28–39.

Adelaja, Oluwaseun, Tajalli Keshavarz, and Godfrey Kyazze. 2015. "The effect of salinity, redox mediators and temperature on anaerobic biodegradation of petroleum hydrocarbons in microbial fuel cells." *Journal of Hazardous Materials* 283:211–217. doi: 10.1016/j.jhazmat.2014.08.066.

Agarry, Samuel E., and Chiedu N. Owabor. 2011. "Anaerobic bioremediation of marine sediment artificially contaminated with anthracene and naphthalene." *Environmental Technology* 32 (12):1375–1381.

Agarry, Samuel, and Ganiyu K. Latinwo. 2015. "Biodegradation of diesel oil in soil and its enhancement by application of bioventing and amendment with brewery waste effluents as biostimulation-bioaugmentation agents." *Journal of Ecological Engineering* 16 (2):82–91.

Ahmed, Fowzia, and A.N.M. Fakhruddin. 2018. "A review on environmental contamination of petroleum hydrocarbons and its biodegradation." *International Journal of Environmental Sciences & Natural Resources* 11(3):63–69.

Ajona, M., and P. Vasanthi. 2021. "Bioremediation of petroleum contaminated soils–A review." *Materials Today: Proceedings* 45:7117–7122.

Al-Hawash, Adnan B., Maytham A. Dragh, Shue Li, Ahmad Alhujaily, Hayder A. Abbood, Xiaoyu Zhang, and Fuying Ma. 2018. "Principles of microbial degradation of petroleum hydrocarbons in the environment." *The Egyptian Journal of Aquatic Research* 44 (2):71–76.

Alegbeleye, Oluwadara Oluwaseun, Beatrice Oluwatoyin Opeolu, and Vanessa Angela Jackson. 2017. "Polycyclic aromatic hydrocarbons: A critical review of environmental occurrence and bioremediation." *Environmental Management* 60 (4):758–783.

Aleruchi, Owhonka, and O. Abu Gideon. 2015. "Aerobic biodegradation of petroleum hydrocarbons in laboratory contaminated groundwater." *Microbiology Research Journal International* 7:313–321.

Ambrosoli, Roberto, Laura Petruzzelli, José Luis Minati, and Franco Ajmone Marsan. 2005. "Anaerobic PAH degradation in soil by a mixed bacterial consortium under denitrifying conditions." *Chemosphere* 60 (9):1231–1236. doi: 10.1016/j.chemosphere.2005.02.030.

Atoufi, Hossein D., and David J. Lampert. 2020. "Impacts of oil and gas production on contaminant levels in sediments." *Current Pollution Reports* 6 (2):43–53.

Bach, Quang-Dung, Sang-Jin Kim, Sung-Chan Choi, and Young-Sook Oh. 2005. "Enhancing the intrinsic bioremediation of PAH-contaminated anoxic estuarine sediments with biostimulating agents." *Journal of Microbiology - Seoul* 43 (4):319.

Bamforth, Selina M., and Ian Singleton. 2005. "Bioremediation of polycyclic aromatic hydrocarbons: Current knowledge and future directions." *Journal of Chemical Technology and Biotechnology* 80 (7):723–736.

Basha, Khazi Mahammedilyas, Aravindan Rajendran, and Viruthagiri Thangavelu. 2010. "Recent advances in the biodegradation of phenol: A review." *Asian Journal of Experimental Biological Sciences* 1 (2):219–234.

Bezza, Fisseha Andualem, and Evans M. Nkhalambayausi Chirwa. 2017. "The role of lipopeptide biosurfactant on microbial remediation of aged polycyclic aromatic hydrocarbons (PAHs)-contaminated soil." *Chemical Engineering Journal* 309:563–576.

Bianco, F., M. Race, S. Papirio, and G. Esposito. 2019. "Removal of polycyclic aromatic hydrocarbons during anaerobic biostimulation of marine sediments." *Science of the Total Environment* 709:136141.

Boufadel, Michel, Xiaolong Geng, Chunjiang An, Edward Owens, Zhi Chen, Kenneth Lee, Elliott Taylor, and Roger C. Prince. 2019. "A review on the factors affecting the deposition, retention, and biodegradation of oil stranded on beaches and guidelines for designing laboratory experiments." *Current Pollution Reports* 5 (4):407–423.

Brown, David M, Louise Camenzuli, Aaron D Redman, Chris Hughes, Neil Wang, Eleni Vaiopoulou, David Saunders, Alex Villalobos, and Susannah Linington. 2020. "Is the Arrhenius-correction of biodegradation rates, as recommended through REACH guidance, fit for environmentally relevant conditions? An example from petroleum biodegradation in environmental systems." *Science of the Total Environment* 732:139293.

Brown, David M., Samson Okoro, Juami van Gils, Rob van Spanning, Matthijs Bonte, Tony Hutchings, Olof Linden, Uzoamaka Egbuche, Kim Bye Bruun, and Jonathan W.N. Smith. 2017. "Comparison of landfarming amendments to improve bioremediation of petroleum hydrocarbons in Niger Delta soils." *Science of the Total Environment* 596:284–292.

Brzeszcz, Joanna, and Paweł Kaszycki. 2018. "Aerobic bacteria degrading both n-alkanes and aromatic hydrocarbons: An undervalued strategy for metabolic diversity and flexibility." *Biodegradation* 29 (4):359–407.

Chen, Shuisen, and Ming Zhong. 2019. "Bioremediation of petroleum-contaminated soil." In *Environmental Chemistry and Recent Pollution Control Approaches*, 161. IntechOpen.

Cheng, Lei, Shengbao Shi, Qiang Li, Jianfa Chen, Hui Zhang, and Yahai Lu. 2014. "Progressive degradation of crude oil n-alkanes coupled to methane production under mesophilic and thermophilic conditions." *PloS One* 9 (11):e113253.

Coates, John D., Kimberly A. Cole, Romy Chakraborty, Susan M. O'Connor, and Laurie A. Achenbach. 2002. "Diversity and ubiquity of bacteria capable of utilizing humic substances as electron donors for anaerobic respiration." *Applied and Environmental Microbiology* 68 (5):2445–2452. doi: 10.1128/aem.68.5.2445-2452.2002.

Cui, Jia-qi, Ya-qi Li, Qing-sheng He, Bing-zhi Li, Ying-jin Yuan, and Jianping Wen. 2021. "Effects of Different Surfactants to Petroleum Hydrocarbons Degradation of Mixed-bacteria." doi: 10.21203/rs.3.rs-540943/v1.

Dadrasnia, Arezoo, and Periathamby Agamuthu. 2014. "Biostimulation and monitoring of diesel fuel polluted soil amended with biowaste." *Petroleum Science and Technology* 32 (23):2822–2828.

Decesaro, Andressa, Alan Rampel, Thaís Strieder Machado, Antônio Thomé, Krishna Reddy, Ana Claudia Margarites, and Luciane Maria Colla. 2017. "Bioremediation of soil contaminated with diesel and bio-diesel fuel using biostimulation with microalgae biomass." *Journal of Environmental Engineering* 143 (4):04016091.

Dev, Subhabrata, Aditya Kumar Patra, Abhijit Mukherjee, and Jayanta Bhattacharya. 2015. "Suitability of different growth substrates as source of nitrogen for sulfate reducing bacteria." *Biodegradation* 26 (6):415–430.

Dhar, Kartik, Suresh R. Subashchandrabose, Kadiyala Venkateswarlu, Kannan Krishnan, and Mallavarapu Megharaj. 2020. "Anaerobic microbial degradation of polycyclic aromatic hydrocarbons: A comprehensive review." *Reviews of Environmental Contamination and Toxicology* 251:25–108.

Emoyan, Onoriode O., Patience O. Agbaire, Efe Ohwo, and Godswill O. Tesi. 2021. "Priority mono-aromatics measured in anthropogenic impacted soils from Delta, Nigeria: Concentrations, origin, and human health risk." *Environmental Forensics* 23 (1–2):1–13.

Fan, Mei-Ying, Rui-Jie Xie, and Gang Qin. 2014. "Bioremediation of petroleum-contaminated soil by a combined system of biostimulation–bioaugmentation with yeast." *Environmental Technology* 35 (4):391–399.

Fitch, Lee Ann Renee. 2016. "Biodegradation of Buried MC252 Oil in Coastal Beach Sands by PAH Degraders in Response to Oxygen Biostimulation." *LSU Master's Theses*, 114.

Ghattas, Ann-Kathrin, Ferdinand Fischer, Arne Wick, and Thomas A Ternes. 2017. "Anaerobic biodegradation of (emerging) organic contaminants in the aquatic environment." *Water Research* 116:268–295.

Ghosal, Debajyoti, Shreya Ghosh, Tapan K. Dutta, and Youngho Ahn. 2016. "Current state of knowledge in microbial degradation of polycyclic aromatic hydrocarbons (PAHs): A review." *Frontiers in Microbiology* 7:1369.

Haller, Henrik, Anders Jonsson, Joel Ljunggren, and Erik Hedenström. 2020. "Appropriate technology for soil remediation in tropical low-income countries-a pilot scale test of three different amendments for accelerated biodegradation of diesel fuel in Ultisol." *Cogent Environmental Science* 6 (1):1754107.

Horel, Agota, Behzad Mortazavi, and Patricia A. Sobecky. 2012. "Seasonal monitoring of hydrocarbon degraders in Alabama marine ecosystems following the Deepwater Horizon oil spill." *Water, Air, & Soil Pollution* 223 (6):3145–3154.

Horel, Agota, Behzad Mortazavi, and Patricia A. Sobecky. 2014. "Biostimulation of weathered MC252 crude oil in northern Gulf of Mexico sandy sediments." *International Biodeterioration & Biodegradation* 93:1–9.

Iskander, Lauren, Charbel Abou Khalil, and Michel C. Boufadel. 2021. "Fate of crude oil in the environment and remediation of oil spills." *STEM Fellowship Journal* 6 (1):1–7.

Jednak, Tanja, Jelena Avdalović, Srđan Miletić, Latinka Slavković-Beškoski, Dalibor Stanković, Jelena Milić, Mila Ilić, Vladimir Beškoski, Gordana Gojgić-Cvijović, and Miroslav M Vrvić. 2017. "Transformation and synthesis of humic substances during bioremediation of petroleum hydrocarbons." *International Biodeterioration & Biodegradation* 122:47–52.

Jiménez, Núria, Brandon E. L. Morris, Minmin Cai, Friederike Gründger, Jun Yao, Hans H. Richnow, and Martin Krüger. 2012. "Evidence for in situ methanogenic oil degradation in the Dagang oil field." *Organic Geochemistry* 52:44–54 doi: 10.1016/j.orggeochem.2012.08.009.

Kappler, Andreas, Marina Lisa Wuestner, Alexander Ruecker, Johannes Harter, Maximilian Halama, and Sebastian Behrens. 2014. "Biochar as an electron shuttle between bacteria and Fe (III) minerals." *Environmental Science & Technology Letters* 1 (8):339–344.

Karr, Michael and D. ARCPACs. "Using Humic Substances in the Bioremediation of Petroleum Polluted Soils." (2011). https://www.semanticscholar.org/paper/Using-Humic-Substances-in-the-Bioremediation-of-Karr-ARCPACs/95b27468a47d6d6b89b80f8ee4db995eb31192da.

Kong, Lulu, Yuanyuan Gao, Qixing Zhou, Xuyang Zhao, and Zhongwei Sun. 2018. "Biochar accelerates PAHs biodegradation in petroleum-polluted soil by biostimulation strategy." *Journal of Hazardous Materials* 343:276–284.

Kraft, Beate. 2014. *"Competition in Nitrate-Reducing Microbial Communities."* University of Bremen, Bremen.

Kronenberg, Maria, Eric Trably, Nicolas Bernet, and Dominique Patureau. 2017. "Biodegradation of polycyclic aromatic hydrocarbons: Using microbial bioelectrochemical systems to overcome an impasse." *Environmental Pollution* 231:509–523.

Lang, Firmin Semboung, Jacqueline Destain, Frank Delvigne, Philippe Druart, Marc Ongena, and Philippe Thonart. 2016. "Biodegradation of polycyclic aromatic hydrocarbons in mangrove sediments under different strategies: Natural attenuation, biostimulation, and bioaugmentation with Rhodococcus erythropolis T902. 1." *Water, Air, & Soil Pollution* 227 (9):1–15.

Langenhoff, Alette A.M., Alexander J.B. Zehnder, and Gosse Schraa. 1996. "Behaviour of toluene, benzene and naphthalene under anaerobic conditions in sediment columns." *Biodegradation* 7 (3):267–274.

Ławniczak, Łukasz, Marta Woźniak-Karczewska, Andreas P. Loibner, Hermann J. Heipieper, and Łukasz Chrzanowski. 2020. "Microbial degradation of hydrocarbons—basic principles for bioremediation: A review." *Molecules* 25 (4):856.

Lei, Li, Amid P. Khodadoust, Makram T. Suidan, and H.H. Tabak. 2005. "Biodegradation of sediment-bound PAHs in field-contaminated sediment." *Water Research* 39 (2):349–361.

Li, Jiang, Xin Li, Zhao Yang, and Tao Tao Tang. 2020. "Effects of cellulose on polycyclic aromatic hydrocarbons removal and microbial community structure variation during anaerobic digestion of sewage sludge." *Journal of Environmental Science and Health, Part A* 55 (9):1104–1110.

Li, Xiaojing, Xin Wang, Lili Wan, Yueyong Zhang, Nan Li, Desheng Li, and Qixing Zhou. 2016. "Enhanced biodegradation of aged petroleum hydrocarbons in soils by glucose addition in microbial fuel cells." *Journal of Chemical Technology and Biotechnology* 91 (1):267–275.

Li, Xiaoling, Ruiyu Zheng, Xuwu Zhang, Zhiwei Liu, Ruiyan Zhu, Xiaoyu Zhang, and Dawei Gao. 2019. "A novel exoelectrogen from microbial fuel cell: Bioremediation of marine petroleum hydrocarbon pollutants." *Journal of Environmental Management* 235:70–76.

Liu, Xiang, Zhengwen Li, Chen Zhang, Xuejun Tan, Xue Yang, Chunli Wan, and Duu-Jong Lee. 2020. "Enhancement of anaerobic degradation of petroleum hydrocarbons by electron intermediate: Performance and mechanism." *Bioresource Technology* 295:122305.

Lovley, Derek R., John D. Coates, Elizabeth L. Blunt-Harris, Elizabeth J.P. Phillips, and Joan C. Woodward. 1996. "Humic substances as electron acceptors for microbial respiration." *Nature* 382 (6590): 445–448.

Lovley, Derek R., Stephen J. Giovannoni, David C. White, James E. Champine, E.J.P. Phillips, Yuri A. Gorby, and Steve Goodwin. 1993. "Geobacter metallireducens gen. nov. sp. nov., a microorganism capable of coupling the complete oxidation of organic compounds to the reduction of iron and other metals." *Archives of Microbiology* 159 (4):336–344.

Ma, Chen, Yueqiang Wang, Li Zhuang, Deyin Huang, Shungui Zhou, and Fangbai Li. 2011. "Anaerobic degradation of phenanthrene by a newly isolated humus-reducing bacterium, Pseudomonas aeruginosa strain PAH-1." *Journal of Soils and Sediments* 11 (6):923–929.

Mahjoubi, Mouna, Simone Cappello, Yasmine Souissi, Atef Jaouani, and Ameur Cherif. 2018. "Microbial bioremediation of petroleum hydrocarbon–contaminated marine environments." *Recent Insights in Petroleum Science and Engineering* 325:325–350.

Mapelli, Francesca, Alberto Scoma, Grégoire Michoud, Federico Aulenta, Nico Boon, Sara Borin, Nicolas Kalogerakis, and Daniele Daffonchio. 2017. "Biotechnologies for marine oil spill cleanup: Indissoluble ties with microorganisms." *Trends in Biotechnology* 35 (9):860–870.

Moliterni, Elena, Lourdes Rodriguez, Francisco Jesus Fernández, and Jose Villaseñor. 2012. "Feasibility of different bioremediation strategies for treatment of clayey and silty soils recently polluted with diesel hydrocarbons." *Water, Air, & Soil Pollution* 223 (5):2473–2482.

Mukome, Fungai N.D., Maya C. Buelow, Junteng Shang, Juan Peng, Michael Rodriguez, Douglas M. Mackay, Joseph J. Pignatello, Natasha Sihota, Thomas P. Hoelen, and Sanjai J. Parikh. 2020. "Biochar amendment as a remediation strategy for surface soils impacted by crude oil." *Environmental Pollution* 265:115006.

Müller, Juliana B., Débora T. Ramos, Catherine Larose, Marilda Fernandes, Helen S.C. Lazzarin, Timothy M. Vogel, and Henry X. Corseuil. 2017. "Combined iron and sulfate reduction biostimulation as a novel approach to enhance BTEX and PAH source-zone biodegradation in biodiesel blend-contaminated groundwater." *Journal of Hazardous Materials* 326:229–236.

Ni, Zheng, Chi Zhang, Zhiqiang Wang, Song Zhao, Xiaoyun Fan, and Hanzhong Jia. 2021. "Performance and potential mechanism of transformation of polycyclic aromatic hydrocarbons (PAHs) on various iron oxides." *Journal of Hazardous Materials* 403:123993.

Nikolopoulou, Maria, and Nicolas Kalogerakis. 2009. "Biostimulation strategies for fresh and chronically polluted marine environments with petroleum hydrocarbons." *Journal of Chemical Technology & Biotechnology: International Research in Process, Environmental & Clean Technology* 84 (6):802–807.

Nikolopoulou, Maria, Nikos Pasadakis, and Nicolas Kalogerakis. 2013. "Evaluation of autochthonous bioaugmentation and biostimulation during microcosm-simulated oil spills." *Marine Pollution Bulletin* 72 (1):165–173.

Nzila, Alexis. 2018. "Biodegradation of high-molecular-weight polycyclic aromatic hydrocarbons under anaerobic conditions: Overview of studies, proposed pathways and future perspectives." *Environmental Pollution* 239:788–802.

Obiakalaije, U.M., O.A. Makinde, and E.R. Amakoromo. 2015. "Bioremediation of crude oil polluted soil using animal waste." *International Journal of Environmental Bioremediation & Biodegradation* 3 (3):79–85.

Ossai, Innocent Chukwunonso, Aziz Ahmed, Auwalu Hassan, and Fauziah Shahul Hamid. 2020. "Remediation of soil and water contaminated with petroleum hydrocarbon: A review." *Environmental Technology & Innovation* 17:100526.

Overton, Edward B., Terry L. Wade, Jagoš R. Radović, Buffy M. Meyer, M. Scott Miles, and Stephen R. Larter. 2016. "Chemical composition of Macondo and other crude oils and compositional alterations during oil spills." *Oceanography* 29 (3):50–63.

Paulo, A.M.S., Andreia F. Salvador, Joana I. Alves, Rita Castro, A.A.M. Langenhoff, A.J.M. Stams, and Ana J. Cavaleiro. 2017. "Enhancement of methane production from 1-hexadecene by additional electron donors." *Microbial Biotechnology* 11:657–666.

Prince, Roger C., and Clifford C. Walters. 2016. "Biodegradation of oil hydrocarbons and its implications for source identification." In: Scott Stout, Zhendi Wang (eds.), *Standard Handbook Oil Spill Environmental Forensics*, pp. 869–916. Elsevier, Amsterdam.

Ramadass, Kavitha, Mallavarapu Megharaj, Kadiyala Venkateswarlu, and Ravi Naidu. 2018. "Bioavailability of weathered hydrocarbons in engine oil-contaminated soil: Impact of bioaugmentation mediated by Pseudomonas spp. on bioremediation." *Science of the Total Environment* 636:968–974.

Ramírez-García, Robert, Nisarg Gohil, and Vijai Singh. 2019. "Recent advances, challenges, and opportunities in bioremediation of hazardous materials." In Kuldeep Bauddh, Vimal Chandra Pandey (eds.), *Phytomanagement of Polluted Sites*, pp. 517–568. Elsevier, Amsterdam.

Robinson-Lora, Mary Ann, and Rachel A. Brennan. 2009. "The use of crab-shell chitin for biological denitrification: Batch and column tests." *Bioresource Technology* 100 (2):534–541.

Ron, Eliora Z., and Eugene Rosenberg. 2014. "Enhanced bioremediation of oil spills in the sea." *Current Opinion in Biotechnology* 27:191–194.

Roy, Ajoy, Avishek Dutta, Siddhartha Pal, Abhishek Gupta, Jayeeta Sarkar, Ananya Chatterjee, Anumeha Saha, Poulomi Sarkar, Pinaki Sar, and Sufia K Kazy. 2018. "Biostimulation and bioaugmentation of native microbial community accelerated bioremediation of oil refinery sludge." *Bioresource Technology* 253:22–32.

Sayara, Tahseen, Eduard Borràs, Gloria Caminal, Montserrat Sarrà, and Antoni Sánchez. 2011. "Bioremediation of PAHs-contaminated soil through composting: Influence of bioaugmentation and biostimulation on contaminant biodegradation." *International Biodeterioration & Biodegradation* 65 (6):859–865.

Schaechter, Moselio. 2009. *Encyclopedia of Microbiology*. Academic Press, Cambridge, MA.

Sharma, Jot. 2019. "Advantages and limitations of in situ methods of bioremediation." *Recent Advances in Biology and Medicine* 5:10941.

Sharma, Ranju, Ngangbam Sarat Singh, Neha Dhingra, and Talat Parween. 2020. "Bioremediation of oil-spills from shoreline environment." In: Mohammad Oves, Mohammad Omaish Ansari, Mohammad Zain Khan, Mohammad Shahadat, Iqbal M.I. Ismail (eds.), *Modern Age Waste Water Problems*, pp. 275–291. Springer, Cham.

Singh, Om V. 2017. *Bio-pigmentation and Biotechnological Implementations*. John Wiley & Sons, Hoboken, NJ.

Su, Lianghu, Haiyan Zhou, Guangzhai Guo, Aihua Zhao, and Youcai Zhao. 2012. "Anaerobic biodegradation of PAH in river sediment treated with different additives." *Procedia Environmental Sciences* 16:311–319.

Talley, Jeffrey. 2016. *Bioremediation of Recalcitrant Compounds*. CRC Press, Boca Raton, FL.

Thapa, Bijay, Ajay Kumar Kc, and Anish Ghimire. 2012. "A review on bioremediation of petroleum hydrocarbon contaminants in soil." *Kathmandu University Journal of Science, Engineering and Technology* 8 (1):164–170.

Tremblay, Pier-Luc, and Tian Zhang. 2020. "Functional genomics of metal-reducing microbes degrading hydrocarbons." In: Matthias Boll (ed.), *Anaerobic Utilization of Hydrocarbons, Oils, and Lipids*, pp. 233–253. Springer, Cham.

Varjani, Sunita J. 2017. "Microbial degradation of petroleum hydrocarbons." *Bioresource Technology* 223:277–286.

Wartell, Brian, Michel Boufadel, and Lucia Rodriguez-Freire. 2021. "An effort to understand and improve the anaerobic biodegradation of petroleum hydrocarbons: A literature review." *International Biodeterioration & Biodegradation* 157:105156.

Wu, Lin, Binghua Li, and Mingzhu Liu. 2018. "Influence of aromatic structure and substitution of carboxyl groups of aromatic acids on their sorption to biochars." *Chemosphere* 210:239–246.

Wu, Manli, Warren A. Dick, Wei Li, Xiaochang Wang, Qian Yang, Tingting Wang, Limei Xu, Minghui Zhang, and Liming Chen. 2016. "Bioaugmentation and biostimulation of hydrocarbon degradation and the microbial community in a petroleum-contaminated soil." *International Biodeterioration & Biodegradation* 107:158–164.

Xia, Yuqiang, and Michel C. Boufadel. 2011. "Beach geomorphic factors for the persistence of subsurface oil from the Exxon Valdez spill in Alaska." *Environmental Monitoring and Assessment* 183 (1–4):5–21.

Xu, Xingjian, Wenming Liu, Shuhua Tian, Wei Wang, Qige Qi, Pan Jiang, Xinmei Gao, Fengjiao Li, Haiyan Li, and Hongwen Yu. 2018. "Petroleum hydrocarbon-degrading bacteria for the remediation of oil pollution under aerobic conditions: A perspective analysis." *Frontiers in Microbiology* 9:2885.

Yan, Zaisheng, Na Song, Haiyuan Cai, Joo-Hwa Tay, and Helong Jiang. 2012. "Enhanced degradation of phenanthrene and pyrene in freshwater sediments by combined employment of sediment microbial fuel cell and amorphous ferric hydroxide." *Journal of Hazardous Materials* 199:217–225.

Yang, Xunan, Zefang Chen, Qunhe Wu, and Meiying Xu. 2018. "Enhanced phenanthrene degradation in river sediments using a combination of biochar and nitrate." *Science of the Total Environment* 619:600–605.

Yang, Xunan, Jiaxin Ye, Limei Lyu, Qunhe Wu, and Renduo Zhang. 2013. "Anaerobic biodegradation of pyrene by Paracoccus denitrificans under various nitrate/nitrite-reducing conditions." *Water, Air, & Soil Pollution* 224 (5):1–10.

Yao, Lunfang, Ying Teng, Yongming Luo, Peter Christie, Wenting Ma, Fang Liu, Yonggui Wu, Yang Luo, and Zhengao Li. 2015. "Biodegradation of polycyclic aromatic hydrocarbons (PAHs) by Trichoderma reesei FS10-C and effect of bioaugmentation on an aged PAH-contaminated soil." *Bioremediation Journal* 19 (1):9–17.

Yuan, Shaw Y., and Bea V. Chang. 2007. "Anaerobic degradation of five polycyclic aromatic hydrocarbons from river sediment in Taiwan." *Journal of Environmental Science and Health Part B* 42 (1):63–69.

Yuan, Sarah Y., S.H. Wei, and Bv Chang. 2000. "Biodegradation of polycyclic aromatic hydrocarbons by a mixed culture." *Chemosphere* 41 (9):1463–1468.

Yuniati, M.D. 2018. "Bioremediation of petroleum-contaminated soil: A Review." *IOP Conference Series: Earth and Environmental Science* 118:012063.

Zawierucha, Iwona, and Grzegorz Malina. 2011. "Bioremediation of contaminated soils: Effects of bioaugmentation and biostimulation on enhancing biodegradation of oil hydrocarbons." In: Ajay Singh, Nagina Parmar, Ramesh C. Kuhad (eds.), *Bioaugmentation, Biostimulation and Biocontrol*, pp. 187–201. Springer, Berlin.

Zedelius, Johannes, Ralf Rabus, Olav Grundmann, Insa Werner, Danny Brodkorb, Frank Schreiber, Petra Ehrenreich, Astrid Behrends, Heinz Wilkes, and Michael Kube. 2011. "Alkane degradation under anoxic conditions by a nitrate-reducing bacterium with possible involvement of the electron acceptor in substrate activation." *Environmental Microbiology Reports* 3 (1):125–135.

Zhang, Zhen, and Irene M.C. Lo. 2015. "Biostimulation of petroleum-hydrocarbon-contaminated marine sediment with co-substrate: Involved metabolic process and microbial community." *Applied Microbiology and Biotechnology* 99 (13):5683–5696.

Zhang, Zhen, Irene M.C. Lo, and Dickson Y.S. Yan. 2015. "An integrated bioremediation process for petroleum hydrocarbons removal and odor mitigation from contaminated marine sediment." *Water Research* 83:21–30.

Zhong, Yin, Tiangang Luan, Xiaowei Wang, Chongyu Lan, and Nora F.Y. Tam. 2007. "Influence of growth medium on cometabolic degradation of polycyclic aromatic hydrocarbons by Sphingomonas sp. strain PheB4." *Applied Microbiology and Biotechnology* 75 (1):175–186.

12 Commercial Bacterial and Fungal Microbial Biostimulants Used for Agriculture in India
An Overview

Aruna Jyothi Kora
Bhabha Atomic Research Centre (BARC) and
Homi Bhabha National Institute (HBNI)

CONTENTS

12.1 INTRODUCTION

The microbial biostimulants are a subgroup under the diverse category of biostimulants, such as humic substances, protein hydrolysates, amino acids, seaweed extracts, biopolymers, inorganics and microbes. The microbial biostimulants include an array of beneficial microorganisms, including bacteria, fungi and algae, which can be applied to the rhizosphere, phyllosphere, seeds and various plant parts. They exhibit unique characteristics through stimulation of different natural process towards improvement/enhancement of plant nutrition, growth, productivity, crop quality, tolerance/resistance to abiotic and biotic stresses, tolerance/protection against various plant diseases, etc. (Sangiorgio et al. 2020, García-Seco 2019, Madeiras and Lanier 2020, Rouphael and Colla 2020, du Jardin 2015).

It is noteworthy that the plants coevolved with microbial symbionts, which are less toxic and environment-friendly. They are an integral part of the ecosystem and involve in carbon, nitrogen and mineral biochemical cycles. The biochemical, physiological and molecular studies on plant–microbe interaction indicated plant response towards abiotic and biotic stresses in terms of microbe-induced systemically acquired tolerance and resistance, respectively. Microbial biostimulant application depends upon soil properties, specific crop requirements, competition with indigenous microflora, mode of biostimulant application, plant–microbe symbiotic relationship, multiple/multifarious traits of microbial biostimulants, etc. (Sangiorgio et al. 2020, Rouphael and Colla 2020, Rouphael et al. 2017).

The term microbial biostimulant is broader, and it can accommodate a vast number of active microbes of various origins; they need not be from the rhizosphere only. In contrast, the plant growth-promoting rhizobacteria (PGPR) are microbial biostimulants of rhizospheric origin/colonizing rhizosphere. The PGPR are characterized by rhizosphere competence, compatibility with rhizospheric microbes, root colonization potential, plant growth promotion, the broad spectrum of action, physiochemical tolerance (heat, desiccation, radiation and salinity), etc. The microbial biostimulants perform diverse activities, including biofertilization, biocontrol, induction of plant defence, and plant growth promotion via phytohormones production and nutrient release. The application of microbial biostimulants depends on many factors, including soil properties, specific crop requirements, competition with indigenous microflora, mode of biostimulant application, plant–microbe symbiotic relationship and multiple/multifarious traits of microbial biostimulants (Sangiorgio et al. 2020, du Jardin 2015, Basu et al. 2021, Rajput et al. 2019).

This chapter mainly focuses on commercially available bacterial and fungal microbial biostimulants, which serve as biofertilizers and biocontrol agents for agriculture in India. The biofertilizers include nitrogen-fixing bacteria (*Rhizobium*, *Azotobacter*, *Azospirillum* and *Acetobacter*); phosphate-, zinc-, potassium-, silicate- and iron-solubilizing bacteria; and lignocellulolytic bacteria. The biocontrol agents including *Pseudomonas fluorescens*, *Bacillus* and *Trichoderma* species that serve the function of bactericide and fungicide are described in this chapter. Further, the Indian guidelines and regulations governing their production and marketing are covered.

12.2   REGULATIONS FOR BIOSTIMULANTS IN INDIA

According to the Fertilizer (inorganic, organic or mixed) Control Amendment Order, 2021, issued by the Ministry of Agriculture and Farmers' Welfare, Government of India, and as published in the Gazette of India, a biostimulant means a "substance or microorganism or a combination of both whose primary function when applied to plants, seeds or rhizosphere is to stimulate physiological processes in plants and to enhance its nutrient uptake, growth, yield, nutrition efficiency, crop quality and tolerance to stress, regardless of its nutrient content, but does not include pesticides or plant growth regulators". Under Schedule VI, data on chemistry such as source (natural extracts of plant/microbe/animal/synthetic), product specification (analysis report from Good Laboratory Practice or National Accreditation Board for Testing and Calibration Laboratories), method of analysis conforming to the specifications, physical and chemical properties of active ingredients and adjuvants and shelf life should be provided. Further, product data on agronomic bio-efficiency trials conducted at the Indian Council of Agricultural Research and state agricultural universities at a minimum of three different doses for one season at three agroecological locations should be provided. Also, test report and recommendations by a GLP accredited laboratory on five acute toxicity tests (oral—rat, dermal—rat, inhalation—rat, primary skin irritation—rabbit and eye irritation—rabbit) and four ecotoxicity tests (bird, freshwater fish, honeybee and earthworm) should be submitted. The biostimulants should not exceed the maximum limit prescribed for various heavy metals, including cadmium (5 mg/kg), arsenic (10 mg/kg), chromium (50 mg/kg), lead (100 mg/kg), copper (300 mg/kg) and zinc (1,000 mg/kg), and any pesticide (0.01 mg/kg). A Central Biostimulant Committee, constituted by the Central Government, frames the toxicology and other related test guidelines and advises on new biostimulant inclusion, various biostimulants specifications, sample drawing and test analysis methods, minimum laboratory requirements, etc. (Eenadu 2021, Adidam 2021).

12.3 BIOFERTILIZERS

The excessive use of chemical fertilizers leads to nutrient deficiency, nutrient imbalance and fertility loss in soils, in addition to water and soil pollution. In this scenario, biofertilizers are gaining importance as an integral part of cost-effective, renewable, integrated nutrient management, sustainable agriculture and organic farming in terms of enhancing soil fertility and health, crop productivity and crop quality (Bindu Madhavi and Naveen Kumar 2018, Rajabi Hamedani et al. 2020).

The biofertilizers are composed of a diverse category of bacteria and fungi and are broadly classified as nitrogen-fixing; phosphate-, zinc-, potassium-, silicate- and iron-solubilizing; and lignocellulolytic microbes (Sangiorgio et al. 2020). The present subsection mainly concentrates on bacterial biofertilizers. The mass-produced biofertilizers are dispersed in a carrier material and available in moist/dry powder, granules or liquid form. The commonly used carrier materials are peat, lignite, talc, vermiculite, charcoal, press mud, peat soil, humus, farmyard manure, etc. The ideal carrier should be inexpensive; locally available; rich in organic matter; amenable for sterilization, processing and friability; and non-toxic to microbial inoculant and plants, and they should exhibit superior buffering, moisture absorption and seed adhesion capabilities. In India, the quality of various carrier-based biofertilizers is monitored through standards set by the national organization, Bureau of Indian Standards (BIS). The parameters such as carrier type, particle size, and moisture, viable count, contamination level, pH, shelf life and efficiency character are regulated by stipulated limits under Fertilizer Control Order (FCO), 1985 (Table 12.1) (Bashan and de-Bashan 2015, Kosaraju 2018, Karthikeyan and Sivasakthivelan 2021). In comparison with carrier-based biofertilizers, the liquid biofertilizers are preferred, for example *Rhizobium* (IS 17134: 2020), *Azotobacter* (IS 17135: 2019), *Azospirillum*

TABLE 12.1
The Specifications for Various Carrier-Based Biofertilizers as per the Bureau of Indian Standards (BIS)

Characteristic	*Rhizobium* (IS 8268:2001)	*Azotobacter* (IS 9138:2009)	*Azospirillum* (IS 14806:2000)	Phosphate-Solubilizing Bacteria (IS 14807:2000)
		Minimum Required Limit		
Carrier form	Powder, granules	Powder, granules	Powder, granules	Powder, granules
Viable count in carrier (CFU/g)	10^7	10^8	10^7	10^7
Carrier particle size (μm)	<150–212	<150–212	<150–212	<150–212
Carrier moisture (%)	30–40	35–40	30–40	30–40
pH	6.5–7.5	6.5–7.5	6.5–7.5	6.5–7.5
Contamination at 10^5 dilution	0	0	0	0
Efficiency character	Effective nodulation in intended crops	10 mg N fixation/g of sucrose	Effective root development in intended crops	30% Phosphate solubilization by spectrophotometry or 10 mm solubilization zone
Shelf life (months)	6	6	6	6

(IS 17136: 2019) and phosphate-solubilizing bacteria (IS 17137: 2019) due to many virtues. They include high population density (> 10^9 CFU/mL), viability at storage up to 45°C, extended shelf life (1–2 years), being contamination-free, high enzymatic activity, high survival rate against native population on seeds and soil, fermented smell-based easier identification, facile and rapid quality control protocols, ease of application by farmers, lower application dosage, cost savings on carrier material and processing, and high revenue and wider export potentials (Karthikeyan and Sivasakthivelan 2021, Kosaraju 2018). The biofertilizer formulations can be applied through seed inoculation, seedling root dipping and soil application. The biofertilizers should not be mixed with other chemical fertilizers, fungicides, herbicides and insecticides (Vijay and Tirumal 2018, Raja Sekhar et al. 2018).

12.3.1 Nitrogen-Fixing Bacteria

The atmospheric air contains 78% of nitrogen, and the nitrogen-fixing microbes (bacteria, cyanobacteria, fungi and algae) convert the plant-inaccessible, atmospheric, gaseous nitrogen into plant-usable ammonia, catalysed by nitrogenase enzyme system. The important nitrogen-fixing bacteria are *Rhizobium, Azotobacter, Azospirillum, Acetobacter*, cyanobacteria, etc. (Bindu Madhavi and Naveen Kumar 2018, Vijay and Tirumal 2018).

12.3.1.1 *Rhizobium*

These are the most significant Gram-negative, symbiotic, organic acid-producing, nitrogen-fixing bacteria that colonize the roots of various leguminous crops and form root nodules. The host-specific *Rhizobium* sp. fix nitrogen (40–300 kg/ha/year) in root nodules of different leguminous plants such as groundnut, soya bean, black gram, green gram, red gram, Bengal gram, horse gram, pea, cowpea, beans and berseem clover. The nitrogen fixation by *Rhizobium* is host specific, and the important species are *R. leguminosarum, R. japonicum, R. phaseoli, R. trifolii, R. meliloti, R. lupini*, etc. Besides nitrogen supply to crops, legume–*Rhizobium* association leaves residual nitrogen for the subsequent crops (Bindu Madhavi and Naveen Kumar 2018, Raja Sekhar et al. 2018, Ram Babu 2018, Vijay and Tirumal 2018, Kosaraju 2018). The yeast extract mannitol Congo red agar is utilized as a differential medium for the isolation and propagation of *Rhizobium* species (Figure 12.1a). They are commercialized under the brand names Rhizoteeka, Jawahar *Rhizobium*, Rhizopower, Well Rhyzo, Polarhizo, Rhizolife, Premium Rhizo, etc., in India.

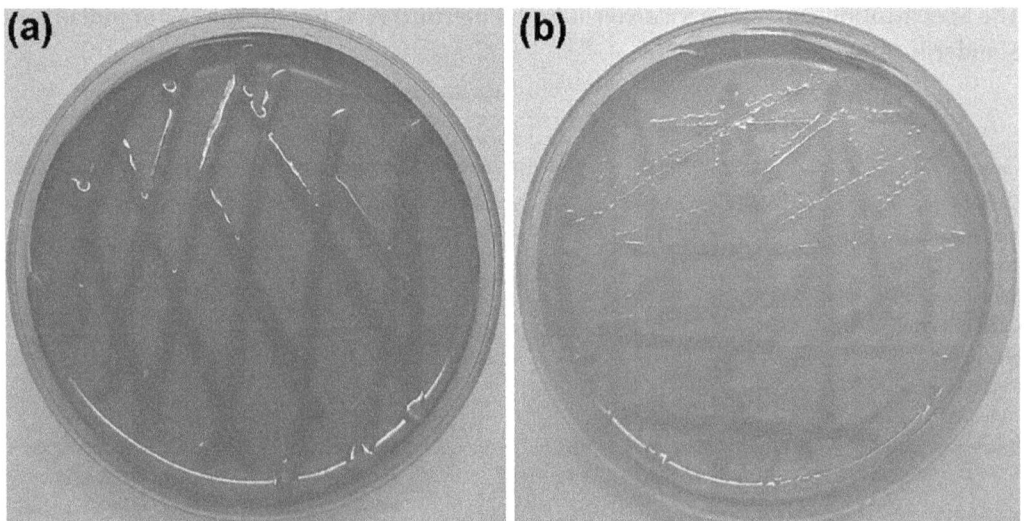

FIGURE 12.1   The growth of (a) *Rhizobium* on yeast extract mannitol Congo red agar and (b) *Azotobacter* on Ashby's mannitol agar.

12.3.1.2 *Azotobacter*

The *Azotobacter* sp. are Gram-negative, free-living, soil-dwelling, pigment-producing, diazotrophic, nitrogen-fixing bacteria (Sumbul et al. 2020, Dhevendaran, Preetha, and Hari 2013). The commercially important *Azotobacter* species are *A. chroococcum, A. vinelandii, A. beijerinckii, A. insignis, A. macrocytogenes* and *A. paspali* (Karthikeyan and Sivasakthivelan 2021). As they fix 10–20 kg N/ha/per year, they are widely utilized for cereal (rice, wheat and sorghum), millet (pearl millet), commercial (sugarcane, cotton, chilli, sunflower, mustard, safflower, tobacco, tea and coffee), leafy, ornamental, olericulture (tomato and cabbage), horticulture and floriculture crops. A selective Ashby's mannitol agar is used for the isolation, identification and cultivation of *Azotobacter*, and the brown-coloured pigmentation is visible (Figure 12.1b). As they flourish well in slightly alkaline soils with high organic matter, the PGPR are mixed with organic manure and applied to the soil at a dosage of 5 kg/ha. In addition, they are known to produce phytohormones, including auxins (indole acetic acid), cytokinins, gibberellins, vitamins, polysaccharides, siderophores, phosphate-solubilizing compounds, antimicrobial substances, phytopathogenic toxins, thereby stimulating seed germination and plant growth, ameliorating soil fertility, tolerating salinity stress and protecting against an array of plant fungal diseases (Bindu Madhavi and Naveen Kumar 2018, Raja Sekhar et al. 2018, Ram Babu 2018, Sumbul et al. 2020, Vijay and Tirumal 2018). They are commercially available under the names Azoteeka, Jawahar *Azotobacter*, Azonik, Azo-Power, K-Azo Power, Azoto, Azoto Plus, AzoVita, Bio Azo, etc., in India.

12.3.1.3 *Azospirillum*

It is another category of Gram-negative, soil-dwelling, pellicle-forming, polyhydroxybutyrate-accumulating, associative symbiotic, diazotrophic, nitrogen-fixing bacteria under the PGPR category. The commercially important *Azospirillum* species are *A. brasilense, A. lipoferum, A. amazonense, A. halopraeferens* and *A. irakense* (Karthikeyan and Sivasakthivelan 2021). The growth of *Azospirillum* on nutrient agar medium is shown in Figure 12.2a. As the bacteria can fix 20–40 kg N/ha/year, they are extensively exploited for rice, wheat, maize, sorghum, sugarcane, pearl millet, finger millet, cotton, chilli, sunflower, banana, and forage and horticulture crops. It also flourishes well in soils with low organic matter and commercially utilized in various countries, including India, Europe, Argentina, Mexico and South Africa (Bindu Madhavi and Naveen Kumar 2018, Raja Sekhar et al. 2018, Ram Babu 2018, Vijay and Tirumal 2018, Bashan, Trejo, and de-Bashan 2011). They produce phytohormones,

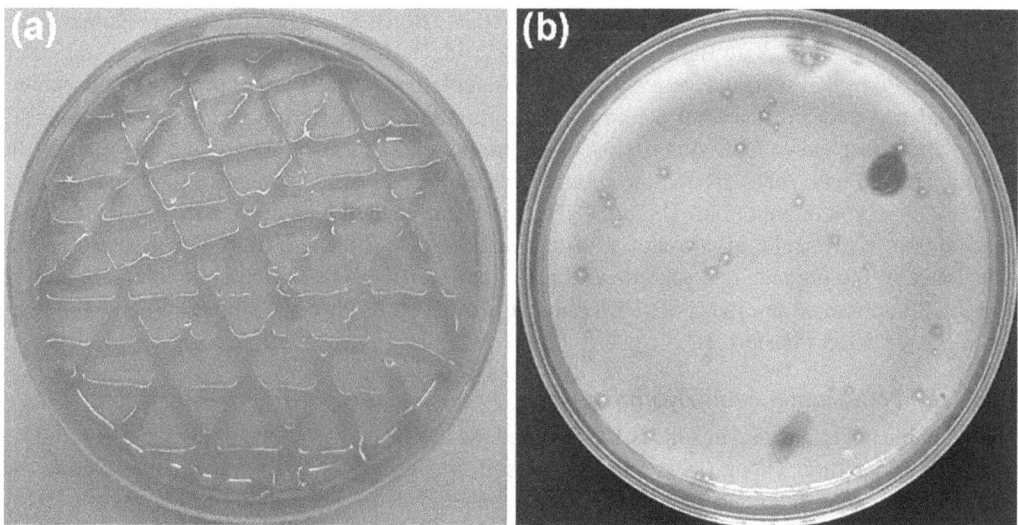

FIGURE 12.2 The growth of (a) *Azospirillum* on nutrient agar and (b) *Acetobacter* demonstrated from the clear halo zones around the colonies.

including indole acetic acid, gibberellins, siderophores and antioxidants, and induce systemic acquired resistance (SAR) in plants (Raffi and Charyulu 2012, Bashan and de-Bashan 2015). The Indian commercial brands of *Azospirillum* are Niromax, Bioteeka, Jawahar *Azospirillum*, Spironik, Azofix, Azospir, etc.

12.3.1.4   *Acetobacter*

The *Acetobacter* sp. are saccharophilic, organic acid-producing, symbiotic, diazotrophic, nitrogen-fixing bacteria living in the roots, stems and leaves of sugarcane, sugar beet, sweet corn and coffee plants and used a biofertilizer for crops, sugarcane, sugar beet, rice, wheat and sorghum. They fix around 30 kg N/ha/year and are known to increase the yield (5–20 t/ha) and sugar content (5%–15%) in sugarcane. The most commonly used strain is *A. diazotrophicus*, and the *Acetobacter* species are identified in the clear halo zones around the colonies because of the acid production in the medium (Figure 12.2b). They flourish well in soils with pH < 6, secrete phytohormones, including indole acetic acid, gibberellins and phosphate solubilizers, and exhibit tolerance to low pH and high salinity. They are particularly recommended for sugarcane via set treatment, soil, application, drip irrigation and foliar spray (Raja Sekhar et al. 2018, Ram Babu 2018, Vijay and Tirumal 2018, Kosaraju 2018). The Indian commercial brands of *Acetobacter* are Aceto Power, Bio Aceto, Bheeshma *Acetobacter*, Acetoz, etc.

12.3.2   Mineral-Solubilizing Bacteria

The minerals including phosphorous, potassium, zinc, silicon and iron are either deficient or tightly bound in soils and are unavailable for the plants to uptake. Because of the secretion of different organic acids and enzymes by soil microbes, the minerals are bioavailable for plants (Nagaraju et al. 2017, Mumtaz et al. 2017, Singh, Maurya, and Bahadur 2018).

12.3.2.1   Phosphate-Solubilizing Bacteria

The phosphate-solubilizing bacteria (PSB) are free-living, organic acid- and enzyme-producing, rhizospheric bacteria that convert, solubilize and mobilize the insoluble, unavailable soil phosphorous into more bioavailable form for the plants. They flourish well in soils with more organic matter, provide around 25 kg of phosphorous/ha and are suitable for all types of crops. The typical examples of PSB are *Pseudomonas* sp. (*P. fluorescens* and *P. striata*), *Bacillus* species (*B. subtilis*, *B. megaterium*, *B. polymyxa* and *B. circulans*) and *Azospirillum* (Bindu Madhavi and Naveen Kumar 2018, Raja Sekhar et al. 2018, Ram Babu 2018, Kosaraju 2018, Hashem, Tabassum, and Fathi Abd Allah 2019). The Pikovskayas agar (PA) contains an insoluble phosphate source, tricalcium phosphate, and is employed as a differential medium for the isolation and detection of PSB (Nagaraju et al. 2017). The phosphate-solubilizing action of *P. fluorescens* and *Bacillus* sp. is evident from the yellow-coloured halo zones around the spot-inoculated PA plates, supplemented with an acid–base indicator dye, bromothymol blue (Figure 12.3). The bromothymol blue used in the media indicates the decrease in pH/acidic pH through a visual medium colour change from blue/green to yellow orange due to the organic acid secretion (Rajawat et al. 2016). The Indian commercial brands of PSB are Phosphomax, Jawahar PSB, Phsophonive, Power Phos, PhoSol, Oshphos, Phospho-Power, Paspo Bacteria, Liv Phosfert, etc.

12.3.2.2   Potassium-Solubilizing Bacteria

The available potassium (K) in soils is 1%–2%, and the low level K of is one of the yield-controlling factors in agriculture. Though K is abundant in soil, it is tightly bound to other minerals and not accessible for plant uptake (Singh, Maurya, and Bahadur 2018, Biswas and Shivaprakash 2020). The potassium-solubilizing bacteria (KSB) such as *Bacillus* species (*B. circulans*, *B. edaphicus*, *B. glucanolyticus*, *B. proteolyticus*, *B. sporothermodurans* and *B. mucilaginous*), *Pseudomonas* species, *Serratia liquefaciens,* etc., are used for increasing the bioavailable potassium in soils due to

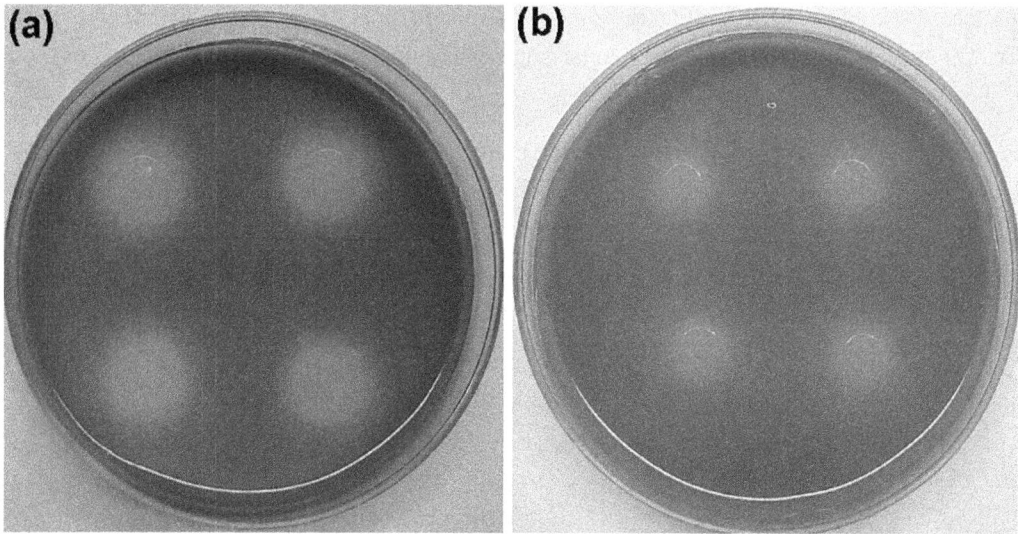

FIGURE 12.3 The phosphate-solubilizing action of (a) *P. fluorescens* and (b) *Bacillus* shown from the yellow-coloured halo zones in Pikovskayas agar medium supplemented with bromothymol blue.

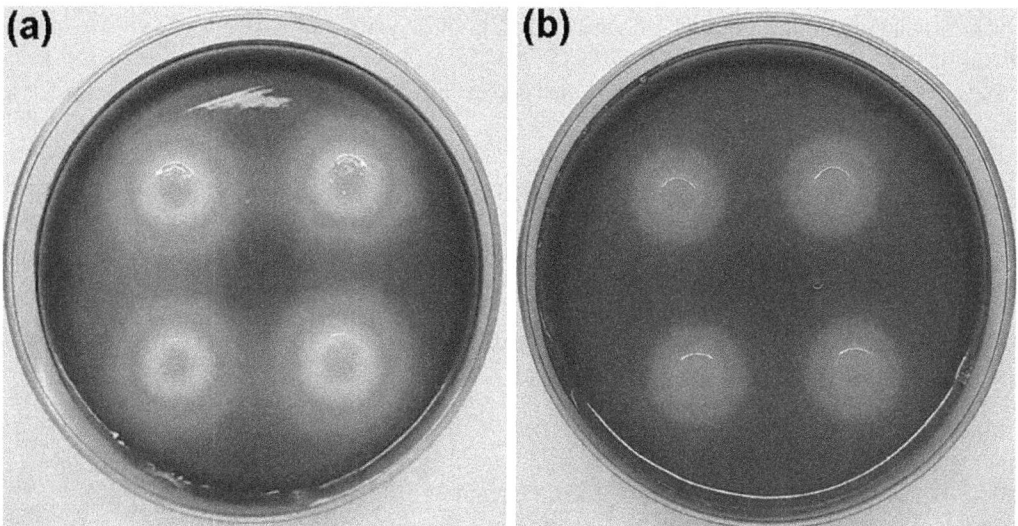

FIGURE 12.4 The potassium-solubilizing potentials of (a) *P. fluorescens* and (b) *Bacillus* species demonstrated from the yellow-coloured halo zones in Aleksandrow agar medium supplemented with bromothymol blue.

their ability to produce organic acids, including formic, acetic, lactic, propionic, glycolic, fumaric and succinic acids (Nagaraju et al. 2017, Raja Sekhar et al. 2018). The Aleksandrow agar (AA) contains insoluble potassium mineral sources such as potassium aluminosilicate (mica), kaolinite, bentonite, K-feldspar and wood ash and is utilized as a differential medium for the isolation and detection of potassium-solubilizing microorganisms (Singh, Maurya, and Bahadur 2018). The K-solubilizing potentials of *P. fluorescens* and *Bacillus* sp. are demonstrated from the yellow-coloured halo zones around the spot-inoculated AA plates supplemented with bromothymol blue (Figure 12.4). The Indian commercial brands of KSB are Potaz, KSB PEP, K-Potash, Potash Activa, Bhumidhan KMB, K Soluble, Liv Potash, etc.

12.3.2.3 Zinc-Solubilizing Bacteria

The deficiency of zinc in plants leads to an impediment in photosynthesis and nitrogen metabolism and is exemplified in terms of decline in carbohydrate and phytohormones syntheses; reduction in flowering and fruiting; delay in crop maturity; and reduction in crop yield and nutrition quality. Many rhizospheric bacterial species such as *Bacillus, Pseudomonas (P. putida), Serratia, Gluconacetobacter* and *Burkholderia ambifaria* are shown to exhibit zinc-solubilizing activity via the synthesis and secretion of chelating ligands, organic acids (2-ketogluconic and gluconic acids), amino acids, vitamins, phytohormones, etc. (Mumtaz et al. 2017). The zinc-solubilizing bacteria (ZSB) solubilize the unavailable, insoluble sources of soil zinc such as zinc sulphide, zinc carbonate, zinc oxide and zinc phosphate and enhance the growth of various crops when applied (Raja Sekhar et al. 2018, Singh, Maurya, and Bahadur 2018). The zinc solubilization activity of spot-inoculated *P. fluorescens* and *Bacillus* is depicted as orange-coloured zones in modified PA plates supplemented with zinc carbonate and bromothymol blue (Figure 12.5). The Indian commercial brands of ZSB are LivZinc, Zinc-B, ZSB PEP, Zinc Extra, Zinc Active, Bio Zinc, Maxgrow Zinc, Sharad ZSB, Z Freelancer, etc.

12.3.2.4 Silicate-Solubilizing Bacteria

The metalloid silicon is the second most abundant element in the earth crust and exists chiefly as silicates. The application of the metalloid is known to induce resistance against the abiotic (salinity, drought, UV and heavy metal (Fe, Mn and Al) toxicity) and biotic (herbivores, bacteria, fungi and insects) stresses (Luyckx et al. 2017, Swain and Rout 2017). The silicon in the soil is insoluble and present as a bound, polymerized form of polysilicate minerals. The silicon uptake from the soil is around 50–70, 230–270 and 500–700 kg for potato, rice and sugarcane, respectively. Usually, farmers supplement the soils with silica-rich sources such as quartz sand; rice husk; straw of rice, wheat and sugarcane; and ash. The silicate-solubilizing bacteria (SSB) such as *Bacillus* sp. solubilize various silicon-containing minerals such as feldspar, muscovite, biotite, vermiculite and aluminium silicate into more bioavailable monosilicic acid by producing organic acids including citric, oxalic, keto and hydroxycarboxylic acids. In addition, they also solubilize

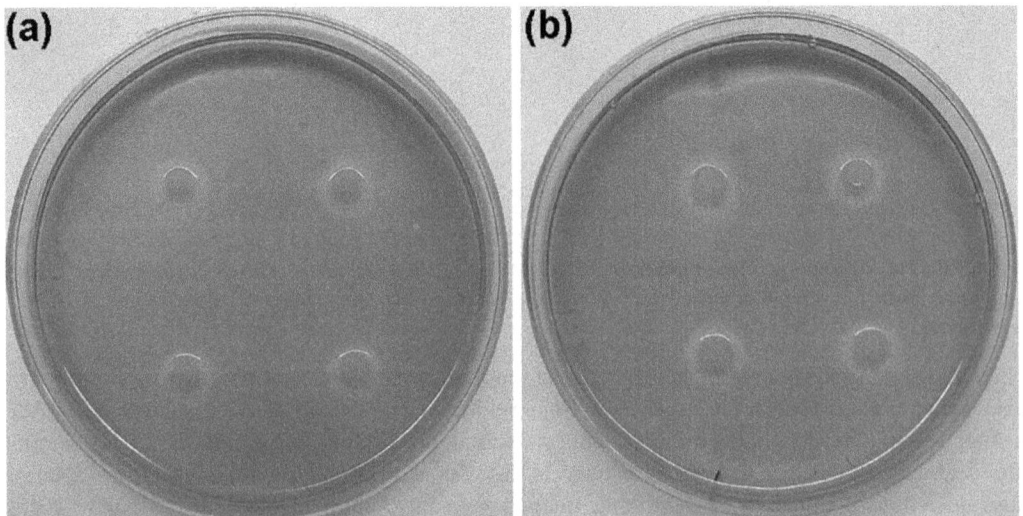

FIGURE 12.5   The zinc solubilization activity of (a) *P. fluorescens* and (b) *Bacillus* species is depicted as orange-coloured zones in modified Pikovskayas agar medium supplemented with zinc carbonate and bromothymol blue.

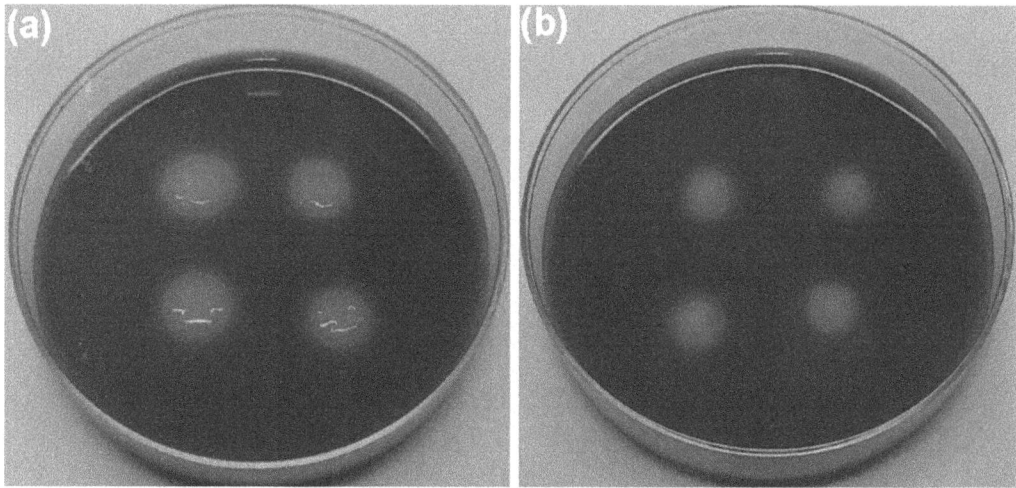

FIGURE 12.6 The siderophore production by (a) *P. fluorescens* and (b) *Bacillus* species confirmed from the yellow orange-coloured halo zones in blue-coloured chrome azurol S agar medium.

potassium and produce auxins (Kosaraju 2018, Vempalli 2021). The Indian commercial brands of SSB are Silica-109, Silica Solubilizer, Silicate Solubilizer, etc.

12.3.2.5 Siderophore-Producing Bacteria

The siderophores are low molecular weight, high-affinity, iron-sequestering biomolecules synthesized by various bacteria (*Pseudomonas*, *Bacillus* and *Acinetobacter*) and fungi (*Trichoderma*). They complex the poorly soluble, non-bioavailable iron (iron oxides and hydroxides) in soil and aid the plants in the iron uptake (Amruta et al. 2018, Rajput et al. 2019). The chrome azurol S (CAS) agar is employed as a selective medium to detect siderophores produced by various microbes (Mumtaz et al. 2017). The siderophore production by *P. fluorescens* and *Bacillus* sp. is confirmed from the yellow orange-coloured halo zones around the spot-inoculated blue-coloured CAS agar plates (Figure 12.6) (Scales et al. 2014, Hashem, Tabassum, and Fathi Abd Allah 2019).

12.3.2.6 Lignocellulolytic Bacteria

The crop residues (wheat, rice and sugarcane residues) are rich in cellulose, hemicellulose and lignin, and their efficient biodegradation increases organic carbon in the soil. In addition, the compost generated by biodegradation provides micronutrients, such as magnesium, sulphur and calcium, and secondary nutrients, including nitrogen, phosphorous and potassium. The common lignocellulolytic bacteria are *Pseudomonas*, *Bacillus amyloliquefaciens*, *Streptomyces*, etc., and the fungi include *Trichoderma*, *Pleurotus* and *Phanerochaete*. The important lignocellulolytic enzymes produced by microbes are glucanase, β-1,3-glucanase, laccase, manganese peroxidase, lignin peroxidase, etc. (Kannan et al. 2013, Prabhu and Thomas George 2002, Ram Babu 2018, Kumar et al. 2017). The carboxymethyl cellulose (CMC) agar medium is employed for the detection of cellulolytic and hydrolytic enzyme (glucanase, xylanase and hemicellulase) production by various microbes (Adlakha et al. 2011, Kasana et al. 2008). The cellulose-degrading activity of *Pseudomonas* and *Trichoderma* is evident from the clear zonation around the spot-inoculated CMC agar plates, detected by iodine staining (Figure 12.7). The most successful, widely used, farmer-friendly, cheaply available, lignocellulolytic bacterial consortium is Waste Decomposer. It was released for farmers by the National Centre of Organic Farming, Ghaziabad, under the Ministry of Agriculture and Farmers' Welfare, Government of India. It is a consortium of few beneficial bacteria isolated from native cow dung and employed for rapid and *in situ* composting of organic wastes and as a biocontrol agent, biofertilizer, soil health reviver, etc. (Kora 2021).

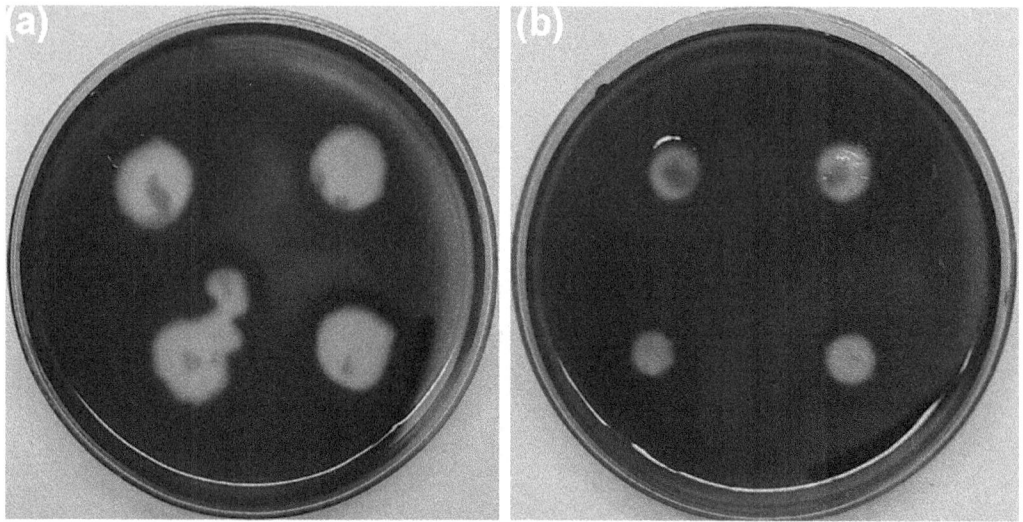

FIGURE 12.7 The cellulose-degrading action of (a) *Pseudomonas* and (b) *Trichoderma* evident from the clear zonation in carboxymethyl cellulose agar medium, detected by iodine staining.

12.4 BIOCONTROL AGENTS

The term biocontrol agent denotes the application of living organisms to suppress, inhibit and control the growth and population of other organisms. In India, the biocontrol agents are regulated under the Section 9(3) of the Insecticides Act of India, 1968, and the information on the systemic and common name; natural occurrence, morphological description, manufacturing process details, mammalian and environmental toxicities; residual analysis, etc., is needed for their registration (Junaid et al. 2013). The current subsection mainly deals with the bactericidal and fungicidal biocontrol agents, especially *Pseudomonas fluorescens* and *Bacillus* and *Trichoderma* species.

12.4.1 PSEUDOMONAS FLUORESCENS

Pseudomonas fluorescens is an aerobic, Gram-negative, fluorescent pigment-producing, metabolically diverse bacterium found in the rhizosphere, water and plant surfaces. The biocontrol agent *P. fluorescens* functions as a bactericide against *Erwinia carotovora, Ralstonia solanacearum* and *Xanthomonas oryzae* pv. *oryzae*, and fungicide against *Magnaporthe oryzae, Sclerotium rolfsii, Botrytis cinerea, Pythium, Fusarium oxysporum, F. solani, F. graminearum, Phytophthora cryptogea*, etc. (Amruta et al. 2018, Hashem, Tabassum, and Fathi Abd Allah 2019, Amruta et al. 2016, Rao et al. 2016, Kumar et al. 2017). It is used for controlling a large variety of plant diseases, including blast, rust, leaf blight, sheath blight, fire blight, damping off, Panama wilt, seedling rot, sheath rot, dry rot, root rot, red rot, bunch rot, and late leaf spot (Junaid et al. 2013, Bindu Madhavi and Naveen Kumar 2018).

It is known to act as a dual-functional biofertilizer and biocontrol agent through diverse mechanisms. It produces phytohormones (auxins, gibberellins and cytokinins), organic acids, antibiotics (pyrrolnitrin, phenazines, pyoluteorin, 2,4-diacetylphloroglucinol and oomycin A), cell wall-degrading enzymes (chitinase and cellulase), biosurfactants, exopolysaccharides, antioxidative enzymes, 1-aminocyclopropane-1-carboxylate (ACC) deaminase, secondary metabolites, volatile compounds (ammonia and hydrogen cyanide), siderophores, etc., and involves in mineral solubilization (phosphate, potassium, zinc and iron), nitrogen fixation, plant SAR induction, etc. (Sangiorgio et al. 2020, Aamir et al. 2020, Basu et al. 2021, Kumar et al. 2017). The application of multifunctional *P. fluorescens* is recommended for cereal (rice, wheat and maize), millet, sugarcane, legume (soya bean, Bengal gram and red gram), tomato, chilli, onion, cucumber, turmeric, mulberry, mustard,

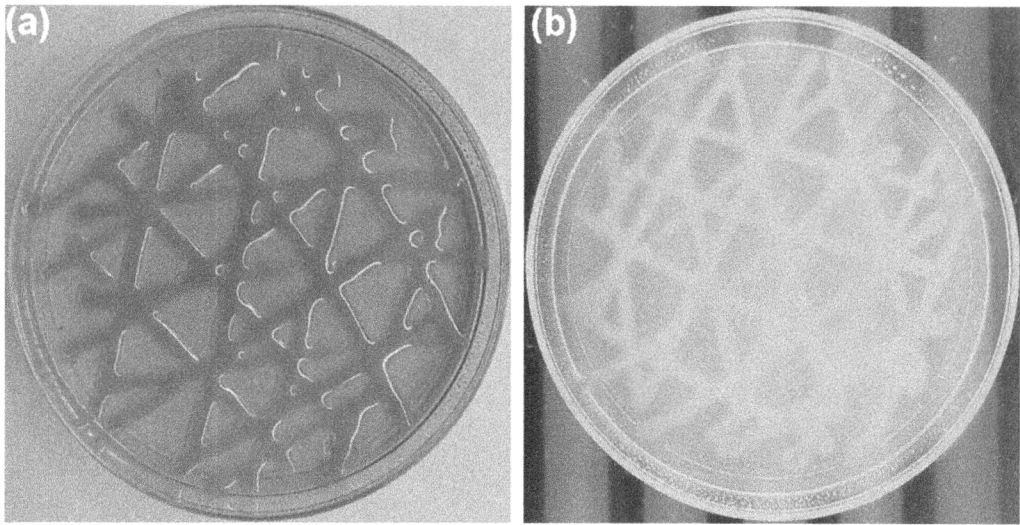

FIGURE 12.8   The *P. fluorescens* culture showing (a) yellow green fluorescein pigment in King's B medium and (b) UV fluorescence at 254 nm.

carrot, and horticulture and floriculture crops (Basu et al. 2021, Rao et al. 2016, Kumar et al. 2017). The King's B medium is used for selective isolation and identification of *P. fluorescens* from different sources (Scales et al. 2014), and the production of UV fluorescent pigment, fluorescein, by *P. fluorescens* is marked from typical UV fluorescence at 254 nm (Figure 12.8). The Indian commercial brands of *P. fluorescens* are Agnee, Monas, Pseudo-FL, Pseudogain, Pseudo, Pseudon-F, Bioshield, Pseudocon, Sudo, Jawahar *Pseudomonas*, Shakti *Pseudomonas*, Sparsha, SudomonRekha, Sudozone, Arka Nehar *Pseudomonas*, Samridhi Mildew, Power All, Bactvipe, etc.

12.4.2   *BACILLUS* SP.

The genus *Bacillus* includes Gram-positive, biofilm-forming, soil bacteria, and some of the *Bacillus* species which function as biofertilizers also exhibit biocontrol action against phytopathogenic bacteria and fungi. The essential qualities of *Bacillus* species are widespread distribution in a diverse environment, sporulation ability, survival capability under adverse conditions (heat and desiccation), biofertilizer action and safety. The important *Bacillus* species with known biocontrol action are *B. subtilis*, *B. amyloliquefaciens*, *B. cereus*, *B. sphaericus* and *B. pumilus* (Figure 12.9) (Amruta et al. 2016, Hashem, Tabassum, and Fathi Abd Allah 2019, Amruta et al. 2018). They are commercialized under the brand names Abacil, Bacil, Tacre Bacillus Plus 2 SC, Stanes Sting, Subwell, Samridhi Bio Downking, Bacilin, Bactus, Baciforte, Mildown, Bio Plus-Bacillus, RhizoPlus, Subilex, etc., in India (Junaid et al. 2013).

The antimicrobial action of *B. subtilis* and *B. amyloliquefaciens* against various phytopathogens such as *X. oryzae, Aspergillus flavus, M. oryzae, Rhizoctonia solani, Sclerotium, Sclerotinia, F. oxysporum, Corticium, Pythium, Oidiopsis, Leveillula, Phakopsora, Alternaria, Botrytis, Phytophthora, Peronospora* and *Rosellinia necatrix* is well known, and they are used for controlling a wide variety of plant diseases including root rot, seedling rot, root wilt, early blight, late blight, leaf spot, leaf blight, downy mildew and rot stalk in many crops. The antagonistic action of various *Bacillus* species against plant pathogens is attributed to diverse mechanisms. They include expression of antimicrobial peptide genes and production of antifungal metabolites (bacillin, bacitracin, bacilysin, bacillomycin, fengycin, iturin, subtilin and surfactin), extracellular enzymes (proteases, cell wall lyases and lipases), siderophores, phytohormones, hydrogen cyanide, antioxidants and secondary metabolites; biofilm formation; nitrogen fixation; mineral solubilization; nutrient and root

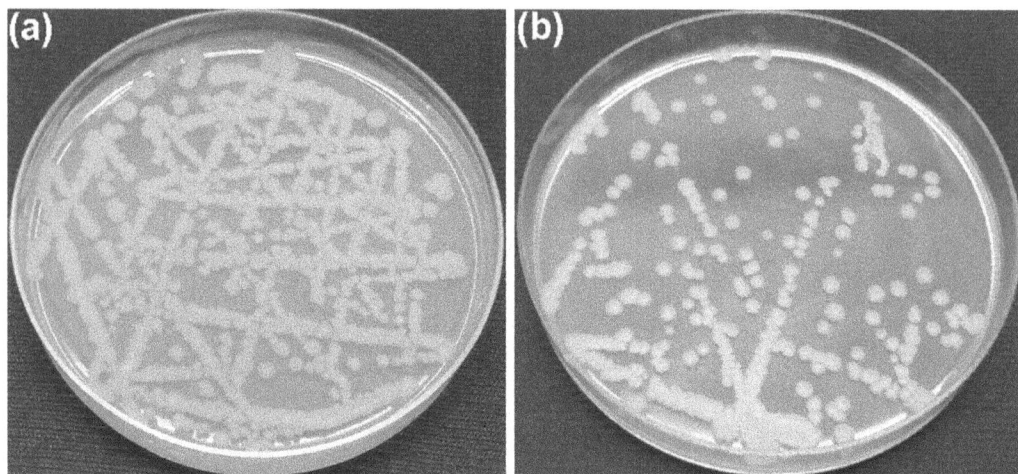

FIGURE 12.9   The colonies of *Bacillus* species in nutrient agar: (a) *B. subtilis* and *(b) B. cereus.*

colonization competition; induction of SAR; stress tolerance enhancement (exopolysaccharide and ACC deaminase production), etc. The application of biofungicide *Bacillus* is suitable for cereal, millet, legume, sugarcane, oilseeds, fibre, forage, plantation, horticulture, olericulture, floriculture, medicinal, aromatic, ornamental crops (Amruta et al. 2018, Hashem, Tabassum, and Fathi Abd Allah 2019, Prasanna Kumar et al. 2017, Amruta et al. 2016, Rajput et al. 2019, Basu et al. 2021, Junaid et al. 2013).

12.4.3 *TRICHODERMA* SP.

Trichoderma is one of the most important, extensively used, antagonistic, filamentous fungi (*Hypocreales* order; Ascomycotina division) in agriculture for enhancing soil fertility and controlling phytopathogens, insects and nematodes, as a part of sustainable agriculture and food production (Ferreira and Musumeci 2021, Srivastava et al. 2016). The biofungicide *Trichoderma* is widely used for controlling seed- and soilborne phytopathogens. In India, the commercially dominant biocontrol species of *Trichoderma* are *T. viride, T. harzianum, T. virens* and *T. hamatum.* For being a successful biocontrol agent, the *Trichoderma* species should exhibit qualities such as high sporulation rate; rapid and facile mass-multiplication; broad antimicrobial spectrum; greater plant growth; superior biocontrol efficiency, rhizospheric competence, saprophytic activity, shelf life, tolerance to high and low temperatures, desiccation, oxidizing environment, field conditions and UV radiation; compatibility with other bioagents; non-toxicity to the environment, plants and non-target organisms, etc. (Kumar, Thakur, and Rani 2014, Bashan, Trejo, and de-Bashan 2011).

It can be mass-multiplied at farmer level or commercially either by solid (inoculating the grains of cereals (wheat, maize and sorghum) and millets) or liquid (potato dextrose broth, molasses yeast medium, molasses soy medium and jaggery soy medium wheat bran) fermentation (Figure 12.10) and formulated with different organic and inorganic (talc) carriers (Kumar, Thakur, and Rani 2014, Kumar et al. 2014, Srivastava et al. 2016, Bindu Madhavi, Naveen Kumar, and Srinivas Rao 2019). In India, the *Trichoderma* formulations are available under different commercial names such as Arka Nehar *Trichoderma*, Biocon, Bioguard, Bip T, Bioderma, Bhoomika, Biosar, Ecoderma, Defense SF, Ecofit, Funginil, F-Stop, Hariz, Jawahar *Trichoderma*, Plant biocontrol agent-1, Soilguard, Trichoguard, Tricho-N, Trico-H and Tricho-X (Junaid et al. 2013, Kumar, Thakur, and Rani 2014). It can be delivered through seed treatment (seed dressing with a dispersing agent), seed biopriming, seedling immersion/dip in diluted solution, soil application either directly or after mass-multiplication in farmyard manure for 10–15 days, and foliar spray. The application of mixed

FIGURE 12.10 The growth of (a) *T. viride* on potato dextrose agar and (b) *T. harzianum* on sorghum grains.

formulations containing different native strains of *Trichoderma* and *Trichoderma* with other bio-control agents showed better control of plant diseases (Kumar, Thakur, and Rani 2014, Srivastava et al. 2016, Naguri et al. 2020).

It is recognized as an antagonistic agent against an array of plant pathogenic fungi such as *A. flavus*, *A. niger*, *Alternaria alternata*, *Armillaria mellea*, *Botrytis theobromae*, *B. cinerea*, *Cercospora moricola*, *Colletotrichum falcatum*, *Elsinoe fawcettii*, *F. moniliforme*, *F. solani*, *F. udum*, *F. oxysporum*, *Gaeumannomyces*, *Macrophomina phaseolina*, *Penicillium*, *P. aphanidermatum*, *P. infestans*, *Pythium ultimum*, *R. solani*, *Sclerotinia sclerotiorum*, *S. rolfsii*, *Thielaviopsis basicola* and *Ustilago segetum* (Gangadharan and Jeyarajan 1990, Kumar, Thakur, and Rani 2014, Srivastava et al. 2016). The biocontrol action of *Trichoderma* towards phytopathogenic fungi includes diverse mechanisms such as cell wall-degrading hydrolytic enzyme secretion, mycoparasitism, competition, antibiotic and phytohormone production, and induction of SAR in plants. The biofungicide *Trichoderma* competes with other plant pathogenic fungi for nutrients due to its rapid root colonization potential and parasitizes on other fungi, known as mycoparasitism. The various species of *Trichoderma* produce an array of cell wall-degrading/lytic enzymes, including chitinases, glucanases, proteases, etc., during the mycoparasitic action towards the pathogenic fungi. After the initial pathogen attack, plants exhibit an immune response against the pathogens known as SAR via the expression and production of β-1,3-glucanase, chitinases and endochitinases. The peptides, proteins and low molecular weight compounds synthesized by *Trichoderma* are known to elicit SAR in plants. The antagonistic fungus also produces various antimicrobial secondary metabolites and antibiotics such as trichodermamides, trichothecenes, viridiofungin A, spironolactone and dermadin; pathogen-inhibiting volatile compounds such as hydrogen cyanide, aldehydes and ketones; and phytohormones. Most significantly, the *Trichoderma* is resistant to many commercially used agricultural fungicides, herbicides and pesticides (Kumar, Thakur, and Rani 2014, Kumar et al. 2014, Kamala et al. 2015, Srivastava et al. 2016, Sangiorgio et al. 2020).

12.5 POPULARIZATION OF MICROBIAL STIMULANTS

Under All India Network Project on Biofertilizers and other programmes, various state agriculture universities, central agriculture research institutes and public sector undertaking companies such as Tamil Nadu Agricultural University, Acharya N. G. Ranga Agricultural University, Professor Jayashankar Telangana State Agricultural University, University of Agricultural Sciences, Chaudhary Charan Singh Haryana Agricultural University, Jawaharlal Nehru Krishi Vishwa Vidyalaya, Odisha University of

Agriculture and Technology, Maharana Pratap University of Agriculture and Technology, Assam Agricultural University, Kerala Agricultural University, Indian Institute of Horticulture Research, National Institute of Plant Health Management, National Fertilizer Ltd, Orissa Agro Industries Corporation Ltd and Brahmaputra Valley Fertilizer Corporation Ltd are promoting and popularizing the use of various microbial biostimulants, biofertilizers and biocontrol agents in agriculture by conducting farmer training, demonstration programmes on their production and mass-multiplication and supplying through the sales counters (Figure 12.11). Also, a large number of private, commercial fertilizer, biotechnology and cooperative companies such as Indian Farmers Fertilizer Cooperative Ltd, Biostadt, Varsha Bioscience and Technology India Pvt Ltd, Niku Bio-Research Lab, Green Agri Biotech and Orgaman R&D Division Ltd are marketing them in large scale (Selva Kumar, Sridhar, and Jaya Lakshmi 2019, Srinivas Reddy 2019, Ram Babu 2018, Rao et al. 2016).

The crucial constraints in the commercialization of biocontrol agents are scaling up, quality control (contamination-free biocontrol agent), long-term stability or shelf life, field persistence, efficacy at farmer field demonstration, development of aerial/foliar formulations and registration process (Ferreira and Musumeci 2021, Kumar, Thakur, and Rani 2014).

FIGURE 12.11 The talc-based biofertilizers and biocontrol agents produced by the University of Agricultural Sciences, Bangalore: (a) *Rhizobium*, (b) *Azotobacter*, (c) *Azospirillum*, (d) phosphate-solubilizing bacteria, (e) potassium-solubilizing bacteria and (f) *T. viride*.

12.6 CONCLUSIONS

The current review provides concise information on commercially used microbial biostimulants for agriculture in India, in terms of biofertilizers and biocontrol agents. It is important to note that some of the microbial biostimulants serve as multifunctional agents through diverse mechanisms. As the demand for microbial biostimulants is burgeoning, many fake and spurious products are flooding the market. The application of microbial biostimulants by farmers are limited by the availability of genuine/quality products, high cost, colonization potential, variable field efficacy, real-time monitoring, lack of knowledge on dosage, application, etc. (García-Seco 2019). The authorized regulatory bodies should concentrate on legislation enforcement, quality control inspections, continuous monitoring, pricing, etc. Further studies are envisaged on multistrain formulation design, multilocational and multicrop field trails, etc.

ACKNOWLEDGEMENTS

The author would like to acknowledge Dr. Mullapudi Venkata Balarama Krishna, Head, Environmental Science and Nanomaterials Section, and Dr. Sanjiv Kumar, Head, NCCCM, BARC, Hyderabad, for their constant support and encouragement throughout the study.

REFERENCES

Aamir, M., K.K. Rai, A. Zehra, M.K. Dubey, S. Kumar, V. Shukla, and R. S. Upadhyay. 2020. "8- Microbial bioformulation-based plant biostimulants: A plausible approach toward next generation of sustainable agriculture." In *Microbial Endophytes*, edited by A. Kumar and E.K. Radhakrishnan, 195–225. Sawston: Woodhead Publishing.

Adidam, N. 2021. The Gazette of India. edited by Cooperation and Farmers Welfare Department of Agriculture. New Delhi: Controller of Publications.

Adlakha, N., R. Rajagopal, S. Kumar, V. S. Reddy, and S. S. Yazdani. 2011. "Synthesis and characterization of chimeric proteins based on cellulase and xylanase from an insect gut bacterium." *Applied and Environmental Microbiology* 77 (14):4859–66. doi: 10.1128/AEM.02808-10.

Amruta, N., M. K. Prasanna Kumar, S. Narayanaswamy, B. Malali Gowda, C. Channakeshava, K. Vishwanath, M. E. Puneeth, and H. P. Ranjitha. 2016. "Isolation and identification of rice blast disease-suppressing antagonistic bacterial strains from the rhizosphere of rice." *Journal of Pure and Applied Microbiology* 10 (2):1043–1054.

Amruta, N., M. K. Prasanna Kumar, M. E. Puneeth, G. Sarika, H. K. Kandikattu, K. Vishwanath, and S. Narayanaswamy. 2018. "Exploring the potentiality of novel rhizospheric bacterial strains against the rice blast fungus *Magnaporthe oryzae*." *Plant Pathology Journal* 34 (2):126–138. doi: 10.5423/PPJ. OA.11.2017.0242.

Bashan, Y., and L. E. de-Bashan. 2015. "Inoculant preparation and formulations for *Azospirillum* spp." In *Handbook for Azospirillum: Technical Issues and Protocols*, edited by F. D. Cassán, Y. Okon and C. M. Creus, 469–485. Cham: Springer International Publishing.

Bashan, Y., A. Trejo, and L. E. de-Bashan. 2011. "Development of two culture media for mass cultivation of *Azospirillum* spp. and for production of inoculants to enhance plant growth." *Biology and Fertility of Soils* 47 (8):963–969. doi: 10.1007/s00374-011-0555-3.

Basu, A., P. Prasad, S. N. Das, S. Kalam, R. Z. Sayyed, M. S. Reddy, and H. E. Enshasy. 2021. "Plant growth promoting rhizobacteria (PGPR) as green bioinoculants: Recent developments, constraints, and prospects." *Sustainability* 13 (3):1140. doi: 10.3390/su13031140.

Bindu Madhavi, G., and P. Naveen Kumar. 2018. "The role of biofertilizers in crop cultivation." *Prakruti Nestam* 5 (10):11–12.

Bindu Madhavi, G., P. Naveen Kumar, and Ch. S. Rao. 2019. "*Trichoderma* production at farmer's level." *Annadata* 5 (3):14–15.

Biswas, S., and M. K. Shivaprakash. 2020. "Comparative evaluation of growth parameters, germination percentage and seedling vigour of tomato and potato sedlings co inoculated with PSB, KMB and KSB isolates under green house condition." *International Journal of Agriculture Science and Research* 10 (2):51–58.

Dhevendaran, K., G. Preetha, and B. N. Vedha Hari. 2013. "Studies on nitrogen fixing bacteria and their application on the growth of seedling of *Ocimum sanctum*." *Pharmacognosy Journal* 5 (2):60–65.

du Jardin, P. 2015. "Plant biostimulants: Definition, concept, main categories and regulation." *Scientia Horticulturae* 196:3–14.

Eenadu. 2021. "Now bistimulants are banned!" *Eenadu*, 25th February 2021, 13.

Ferreira, F. V., and M. A. Musumeci. 2021. "*Trichoderma* as biological control agent: Scope and prospects to improve efficacy." *World Journal of Microbiology and Biotechnology* 37 (90). doi: 10.1007/s11274-021-03058-7.

Gangadharan, K., and R. Jeyarajan. 1990. "Mass multiplication of *Trichoderma* spp." *Journal of Biological Control* 4 (1):70–71.

García-Seco, D. 2019. "The importance of applying a high quality microbial biostimulant." *AgriBusiness Global*, 29th March 2019. Accessed 12th January 2021.

Hashem, A., B. Tabassum, and E. Fathi Abd Allah. 2019. "*Bacillus subtilis*: A plant-growth promoting rhizobacterium that also impacts biotic stress." *Saudi Journal of Biological Sciences* 26 (6):1291–1297. doi: 10.1016/j.sjbs.2019.05.004.

Junaid, J. M., N. A. Dar, T. Bhat, and A. H. Bhat. 2013. "Commercial biocontrol agents and their mechanism of action in the management of plant pathogens." *International Journal of Modern Plant & Animal Sciences* 1 (2):39–57.

Kamala, Th., S. Indira Devi, K. Chandradev Sharma, and K. Kennedy. 2015. "Phylogeny and taxonomical investigation of *Trichoderma* spp. from Indian region of Indo-Burma biodiversity hot spot region with special reference to Manipur." *BioMed Research International* 2015:285261. doi: 10.1155/2015/285261.

Kannan, K., V. Selvi, D.V. Singh, O.P.S. Khola, R. Mohanraj, and A. Murugesan. 2013. Coir pith composting-An alternate source of organic manure for rainfed maize. edited by Central Soil & Water Conservation Research & and Research Centre Training Institute. Udhagamandalam: Indian Council of Agricultural Research.

Karthikeyan, B., and P. Sivasakthivelan. 2021. "Practical manual cum record." *Biofertilizer Technology* 227:1–40. Department of Agricultural Microbiology, Faculty of Agriculture, Annamalai University CAG – AGM.

Kasana, R.C., R. Salwan, H. Dhar, S. Dutt, and A. Gulati. 2008. "A rapid and easy method for the detection of microbial cellulases on agar plates using Gram's iodine." *Current Microbiology* 57 (5):503–507. doi: 10.1007/s00284-008-9276-8.

Kora, A.J. 2021. "Applications of waste decomposer in plant health protection, crop productivity and soil health management." In *Application of Microbes in Environmental and Microbial Biotechnology*, edited by Inamuddin, A.M. Imran and P. Ram. Springer, Singapore.

Kosaraju, C.R. 2018. "Types of biofertilizers-Usage methods." *Raitu Nestam* 13 (6):47–50.

Kumar, A., H. Verma, V. K. Singh, P. P. Singh, S. K. Singh, W. A. Ansari, A. Yadav, P. K. Singh, and K. D. Pandey. 2017. "Role of *Pseudomonas* sp. in sustainable agriculture and disease management." In *Agriculturally Important Microbes for Sustainable Agriculture: Volume 2: Applications in Crop Production and Protection*, edited by V. S. Meena, P. K. Mishra, J. K. Bisht, and A. Pattanayak, pp. 195–215. Singapore: Springer.

Kumar, S., M. Thakur, and A. Rani. 2014. "*Trichoderma*: Mass production, formulation, quality control, delivery and its scope in commercialization in India for the management of plant diseases." *African Journal of Agricultural Research* 9 (53):3838–3852.

Kumar, S., P. D. Roy, M. Lal, G. Chand, and V. Singh. 2014. "Mass multiplication and shelf life of *Trichoderma* species using various agroproducts." *The Bioscan* 9 (3):1143–1145.

Luyckx, M., J.-F. Hausman, S. Lutts, and G. Guerriero. 2017. "Silicon and plants: Current knowledge and technological perspectives." *Frontiers in Plant Science* 8 (411). doi: 10.3389/fpls.2017.00411.

Madeiras, A., and J. Lanier. 2020. "What are biostimulants?". University of Massachusetts Amherst Last Modified 10th October 2020. Accessed 12th January 2021.

Mumtaz, M. Z., M. Ahmad, M. Jamil, and T. Hussain. 2017. "Zinc solubilizing *Bacillus* spp. potential candidates for biofortification in maize." *Microbiological Research* 202:51–60 doi: 10.1016/j.micres.2017.06.001.

Nagaraju, Y., S. Triveni, A. Vijaya Gopal, G. Thirumal, B. Prasanna Kumar, and P. Jhansi. 2017. "*In vitro* screening of Zn solubilizing and potassium releasing isolates for plant growth promoting (PGP) characters." *Bulletin of Environment, Pharmacology and Life Sciences* 6 (SI 3):590–597.

Naguri, S., S. Firdouse, J. Nayak, S. Swathi, and Sridhar. 2020. "Benefits of sedd dressing treatment." *Annadata* 52 (6):14–15.

Prabhu, S.R., and V. Thomas George. 2002. "Biological conversion of coir pith into a value-added organic resource and its application in Agri-Horticulture: Current status, prospects and perspective." *Journal of Plantaion Crops* 30 (1):1–17.

Prasanna Kumar, M. K., N. Amruta, C. P. Manjula, M. E. Puneeth, and K. Teli. 2017. "Characterisation, screening and selection of *Bacillus subtilis* isolates for its biocontrol efficiency against major rice diseases." *Biocontrol Science and Technology* 27 (4):581–599. doi: 10.1080/09583157.2017.1323323.

Raffi, M. Md., and P. B. B. N. Charyulu. 2012. "Nitrogen fixation by the native *Azospirillum* spp. isolated from rhizosphere and non-rhizosphere of foxtail millet." *Asian Journal of Biological and Life Sciences* 1 (3):213–218.

Raja Sekhar, P., A. Srinivas, S. Adarsha, K. C. Bhanu Murty, and S. V. Rani. 2018. "Biofertilizers-their main role in the soil." *Prakruti Nestam* 5 (1):9–12.

Rajabi Hamedani, S., Y. Rouphael, G. Colla, A. Colantoni, and M. Cardarelli. 2020. "Biostimulants as a tool for improving environmental sustainability of greenhouse vegetable crops." *Sustainability* 12 (12):5101. doi: 10.3390/su12125101.

Rajawat, M. V. S., S. Singh, S. P. Tyagi, and A. K. Saxena. 2016. "A modied plate assay for rapid screening of potassium-solubilizing bacteria." *Pedosphere* 26 (5):768–773.

Rajput, R. S., R. M. Ram, A. Vaishnav, and H. B. Singh. 2019. "Microbe-based novel biostimulants for sustainable crop production." In *Microbial Diversity in Ecosystem Sustainability and Biotechnological Applications: Volume 2. Soil & Agroecosystems*, edited by T. Satyanarayana, S. K. Das and B. N. Johri, 109–144. Singapore: Springer.

Ram Babu, E. 2018. "Biofertilizers for crop nutrition and high yields." *Prakruti Nestam* 5 (5):49–50.

Rao, M. S., R. Umamaheswari, R. Rajinikanth, M. K. Chaya, and P. Prabu. 2016. "Microbial consortia based biopesticides for sustainable nematode management in horticutural crops." 2nd International sumposium on new processes & applications for plant and microbial products, New Delhi, India, 1st & 2nd March, 2016.

Rouphael, Y., M. Cardarelli, P. Bonini, and G. Colla. 2017. "Synergistic action of a microbial-based biostimulant and a plant derived-protein hydrolysate enhances lettuce tolerance to alkalinity and salinity." *Frontiers in Plant Science* 8 (131):1–12.

Rouphael, Y., and G. Colla. 2020. "Editorial: Biostimulants in agriculture." *Frontiers in Plant Science* 11 (40):1–7. doi: 10.3389/fpls.2020.00040.

Sangiorgio, D., A. Cellini, I. Donati, C. Pastore, C. Onofrietti, and F. Spinelli. 2020. "Facing climate change: Application of microbial biostimulants to mitigate stress in horticultural crops." *Agronomy* 10 (6):794. doi: 10.3390/agronomy10060794.

Scales, B.S., R. P. Dickson, J. J. LiPuma, and G. B. Huffnagle. 2014. "Microbiology, genomics, and clinical significance of the *Pseudomonas fluorescens* species complex, an unappreciated colonizer of humans." *Clinical Microbiology Reviews* 27 (4):927–948. doi: 10.1128/CMR.00044-14.

Selva Kumar, G., G. Sridhar, and M. Jaya Lakshmi. 2019. "New microbial consortium for horticuture crops." *Annadata* 51 (3):37.

Singh, S., B.R. Maurya, and I. Bahadur. 2018. "Solubilization of potassium containing various K-mineral sources by K-solubilizing bacterial isolates on Aleksandrov medium." *International Journal of Current Microbiology and Applied Sciences* 7 (03):1142–1151. doi: 10.20546/ijcmas.2018.703.136.

Srinivas Reddy, K. 2019. "PGPR conference on sustainable and organic agriculture in Guntur." *Annadata* 51 (4):54–55.

Srivastava, M., V. Kumar, M. Shahid, S. Pandey, and A. Singh. 2016. "*Trichoderma*-a potential and effective bio fungicide and alternative source against notable phytopathogens: A review." *African Journal of Agricultural Research* 11 (5):310–316.

Sumbul, A., R. A. Ansari, R. Rizvi, and I. Mahmood. 2020. "*Azotobacter*: A potential bio-fertilizer for soil and plant health management." *Saudi Journal of Biological Sciences* 27 (12):3634–3640. doi: 10.1016/j.sjbs.2020.08.004.

Swain, R., and G. R. Rout. 2017. "Silicon in agriculture." In *Sustainable Agriculture Reviews*, edited by E. Lichtfouse, 233–260. Cham: Springer International Publishing.

Vempalli, L. R. 2021. "Importance of silicon for crops." *Annadata* 53 (4):48–49.

Vijay, J., and G. Tirumal. 2018. "Biofertilizers-their importance in agriculture." *Prakruti Nestam* 5 (4):43–45.

13 Microbial Biostimulants for Crop Production
Industry Advances, Bottlenecks, and Future Prospects

B.N. Aloo
University of Eldoret

E.R. Mbega
Nelson Mandela African Institution of Science and Technology

J.B. Tumuhairwe
Makerere University

B.A. Makumba
Moi University

CONTENTS

13.1 INTRODUCTION

The number of people in the world is expected to hit approximately 10 billion by 2025 (United Nations, Department of Economic and Social Affairs, Population Division 2015). Providing food for this gigantic population will require that agricultural production be increased by about 60% in

DOI: 10.1201/9781003188032-13

the coming 40 years to match the increasing food demand (FAO 2018). What is worrying is that the size of arable land will not increase in the same proportion because of the rapid globalization and modern industrialization trends that continue to shrink the arable lands (Eickhout, Bouwman, and Van Zeijts 2006). This means that other avenues of increasing food production other than the expansion of arable lands must be pursued.

Agricultural intensification conventionally relies on the use of artificial fertilizers (Aeron et al. 2020). In 2011, the total fertilizer consumption was estimated at 176 million tons and projected to further increase by the year 2050 in efforts to feed the growing world population (FAO 2011; Blanco 2011). Furthermore, the use of nitrogen (N), potassium (K), and phosphorus (P) fertilizers was estimated to have increased from about 18, 26, and 65 kg/ha, respectively, in 2000 to about 20, 33, and 86, 33 kg/ha, respectively, in 2014 after the global population surpassed 7.2 million (FAO 2013). The continuous and indiscriminate use of artificial fertilizers, however, continues to elicit a lot of debate worldwide. Fertilizer runoffs from agricultural fields are the major cause of the pollution and eutrophication of surface water bodies (Gouda et al. 2018). Besides, artificial fertilizers have also been linked to greenhouse gas emissions (Vejan et al. 2016) and acidification of arable lands (Zhu et al. 2018; Han et al. 2015; Goulding 2016). Cognizant of this, the continued use of chemical fertilizers is neither environmentally nor economically feasible and there is need for alternative and sustainable approaches to crop fertilization (Sukul et al. 2021; Aloo, Tripathi et al. 2021).

The past few decades have realized the development of technological tools to promote sustainable agroecosystems (Peter, Amalraj, and Talluri 2020). One such technology is the use of microbial biostimulants, which are majorly preparations or formulations of PGP rhizobacteria (PGPR) for the stimulation of plant growth (Fasusi and Babalola 2021; Aloo, Makumba, and Mbega 2021). Literature advances that these bacteria can improve the growth of plants through several biochemical processes such as increased nutrient availability and solubility, production of plant regulatory hormones, amelioration of abiotic stresses, and phytopathogen control (Fasusi and Babalola 2021; Agustian et al. 2010; Bhat 2019). Research continues to advance the diversity, metabolism, functional attributes, and genetic potential of PGPR, and consequently, their development into microbial products/biostimulants and commercialization for crop production (Novello et al. 2021; Aeron, Dubey, and Maheshwari 2021; Vasseur-Coronado et al. 2021).

Biostimulants are emerging as vital components of sustainable agricultural systems, but there is neither an account of the commercially available microbial biostimulants, nor a consistent definition of these types of products. Besides, several bottlenecks continue to constrain the industry, but have not been articulated for relevant solutions to be sought. This chapter gives a comprehensive discussion on the concepts, definitions, and terminologies regarding microbial biostimulants and provides an overview of these biostimulants. The chapter further gives an account of the formulation and commercialization of these products and outlines the bottlenecks that face the industry. Finally, the chapter critically examines the future perspectives that can together pave the way for the advancement of the industry and application of microbial biostimulants for crop production and the development of sustainable agricultural systems globally.

13.2 CONCEPTS, DEFINITIONS, AND TERMINOLOGIES

The definition of biostimulants has rigorously been debated for several years now, and no single definition is uniformly available. Despite extensive discussions, to date, there is no consolidated definition of biostimulants (du Jardin 2015). The term 'biostimulant' was conceived by horticulture specialists to depict substances other than nutrients, pesticides, or soil improvers that can promote plant growth (du Jardin 2015). It is widely acknowledged that the 'bio' denotes the presence of living organisms, and/or their natural substances.

In general terms, biostimulants can be defined as eco-friendly and biological substances originating from microbial biomass or products and metabolites that can promote nutrient acquisition, plant growth, and/or phytopathogen management (Carletti et al. 2021). Based on this definition, microbial biostimulants may include macroalgae/seaweed extracts and all plant-beneficial microbes such as arbuscular mycorrhizal fungi and PGPR (Ahemad and Kibret 2014). A biostimulant can similarly be described as 'a bio-product that enhances plant growth and productivity with several variants' (Table 13.1).

The concept of 'biostimulants' and the complexity of the term are extensively discussed by du Jardin et al. (2020). The widely accepted definition of microbial plant biostimulants is 'micro-organisms that when administered to plants or plant rhizospheres, induce processes that promote nutrient availability, plant growth, and abiotic stress tolerance' (du Jardin 2015). This is also the definition advanced by the European Biostimulant Industry Council (EBIC) (http://www.biostimulants.eu). In general, the term 'biostimulant' appears to signify any substance other than nutrients and pesticides that is beneficial to plants. Somewhat, biostimulants are primarily described by what they are not, to draw a line between them and other substances such as pesticides and fertilizers that are widely used to improve plant growth (du Jardin 2015).

Regardless of the several definitions and concepts that exist for biostimulants, the actual product often comprises an active ingredient in the form of living microbes, and their spores or other products. Generally, biostimulants do not have any apparent action against pests (Dipak and Aloke 2020; du Jardin 2015). Based on these concepts, we propound a simple definition of a biostimulant in this chapter to be 'a biological product formulated to improve plant productivity as a result of increased nutrient availability through various mechanisms, and plant growth stimulation via plant growth-regulating hormones.'

TABLE 13.1
Diversity of Concepts, Terminologies, and Definitions Related to Biostimulants

Definition	Reference
A product that when applied to plants, promotes growth and/or development without the addition of nutrients	Schmidt, Ervin, and Zhang (2003)
A natural/synthetic product that positively contributes to plant nutrition and health by different modes of action	Dixon and Walsh (2002)
Agents that advance the plant/soil biochemical processes, and thereby plant growth and pest resistance	Basak (2008)
Materials that are neither fertilizers nor pesticides that enhance plant growth, health, and protection when applied	Banks and Percival (2012)
Substances other than pesticides and nutrients that when introduced to soil, seeds, or plants, can modify plant physiological processes and enhance their growth and response to stress	du Jardin (2012)
Substances and/or microbes that when applied to plant rhizospheres, stimulate nutrient uptake and abiotic stress tolerance	EBIC (2012)
Organic materials that when applied, enhance plant growth and development which cannot be attributed to the addition of nutrients	Sharma et al. (2014)
Any substance/microorganism that when introduced to seeds, soil, or plant roots, stimulates abiotic stress tolerance and nutrient use efficiency	Traon et al. (2014)
A substance/microorganism introduced to plants to enhance nutrient use efficiency, and abiotic stress tolerance, and by extension, commercial products comprising mixtures of such	du Jardin (2015)
Substance or materials other than nutrients/pesticides that when applied to plants or seeds, in specific formulations can modify plant physiological processes that potentially benefit the plants in terms of growth and stress tolerance	Halpern et al. (2015)
A material that is neither a pesticide nor a fertilizer but when introduced to a plant enhances its health and growth	Lovatt (2015)
Microorganisms which enhance crop growth/yield by improving the availability of plant nutrients and growth hormones/factors	Vessey (2003)
Biological or biologically derived fertilizer additives that are used in crop production to improve plant growth and productivity	Nori et al. (2019)

In the context of microbial biostimulants, the term 'biostimulants' can be used interchangeably with other terms such as phytostimulators, biofertilizers, bioformulations, or bioinoculants, and today, microbial biostimulants are marketed under these different trade names. To sum up this section, although a uniform definition of biostimulants has so far been lacking, the general consensus is that they are stimulators of plant processes, resulting in enhanced nutrient use efficiency and growth. In this chapter, we focus on microbial biostimulants and, specifically, the PGPR.

13.3   OVERVIEW OF MICROBIAL BIOSTIMULANTS

Previous and current research on microbiomes has broadened our knowledge of the structure and complexity of microbial rhizobacterial communities. The academia and industry are now both intensely interested in biostimulants made of living microbes attributable to their immense abilities and potential to enhance plant growth. Plant-beneficial bacteria or PGPR are considered the most promising microbial biostimulants (Kumari et al. 2019; du Jardin 2015). The PGPR populate plant rhizospheres and execute many direct and indirect ecological functions for plants such as phytopathogen control and improved nutrient solubility and availability (Figure 13.1).

There is voluminous literature on the potential applications of plant rhizobacteria, particularly as PGP agents (Aloo, Mbega, et al. 2021; Bharadwaj et al. 2017; Gholami, Shahsavani, and Nezarat 2009). While many rhizobacteria are documented as PGP agents, the molecular basis of their PGP mechanisms is still not fully comprehended. However, whole-genome sequences of several plant rhizobacteria are now available (Aloo et al. 2020; Belbahri et al. 2017; Chen et al. 2020; Deng et al. 2011; Duan et al. 2013) and these molecular processes are now starting to be understood. With the development of more advanced sequencing technologies such as next-generation sequencing, we can now strive to catalog genome-wide variations among organisms and identify genome loci toward agriculturally important traits (Niazi 2014; Cole et al. 2017; Cheng et al. 2017).

The representative groups of PGPR-based plant biostimulants comprise *Rhizobium*, *Bacillus*, *Azospirillum*, *Pseudomonas*, and *Azotobacter* spp. (Bashir et al. 2021). Nitrogen-fixing PGPR such as rhizobia can accumulate plant-utilizable N through biological N fixation (BNF) in soil and plant roots (Monteiro et al. 2021). Thus, such PGPR greatly contribute to the production of organic fertilizers and biostimulants (Vejan et al. 2016). Some PGPR can solubilize phosphates (P) in soil (Gupta et al. 2021; Yazdani et al. 2009; Hussein and Joo 2015) and potassium (K) (H. S. Han and Lee 2006;

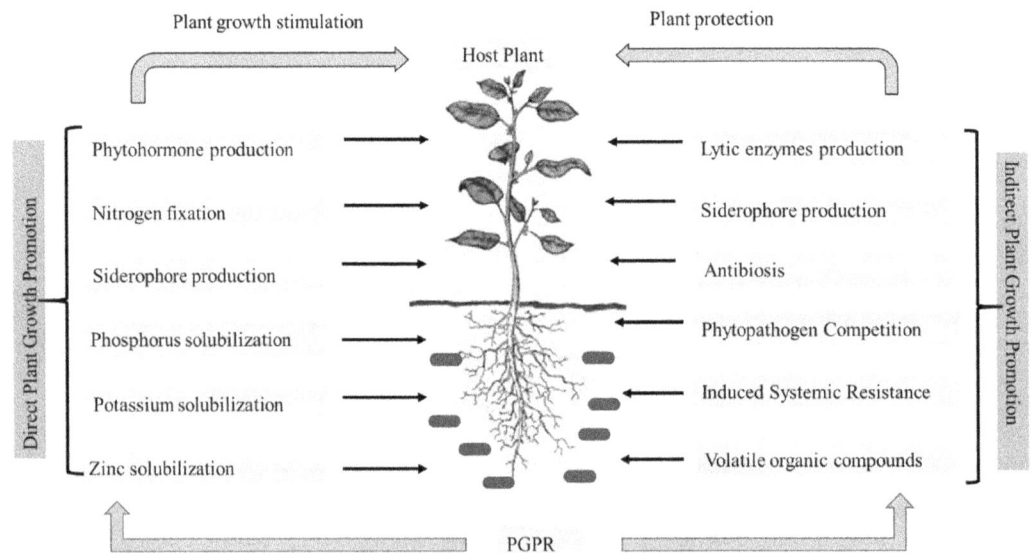

FIGURE 13.1   Beneficial functions of plant growth-promoting rhizobacteria in plant rhizosphere.

Dhaked et al. 2017; Wan et al. 2016) and improve their accessibility to plants. Considering the critical role of these nutrient elements in plants, such PGPR are also an integral part of biostimulants.

Hormones are also key players in plant growth regulation and are thus key candidates for the bioactivity of microbial biostimulants (Niazi et al. 2014; Mushtaq, Faizan, and Hussain 2021). According to Khatoon et al. (2020), phytohormones include auxins, gibberellins, cytokinins, and ethylene that together act as signaling molecules in plants and control several physiological processes such as cell elongation, tissue differentiation, seed germination, and plant defense at very low concentrations. The functions of rhizobacterial phytohormones are summarized in Figure 13.2. Bearing in mind that phytohormones can have significant effects on plant growth even at very low concentrations, rhizobacteria that produce them can be very useful in agriculture. Examples of reports on microbial biostimulants and enhanced crop production are detailed in Table 13.2.

13.4 FORMULATION AND COMMERCIALIZATION OF MICROBIAL BIOSTIMULANTS

The general processes and steps involved in the formulation and commercialization of microbial biostimulants are displayed in Figure 13.3. The formulation and commercialization of microbial biostimulants begin with the selection or screening of PGPR to identify suitable candidates for formulation.

Presently, over 100 companies are involved in the marketing of biostimulant/biofertilizer formulations of *Pseudomonas*, *Klebsiella*, *Bacillus*, *Azotobacter*, *Enterobacter*, and *Azospirillum* (Aloo, Makumba, and Mbega 2021). A summary of the commercially available biostimulants/biofertilizers in different countries is provided in Table 13.3.

The stepwise screening of biostimulants involves (i) the identification of target crop, (ii) growth medium selection, (iii) screening for PGP traits, (iv) screening for product development, (v) PGPR characterization, and (vi) valuation of plant growth efficacy as discussed by Vasseur-Coronado et al. (2021).

While some formulations may include fungal components alongside the PGPR strains, the preparations are majorly PGPR based. The most popular biostimulants are N-fixers, which represented

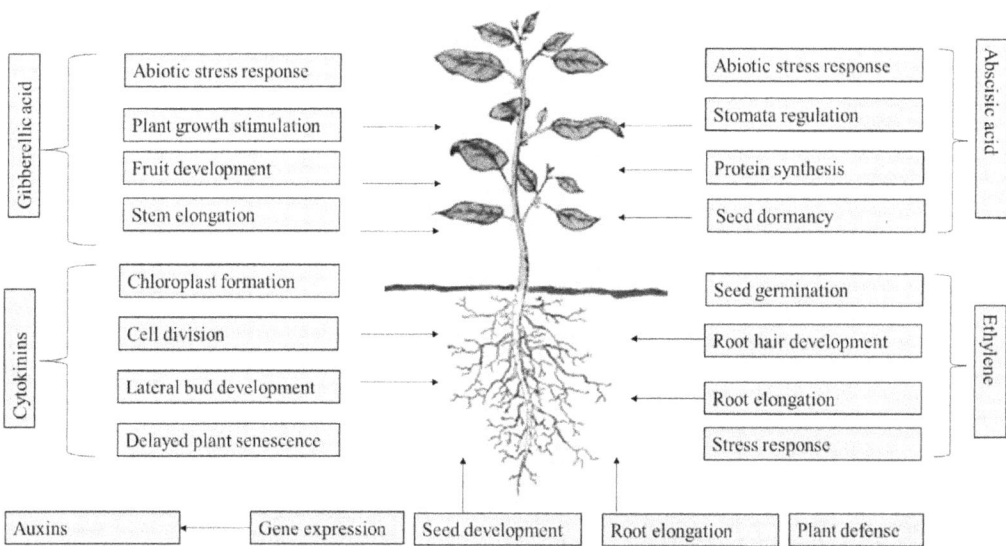

FIGURE 13.2 Effects of rhizobacterial phytohormones in plant development.

TABLE 13.2
Reports on Microbial Biostimulants and Enhanced Crop Production

Crop	Test Biostimulant	Growth Effect	Test Conditions	Reference
Barley (*Hordeum vulgare*)	*B. megaterium M3, Bacillus OSU-142, Azospirillum brasilense Sp.245, B. megaterium RC07, Raoultella terrigena, Paenibacillus polymyxa RC05, Burkholderia cepacia FS, B. licheniformis RC08*	Increased yields by up to 33.7% in combination with N fertilizer	Field	Cakmakci et al. (2014)
Chickpea (*Cicer arietinum*)	*Rhizobium* spp.	Increased yield attribute	Field	Verma and Yadav (2019)
Common Bean (*Phaseolus vulgaris*)	*Pseudomonas korrensis*, Rhizobia	Significantly improved plant height, biomass, nodulation, and macro-/microelements	Greenhouse	El-Nahrawy and Omara (2017)
Cotton (*Gossypium* spp.)	*Azotobacter chroococcum*	Increased growth with reduced fertilizer use	Glasshouse	Romero-Perdomo et al. (2017)
Kikuyu grass (*Pennisetum clandestinum*)	PGPR consortia	Increased growth	Field	Paungfoo-Lonhienne et al. (2019)
Mung bean (*Vigna radiata*)	*Bradyrhizobium, Streptomyces griseoflavus*	Increased growth, nodulation, nitrogen, phosphorus, and potassium uptake	Pots	Htwe, Moh, Soe, et al. (2019)
Rice (*Oryza sativa*)	*Rhizobium leguminosarum*	Increased shoot/root weights	Greenhouse	Kecskés et al. (2016)
Soybean (*Glycine max*)	*Streptomyces griseoflavus P4, Bradyrhizobium elkanii BLY3-8, B. japonicum SAY3-7*	Increased nutrient uptake, plant growth and yield, and nodulation	Pots	Htwe, Moh, Moe, et al. (2019)
		Increased nodulation, growth, and nitrogen, phosphorus, and potassium uptake	Pots	Htwe, Moh, Soe, et al. (2019)
Sugar beet (*Beta vulgaris*)	*Acinetobacter, Bacillus, Burkholderia* spp.	Increased yield and yield components compared with the untreated controls	Field	Hanan and El Sebai (2019)
Tomato (*Solanum lycopersicum*)	*Azospirillum lipoferum, Brevibacillus parabrevis*	Increased nutritional quality	Field	Oancea, Răut, and Zamfiropol-Cristea (2017)
Wheat (*Triticum aestivum*)	*B. velezensis GB03, Azospirillum brasilense 65B, B. megaterium SNj*	Increased plant biomass and nutrient contents	Greenhouse	Nguyen (2018)
	P. aeruginosa	Improved nutrient uptake, plant biomass, and leaf chlorophyll under Zn stress	Greenhouse	Islam et al. (2014)
	Arthrobacter sp., *Bacillus* sp.	Increased dry weight by 26% and 40% under 2 and 6 dS/m of salinity, respectively.	Greenhouse	Upadhyay and Singh (2015)
	Azospirillum spp. + *Azoarcus* spp. + *Azorhizobium* spp.	Enhanced nitrogen accumulation, heading, and stem elongation	Field	Dal Cortivo et al. (2020)
	B. megaterium M3, Raoultella terrigena, Azospirillum brasilense.245, P. polymyxa RC05, B. licheniformis RC08, B. cepacia FS, B. megaterium RC07	Increased root and shoot weights and yields by up to 40.4% field in combination with N fertilizer	Greenhouse and field	Cakmakci et al. (2014)

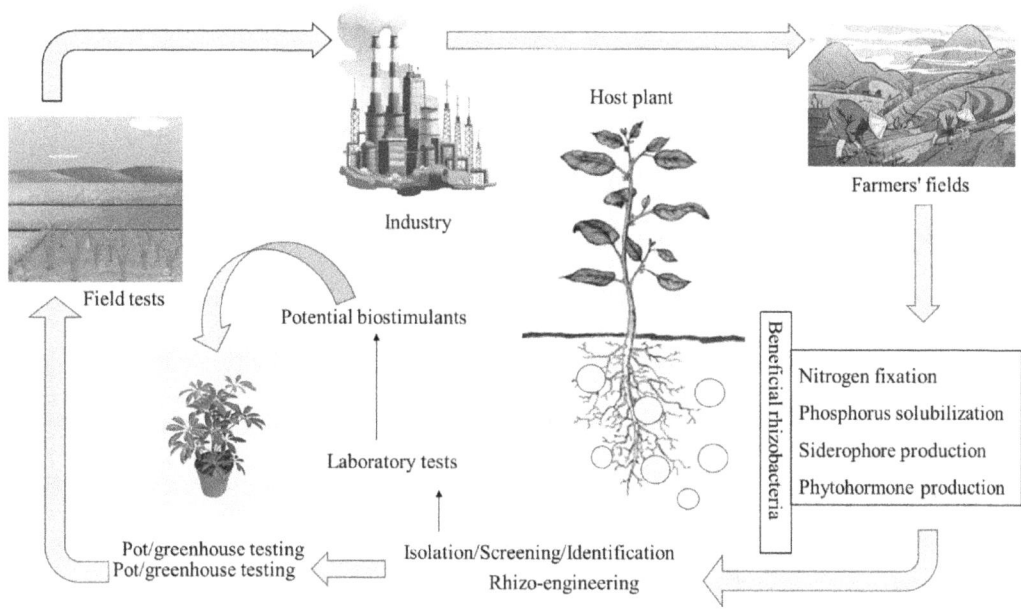

FIGURE 13.3 Processes and steps involved in the formulation and commercialization of microbial formulations.

about 77 and 79% of the world market demand in 2012 and 2017, respectively (Agro News 2014) (Figure 13.4). *Rhizobium*, *Azospirillum*, and *Azotobacter* spp. are the main commercially available N-fixing biostimulants. While these biostimulants are largely used for the production of legumes, they can also be used on cereals and other crops. Biostimulant products can consist of single PGPR strains, PGPR mixtures, or a mixture of PGP fungi and PGPR. Unlike single-strain products, products with microbial consortia can be more efficient due to greater genetic diversity and the ability to colonize roots faster (Ray et al. 2020; Rivett and Bell 2018).

The world biostimulant market was valued at USD 1.7 million about 5 years ago and is predicted to increase at 10.4% annually to reach about USD 3 billion in the current year and USD 4.14 billion by 2025 (Hamidi et al. 2021). Most developed countries employ biofertilizers/biostimulants to enhance the productivity of various crops. The European biostimulant market is not only the fastest growing, but also the most developed. The EBIC propounds that the European biostimulant market hit approximately 500 million about 10 years ago and is now estimated at 3 million ha at a stable annual growth rate of 10% or more. This is largely fueled by the rising demand for organic food in the region (Markets and Markets 2017). Developing countries are also coming up as biostimulant consumers. In India, over 150 biostimulant production units are operated by governmental and non-governmental bodies (Mahajan and Gupta 2009). This significant uptake of microbial biostimulants demonstrates the interest of farmers, academics, and agrochemical companies in them (Caradonia et al. 2019). Nevertheless, the global biofertilizer/biostimulant market still occupies only a small fraction of the market for artificial agrochemicals despite gathering momentum in recent years. Several bottlenecks have been identified as contributors to this low uptake. These are discussed in detail in Section 13.5.

13.5 BOTTLENECKS IN THE INDUSTRY

The successful production and delivery of quality microbial biostimulants to the market is still a daunting task. Despite their extensive diversity, genetic potential, genomic versatility, and multifaceted PGP functions, microbial bioproducts have not dominated agro-markets because of several

TABLE 13.3

Examples of Commercially Available Microbial Biostimulants/Biofertilizers around the World

Country	Product	PGPR	Company	Crop	References
Argentina	Liquid PSA	*P. aurantia*	Laboratorios BioAgro	Wheat	Celador-Lera et al. (2018)
	Nodulest 10, BiAgro	*Bradyrhizobium japonicum*	Not mentioned	Not mentioned	Mehnaz et al. (2016)
	Rhizo Liq	*Bradyrhizobium* sp., *Mesorhizobium cicero*, *Rhizobium* sp.	Rhizobacter	Green gram, groundnut, chickpea	Adeleke et al. (2019)
	Zadspirillum	*Azospirillum brasilense*	Semillera Guasch	Maize, Beans	Celador-Lera et al. (2018)
Australia	Nitorguard® TwinN®	*A. brasilense NAB317, Bacillus* sp., *Azoarcus indigens NAB04, Azorhizobium caulinodans NAB38*	Mapleton AgriBiotec Pty Ltd	Beet, rapeseed, sugarcane, vegetables	Hamidi et al. (2021), Dipak and Aloke (2020)
	Rhizobium N	*Azotoformans, Pseudomonas* sp.	Mapleton AgriBiotec Pty	Wheat	Hamidi et al. (2021)
Brazil	BioAtivo®	PGPR consortia	Embrafros Ltd.	Beans, maize, sugarcane, rice	Odoh et al. (2019)
	Onix®	*B. methylotrophicus*	Farroupilha's group	Carrot	Clemente et al. (2016)
	Quartz®	*B. methylotrophicus*	Farroupilha's group	Carrot	Clemente et al. (2016)
	Rizos®	*B. subtilis*	Farroupilha's group	Carrot	
Canada	Nodulator® PRO, Nodulator® N/T	*B. subtilis, B. japonicum*	BASF Inc.	Not mentioned	Sekar, Raj, and Prabavathy (2016)
	EVL Coating®	PGPR consortia	EVL	Not mentioned	García-Fraile, Menéndez, and Rivas (2015)
Colombia	Rhizocell® GC	*B. amyloliquefaciens IT45*	Lallem & Plant Care	Fruits, vegetables	Dipak and Aloke (2020)
	Dimargon®	*Azotobacter chroococcum*	Biocultivos	Not mentioned	Uribe, Sanchez-Nieves, and Vanegas (2010)
Cuba	Nitrofix Bioenraiz	*P. aurantiaca SR1 +A. brasilense + Rhizobium phytohormones*	Biagro	Not mentioned	http://www.biagrosa.com.ar
Cuba	Fosforina®	*P. fluorescens*	Not mentioned	Not mentioned	Uribe, Sanchez-Nieves, and Vanegas (2010)
France	Ceres®	*P. fluorescens*	Biovitis	Field and horticultural crops	Dipak and Aloke (2020)
Germany	FZB24®fl, Rhizo Vital 42®	*B. amyloliquefaciens, B. velezensis*	ABiTEP	Ornamentals, vegetables, field crops	Hamidi et al. (2021), Dipak and Aloke (2020)
Hungary	BactoFil A10®	*A. brasilense, B. megaterium, P. fluorescens, Azotobacter vinelandii, B. polymyxa*	AGRO.bio	Cereals	Dipak and Aloke (2020)
	BactoFil B10®	*A. vinelandii, P. fluorescens, B. subtilis, B. megaterium, Azospirillum lipoferum, B. circulans*	AGRO.bio	Sunflower, potato, rapeseed	Dipak and Aloke (2020)
	Phylazonit-M Zn Sol B	*B. megaterium, A. chroococcum, Thiobacillus thiooxidans*	Novozymes	Not mentioned	www.novozymes.com

(Continued)

TABLE 13.3 (Continued)
Examples of Commercially Available Microbial Biostimulants/Biofertilizers around the World

Country	Product	PGPR	Company	Crop	References
India	Biozink®	PGPR consortia	BioMax	Not mentioned	García-Fraile et al. (2017)
	Bio Power	Azospirillum, Rhizobium, Acetobacter, Azotobacter	SKS Bioproducts Pvt Ltd.	Not mentioned	Sekar, Raj, and Prabavathy (2016)
	VOTiVO®[a]	B. firmus I-1582	Bayer crop science	Maize, soybean	Mendis et al. (2018)
	Biomix®, Biozink®, Biodine®, Llife®	PGPR consortia	Biomax	Not mentioned	Sekar, Raj, and Prabavathy (2016)
	Calsophiere	PGPR consortia	Camson Bio Technologies	Not mentioned	Sekar, Raj, and Prabavathy (2016)
	Calspiral	Azospirillum + other PGPR consortia	Camson Bio Technologies Ltd	Not mentioned	Sekar, Raj, and Prabavathy (2016)
	CataPult Nodulest 10 Agrilife Nitrofix	B. japonicum, A. vinelandii, Acetobacter diazotrophicus, A. chroococcum, Azospirillum lipoferum, P. striata, B. polymyxa, B. megaterium, Rhizobium japonicum	Phylazonit KftAgrilife	Cereals, millets, pulses, oilseeds	www.agrilife.in
	Gmax® PGPR	P. fluorescens, Azotobacter	Greenmax AgroTech	Field crops	Dipak and Aloke (2020)
	Economas[a]	P. fluorescens®	PJ Margo Pvt Ltd	Rice	Vijay et al. (2009)
	Florezen P[a]	P. fluorescens®	Bab Inda Private Limited Ltd., Hyderabad	Rice	Vijay et al. (2009)
	Inogro Bio Gold	>30 PGPR	Agrilife	Rice	www.agrilife.in Mehnaz et al. (2016)
	Serenade®[a]	B. subtilis	Bayer Crop Science	Cucurbits, hops, peanuts, grapes, pome fruits, stone fruits	www.bayercropscience.com Nicot et al. (2016), Tabassum et al. (2017)
	Fe Sol B	Bacillus sp.	Agri Life Bio Solutions	Not mentioned	Mishra and Arora (2016)
	P Sol B	P. striata, B. polymyxa, B. megaterium	Not mentioned	Not mentioned	Mącik, Gryta, and Frąc (2020), Mehnaz et al. (2016)
	Symbion-K, Symbion-P, Symbion-N	Frateuria aurantia, B. megaterium var. phosphaticum, Acetobacter, Azotobacter, Rhizobium, Azospirillum	T. Stanes & Company Ltd	Field crops, vegetables	http://www.tstanes.com/products.html Dipak and Aloke (2020)
	VitaSoil®	PGPR consortia	Symborg	Not mentioned	Sekar, Raj, and Prabavathy (2016)
	Bio Super	Pseudomonas, Cellulomonas, Bacillus, Rhodococcus	SKS Bioproducts Pvt Ltd.	Not mentioned	Sekar, Raj, and Prabavathy (2016)
	Omega®	Bacillus sp., F. aurantia, Azotobacter, Streptomyces sp.	Varsha Bioscience and Technology	Field crops and vegetables	Le Mire et al. (2016)

(Continued)

TABLE 13.3 (*Continued*)

Examples of Commercially Available Microbial Biostimulants/Biofertilizers around the World

Country	Product	PGPR	Company	Crop	References
Italy	Micosat F® Cereali	*Paenibacillus durus PD 76, Streptomyces spp. ST 60, B. subtilis BR 62*	CCS Aosta Srl	Cereals, tomatoes, sunflowers, beet, soybeans	Dipak and Aloke (2020)
	Micosat F® Uno	*B. subtilis BA 41, Agrobacterium radiobacter AR 39, Streptomyces spp. SB 14*	CCS Aosta Srl	Fruits, flowers, vegetables	Dipak and Aloke (2020)
Japan	Mamezo®	*Rhizobia*	Not mentioned	Not mentioned	García-Fraile, Menéndez, and Rivas (2015), García-Fraile et al. (2017)
Kenya	Biofix	*Rhizobia*	MEA Fertilizer Ltd.	Not mentioned	Adeleke et al. (2019)
Nigeria	Nodumax	*Bradyrhizobium*	IITA	Not mentioned	Adeleke et al. (2019)
Russia	Phosphobacterin	*B. megaterium*	Not mentioned	Not mentioned	Mahajan and Gupta (2009)
South Africa	Mazospirflo-2	*A. brasilense*	Soygrow Ltd	Soybean, maize	Laditi et al. (2012)
	Organico	*Bacillus spp., Pseudomonas, Enterobacter spp., Stenotrophomonas, Rhizobium*	Amka Products (Pty) Ltd	Several	Raimi, Adeleke, and Roopnarain (2017)
	PHC Biopak	*Paenibacillus azotofixans, Bacillus spp.*	Plant Health Product (Pty) Ltd.	Soybean, maize	Laditi et al. (2012)
	Firstbase, Landbac, Biostart, Lifeforce	*Bacillus spp.*	Microbial Solution (Pty)	Not mentioned	Mohammadi and Sohrabi (2012)
Spain	Inomix® Biostimulant	*B. polymyxa, B. subtilis, B. megaterium, A. vinelandii, R. leguminosarum P. fluorescens, B. megaterium*	Iabiotech	Cereals	Dipak and Aloke (2020)
	Inomix® Biofertiliant[b]	*B. megaterium, A. vinelandii, R. leguminosarum*	Iabiotech	Cereals	Dipak and Aloke (2020)
	Inomix® Phosphore[a]	*P. fluorescens, B. megaterium*	Iabiotech	Cereals	Dipak and Aloke (2020)
	Rhizosum PK®[b]	*B. megaterium, Frateuria aurantia*	Biosum Technology	Not mentioned	García-Fraile et al. (2017)
Sri Lanka	Bio-Gold	*P. fluorescens, Azotobacter chroococcum*	BioPower	Horticultural crops	Hamidi et al. (2021)
	Bio Phos®	*B. megaterium*	Not mentioned	Not mentioned	Maçik, Gryta, and Frąc (2020), Mehnaz et al. (2016)
Sweden	Amase®	*P. azotoformans*	Lantmannen Bioagri	Cucumber, lettuce, tomato, pepper	Dipak and Aloke (2020)
Thailand	BioPlant	*Clostridium, Achromobacter, Streptomyces, Aerobacter, Nitrobacter, Nitrosomonas, Bacillus*	Artemis & Angelio Co. Ltd.	Not mentioned	Adeleke et al. (2019)

(Continued)

TABLE 13.3 (Continued)
Examples of Commercially Available Microbial Biostimulants/Biofertilizers around the World

Country	Product	PGPR	Company	Crop	References
The UK	Amnite A 10®	*Azotobacter, Bacillus, Pseudomonas, Rhizobium, Chaetomium*	Cleveland Biotech	Cucumber, lettuce, tomato, pepper	Dipak and Aloke (2020)
	Legume Fix	*Rhizobium* sp., *B. japonicum*	Legume Technology	Beans, soybeans	Adeleke et al. (2019)
	Amnite A100	*Azotobacter, Bacillus, Rhizobium, Rhizobium, Pseudomonas*	Cleveland Biotech	Cucumber, tomato	Odoh et al. (2019)
The USA	BioJet®	*Pseudomonas* sp., *Azospirillum* sp.	Eco Soil Systems Inc.	Not mentioned	Sekar, Raj, and Prabavathy (2016)
	BioYiled	*B. amyloliquefaciens, B. subtilis*	Gustafso Inc.	Not mentioned	Sekar, Raj, and Prabavathy (2016)
	Cell-Tech (N Prove) Nitragin Gold Apron	*Rhizobia*	Flozyme	Alfalfa, sweet clover	http://www.flozyme.com/agriclutre
	Compete ® Plus	*B. azotofixans, B. pumilus, B. polymyxa, B. licheniformis, B. megaterium, B. subtilis*	Plant Health Care	Filed crops and tree nurseries	Dipak and Aloke (2020)
	Nitragin Gold	*Rhizobia*	Not mentioned	Legumes	Aamir et al. (2020)
	PGA®	*Bacillus* sp.	Organica Technologies	Fruits, vegetables	Dipak and Aloke (2020)
	Quickroots	*B. amyloliquefaciens*	Monsanto	Wheat, beans	Celador-Lera et al. (2018)
Vietnam	BioGro®	*P. fluorescens, Klebsiella pneumoniae, P. putida, Citrobacter freundii*	Not mentioned	Not mentioned	Uribe, Sanchez-Nieves, and Vanegas (2010)

[a] Doubles as a biostimulant/biofertilizer and biocontrol product for phytopathogen.
[b] Also contains vesicular-arbuscular mycorrhizal fungus.

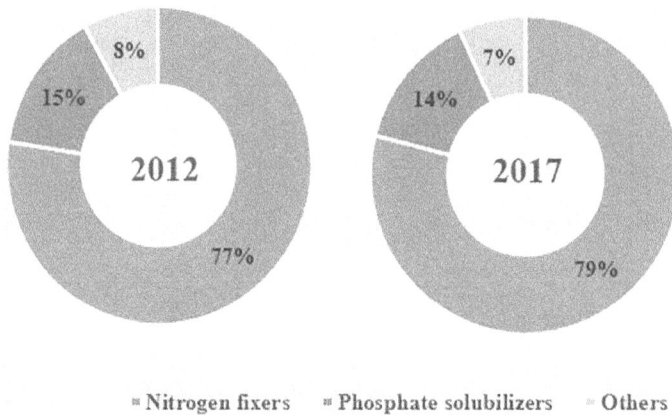

Nitrogen fixers   Phosphate solubilizers   Others

FIGURE 13.4   Global market share of microbial biostimulant/biofertilizer by-products.

bottlenecks such as poor shelf life, inconsistent performance, and poor acceptability, which are further discussed in detail in Subsections 13.5.1–13.5.7.

13.5.1   PRODUCT QUALITY

The quality of microbial biostimulants is an important aspect of formulation and commercialization. Since biostimulants comprise living microbes, microbial loads must be carefully maintained to avoid contamination (Herrmann and Lesueur 2013). During the manufacturing process, production requires sophisticated technologies to maintain high-quality products. The inadequacy of knowledge and sophisticated technologies, coupled with stringent importation laws in developing countries, is one of the factors that lead to the production of substandard microbial biostimulants and, subsequently, the low level of adoption of biostimulants (Herrmann and Lesueur 2013; Arora and Mishra 2016). Besides, potential contaminants cannot be avoided during the manufacture of biostimulants as long as non-sterile carriers are used. There is also the risk of product deterioration due to unplanned mutations (Takors 2012). During transportation and distribution, the microorganisms can be exposed to high temperatures, which may kill them and result in ineffective products (Basu et al. 2021). Consequently, the end products may have a low quality which cannot produce the desired effects on plants which may lessen their acceptability.

13.5.2   PRODUCTION COST

The use of high-tech instruments to produce quality biostimulant products is costly. Besides, microbes are highly sensitive to external conditions such as temperature and require care during manufacturing, transportation/distribution, and application (Basu et al. 2021). This means that huge investments are essential in the acquisition of suitable carrier materials, packaging, and storage. Generally, larger firms are associated with more expenditure and production costs. This way, the inadequacy of financial resources can be a huge drawback in the large-scale production of biostimulants. Besides, the development of appropriate delivery systems is an additional obstacle to the successful commercialization of biostimulants. Even after manufacture, small-scale manufacturers may not have sufficient resources to distribute/market the biostimulants (Basu et al. 2021).

13.5.3 REGULATORY ISSUES

The global scenario of biostimulant regulation has extensively been discussed by Bharti and Suryavanshi (2021) and Caradonia et al. (2019). Although numerous biological products are currently registered, very few are available commercially due to the complex regulatory policies and constraints (Basu et al. 2021). The regulation of biostimulants is very complex, time-consuming, and expensive (Trivedi et al. 2021), particularly due to the lack of acceptance by regulatory authorities (Berg 2009). The product registration and documentation procedures are similarly long and complex, with varying and inconsistent rules between regions and nations (Basu et al. 2021). According to du Jardin (2015), the lack of a standardized and regulatory meaning of 'plant biostimulants' could be the main cause for the lack of a universally uniform regulatory biostimulant policy. The whole process from registration to commercialization of potential biostimulants is extensive and may prolong for years. Although countries have individual registration guidelines, the registering agencies may also ask for additional data and further lengthen the process (Basu et al. 2021). Nevertheless, efforts are underway to develop legal descriptions and universal guidelines, and standards for the commercialization of biostimulants (Trivedi et al. 2021). What awaits is how fast the regulatory regime will be concluded and how individual country regulatory structures will fit within the international regulations.

13.5.4 SHELF LIFE

The short shelf life of microbial products is one of the most pronounced challenges in the development and commercialization of effective PGPR strains as biostimulants. The formulation of microbial biostimulants with prolonged shelf life is especially daunting for non-sporulating Gram-negative bacteria (Berninger et al. 2018). This is because formulations made from spores tend to have a longer shelf life and are easy to handle, both of which are qualities for successful commercialization (Trivedi et al. 2021). Several elements, e.g., inoculant selection, carrier selection, mass multiplication, packaging, distribution, and marketing, dictate the success of biostimulants (Orozco-Mosqueda et al. 2021). Carriers are used to protect the microbes, aid their delivery, and increase their efficacy. According to Shaikh and Sayed (2015), the shelf life of microbial biostimulants is dictated by multiple factors, including the type of carrier/packaging material, production technology, and the distance/mode of transport. Regarding microbial biostimulants that contain bacterial consortia, bacterial species may be incompatible, causing problems with the selection of species for consortia development.

13.5.5 BIOSAFETY OF PRODUCTS

Although most potential biostimulants are usually harmless, a few may potentially be harmful. These risks become quite considerable because common PGPR species belong to genera such as *Stenotrophomonas*, *Enterobacter*, *Serratia*, *Bacillus*, *Burkholderia*, and *Acinetobacter*, which include numerous human-opportunistic pathogens (Martínez-Hidalgo et al. 2019). Although not all species in these genera are pathogenic (Berg et al. 2013), some are opportunistic pathogens and may colonize human tissues and organs and cause infections (Berg et al. 2011; Martínez-Hidalgo et al. 2019). Some PGP strains may also be classified as potential opportunistic pathogens if related phylogenetically to strains isolated in hospital environments (Uzcátegui-Negrón et al. 2011). According to Berg et al. (2005), many rhizobacteria which are closely related to nosocomial pathogens such as *B. cepacia*, *S. marcescens*, *S. maltophilia*, and *P. aeruginosa* are also famous rhizosphere colonizers. Besides, the rhizosphere hosts bacteria that have previously been evidenced to cause urinary tract infectionsw (*B. cepacia* and *P. vulgaris*), cystic fibrosis (*P. aeruginosa* and *B. cepacia*) (Mendes, Garbeva, and Raaijmakers 2013), and wound/skin infections (*Proteus vulgaris*, *B. cereus* and *Pseudomonas* spp.).

Stenotrophomonas maltophilia, for instance, is a well-known species for both plant coloniza-tion and clinical infections (Ribbeck-Busch et al. 2005). Similarly, Burkholderiales (Eberl and Vandamme 2016) and Enterobacteriales can be potent human pathogens that use plants as inter-mediate reservoirs/hosts (Holden, Pritchard, and Toth 2009). *Bacillus cereus* is a plant-beneficial bacterium, but is also classified as an opportunistic human pathogen because it produces many intestinal/non-intestinal tissue-destructive exoenzymes (Bottone 2010). According to Compant et al. (2010), some plant endophytes also exhibit extremely high homology to clinical isolates.

A principal mechanism by which non-pathogenic bacteria evolve into pathogenic ones is through host niche or host alteration, which subsequently exposes their virulence (Berg et al. 2011). Since PGPR are basically assumed as probable prospects in sustainable agriculture, they must not possess any potential harm. Although most strains used as microbial biostimulants are primarily of low risk (Mitter et al. 2021; Martínez-Hidalgo et al. 2019), their secondary negative effects cannot entirely be overlooked. Thus, the formulation and commercialization of microbial biostimulants continue to raise a lot of concern and their use remains a complex discern between pathogenic or opportunistic, and beneficial strains.

13.5.6 Farmer Perceptions and Acceptability

Although microbial biostimulants are environmentally safe products for crop production, their use is still negligible compared to synthetic pesticides because of a lack of awareness and acceptabil-ity among farmers. Most of the time, farmers do not eagerly accept alternative cropping methods. Besides, farmers can only adopt biostimulant products if their success is guaranteed, which is not always the case. The incorporation of biostimulants into agricultural practices is pegged on their economic relevance relative to the existing conventional inputs. Therefore, the major reason for farmer skepticism about biostimulants concerns their inconsistent performance relative to conven-tional chemical inputs. A further discussion on the variability of field performance of biostimulants is provided in Subsection 13.5.7.

13.5.7 Variability of Performance

Perhaps one of the greatest bottlenecks in the biostimulant industry is the inconsistency of results from laboratories to fields (Vessey 2003). Many PGPR tend to perform well *in vitro*, but fail to produce the desired outcomes in crop fields (Vejan et al. 2016). The main reason for such failures is the fast decline in active inoculants/cells to levels that cannot be effective when introduced into soil (Barquero et al. 2019). Soil being a heterogeneous environment, inoculants in microbial bio-stimulants are subjected to competition by native microflora and survival in new locations which is similarly dependent on several factors in the rhizosphere (Barquero et al. 2019). The field per-formance of biostimulants changes with the variety of environmental and climatic conditions such as temperature, rainfall, and soil types, leading to disparity in their potentiality. Unlike the con-trolled environment in laboratories, it is problematic to predict the effectiveness of inoculants in the field. Besides, the insufficient availability of soil-specific microbial strains substantially restricts the extensive use of bioinoculants (Saad, Eida, and Hirt 2020).

13.6 FUTURE PERSPECTIVES

The global market for PGPR biostimulants is increasing annually, probably because they offer eco-friendly alternatives to chemical fertilizers (Brisk Insights 2016). The market is estimated to expand at about 14% annually to generate USD 2 billion globally by 2022 (Allied Market Research 2016). Generally, the existing and emerging microbial technologies present sustainable and new approaches to agriculture productivity. Nonetheless, significant scientific and technological challenges exist and must be prioritized alongside the improvement of the regulatory frameworks for microbial biostimulants to transform sustainable agriculture (Trivedi et al. 2017). A proper comprehension of appropriate application types/modes for specific plants is still essential for more large-scale production and application. Genomic studies can rapidly expand our knowledge regarding various microbial metabolic pathways responsible for novel plant biostimulation traits. Attempts to engineer and design microbial cells to illuminate and explore new metabolic pathways hold a lot of potential for the industry (Trivedi et al. 2017). The genetic engineering of efficient PGPR strains may further heighten their expression of PGP benefits.

Taking into consideration the potential pathogenicity of PGPR biostimulants, new methods need to be incorporated while studying the biosafety of novel biostimulants during product development to guarantee their safety to humans and the ecosystem at large. In this regard, the determination of virulence genes could be a clear-cut way of assessing the safety of novel bacterial isolates relative to their pathogenic relatives (Martínez-Hidalgo et al. 2019). Rather than depending on common conventional methods of strain identification, current interdisciplinary tools need to be incorporated for risk assessment of strains. In this context, the increasing public availability of whole-genome sequences of rhizobacterial strains is invaluable. The identification of efficient and safe microbes, techniques standardization, and information sharing will altogether permit the identification of rhizosphere microbes with no potential risks for biotechnological applications and the development of microbial biostimulants.

A majority of research work on microbial biostimulants is still confined either to laboratories or greenhouses; hence, it should be taken up to fields. Further research is needed to elucidate the suitable field conditions for specific biostimulants. The development of PGPR as reliable microbial biostimulants in agriculture will be accelerated by a comprehensive knowledge of their diversity, action mechanisms, formulation, and proper application, and colonization abilities. In addition to addressing the limitations discussed in Section 13.5, researchers must acquire knowledge regarding the field application/delivery of microbial biostimulants to guarantee utmost performance, which is currently the biggest source of concern for producers and consumers.

Just like other microbial products, increasing confidence and faith among farmers can be the key goal to enhance the demand and use of microbial biostimulants for crop production. This can be achieved by educating farmers about the benefits of bioinoculants and methods for effective application. The success of microbial biostimulant industries is subject to measures such as extensive marketing, research, and education. Besides, the full adoption of PGPR strains in the agricultural industry depends very much on their optimization for improved products and performance (Tabassum et al. 2017).

13.7   CONCLUDING REMARKS

Feeding the rising global population is among the greatest challenges facing mankind today, especially with regard to agricultural and environmental sustainability. Microbial biostimulants continue to attract huge attention as environmentally sustainable agricultural inputs. Although these products have significant abilities to enhance crop growth by increasing nutrient availability and producing PGP hormones, their commercialization and uptake are still in infancy. The major reasons for this include the lack of acceptability among farmers and complex regulatory processes, all of which have been articulated in this chapter. To cope up with these challenges, a holistic approach by governmental, non-governmental, private, and public institutes and universities is required. Broad-spectrum formulations with reliable performance and prolonged shelf life hold a great potential for hastening PGPR commercialization and utilization.

FUNDING

The preparation of this chapter was supported by the African-German Network of Excellence in Science (AGNES), with generous funds from the Federal Ministry of Education and Research (BMBF) and the Alexander von Humboldt Foundation (AvH).

REFERENCES

Aamir, M., K.K. Rai, A. Zehra, M.K. Dubey, S. Kumar, V. Shukla, and R.S. Upadhyay. 2020. Microbial bioformulation-based plant biostimulants: A plausible approach toward next generation of sustainable agriculture. In *Microbial Endophytes*, eds. A. Kumar and E.K. Radhakrishnan, 195–225. Sawston: Woodhead Publishing. https://www.sciencedirect.com/science/article/pii/B9780128196540000089.

Adeleke, R.A., A.R. Raimi, A. Roopnarain, and S.M. Mokubedi. 2019. Status and prospects of bacterial inoculants for sustainable management of agroecosystems. In *Biofertilizers for Sustainable Agriculture and Environment*, eds. B. Giri, R. Prasad, Q.S. Wu, and A. Varma, 137–172. Cham: Springer International Publishing. https://doi.org/10.1007/978-3-030-18933-4_7.

Aeron, A., R.C. Dubey, and D.K. Maheshwari. 2021. Next-generation biofertilizers and novel biostimulants: Documentation and validation of mechanism of endophytic plant growth-promoting rhizobacteria in tomato. *Archives of Microbiology* 203: 1–12.

Aeron, A., E. Khare, C.K. Jha, V.S. Meena, S.M.A. Aziz, M.T. Islam, K. Kim, S.K. Meena, A. Pattanayak, and H. Rajashekara. 2020. Revisiting the plant growth-promoting rhizobacteria: Lessons from the past and objectives for the future. *Archives of Microbiology* 202, no. 4: 665–676.

Agro News. 2014. Biofertilizers Market–Global Industry Analysis, Size, Share, Growth, Trends and Forecast, 2013–2019. http://news.agropages.com/News/NewsDetail11612–e.htm.

Agustian, A., L. Nuriyani, L. Maira, and O. Emalinda. 2010. Phytohormone IAA producing rhizobacteria in rhizospheric plant of karamunting, titonia, and crops. *Indonesia Journal of Solum*, 7: 49–60.

Ahemad, M., and M. Kibret. 2014. Mechanisms and applications of plant growth promoting rhizobacteria: Current perspective. *Journal of King Saud University-Science* 26, no. 1: 1–20.

Allied Market Research. 2016. Agricultural Biologicals Market by Product Type (Biopesticides, Biostimulants, and Biofertilizers) and Application (Cereals & Grains, Oilseeds &Pulses, and Fruits) - Global Opportunity Analysis and Industry Forecast, 2017–2023. https://www.alliedmarketresearch.com/agricultural-biologicals-market.

Aloo, B.N., B. Makumba, and E. Mbega. 2021. Status of biofertilizer research, commercialization, and practical applications: A global perspective. In *Biofertilizers: Advances in Bio-Inoculants*, eds. A. Rakshit, V.S. Meena, M. Parihar, H.B. Singh, and A.K. Singh, vol. 1, 191–208. Elsevier. https://doi.org/10.1016/B978-0-12-821667-5.00017-8.

Aloo, B.N., E.R. Mbega, B.A. Makumba, I. Friedrich, R. Hertel, and R. Daniel. 2020. Whole-genome sequences of three plant growth-promoting rhizobacteria isolated from *Solanum tuberosum* L. Rhizosphere in Tanzania. *Microbiology Resource Announcements* 9, no. 20: e00371-20.

Aloo, B.N., E.R. Mbega, B.A. Makumba, R. Hertel, and R. Danel. 2021. Molecular identification and in vitro plant growth-promoting activities of culturable potato (*Solanum tuberosum* L.) rhizobacteria in Tanzania. *Potato Research* 64: 67–95.

Aloo, B.N., V. Tripathi, E.R. Mbega, and B.A. Makumba. 2021. Endophytic rhizobacteria for mineral nutrients acquisition in plants: Possible functions and ecological advantages. In *Endophytes: Mineral Nutrient Management*, eds. D.K. Maheshwari and S. Dheeman, vol. 3:267–291. Sustainable Development and Biodiversity 26. Cham: Springer.

Arora, N.K., and J. Mishra. 2016. Prospecting the roles of metabolites and additives in future bioformulations for sustainable agriculture. *Applied Soil Ecology* 107: 405–407.

Banks, J.M., and G.C. Percival. 2012. Evaluation of biostimulants to control guignardia leaf blotch (*Guignardia aesculi*) of horsechestnut and black spot (*Diplocarpon rosae*) of roses. *Arboriculture and Urban Forestry* 38, no. 6: 258.

Barquero, M., R. Pastor-Buies, B. Urbano, and F. González-Andrés. 2019. Challenges, regulations and future actions in biofertilizers in the European agriculture: From the lab to the field. In *Microbial Probiotics for Agricultural Systems*, 83–107. https://doi.org/10.1007/978-3-030-17597-9_6.

Basak, A. 2008. Biostimulators: Definitions, classification and legislation. In *Biostimulators in Modern Agriculture. General Aspects*, ed. H. Gawrońska, 7–17. Warszawa: Editorial House Wieś Jutra.

Bashir, M.A., A. Rehim, H.M. Qurat-Ul-Ain Raza, A. Raza, L. Zhai, H. Liu, and H. Wang. 2021. Biostimulants as Plant Growth Stimulators in Modernized Agriculture and Environmental Sustainability. In *Technology in Agriculture*, ed. F. Ahmad and S. Muhammad, 311-322. IntechOpen.

Basu, A., P. Prasad, S.N. Das, S. Kalam, R. Sayyed, M. Reddy, and H. El Enshasy. 2021. Plant growth promoting rhizobacteria (PGPR) as green bioinoculants: Recent developments, constraints, and prospects. *Sustainability* 13, no. 3: 1140.

Belbahri, L., A. Chenari Bouket, I. Rekik, F.N. Alenezi, A. Vallat, L. Luptakova, E. Petrovova, et al. 2017. Comparative genomics of *Bacillus amyloliquefaciens* strains reveals a core genome with traits for habitat adaptation and a secondary metabolites rich accessory genome. *Frontiers in Microbiology* 8: 1438.

Berg, G. 2009. Plant-microbe interactions promoting plant growth and health: Perspectives for controlled use of microorganisms in agriculture. *Applied Microbiology and Biotechnology* 84: 11–18.

Berg, G., M. Alavi, M. Schmid, and A. Hartmann. 2013. The rhizosphere as a reservoir for opportunistic human pathogenic bacteria. In *Molecular Microbial Ecology of the Rhizosphere*, ed. F.J. de Bruijn, 1209–1216. 1st ed. Hoboken, NJ: John Wiley & Sons, Inc.

Berg, G., L. Eberl, and A. Hartmann. 2005. The rhizosphere as a reservoir for opportunistic human pathogenic bacteria. *Environmental Microbiology* 7, no. 11: 1673–1685.

Berg, G., C. Zachow, M. Cardinale, and H. Müller. 2011. Ecology and human pathogenicity of plant-associated bacteria. In *Regulation of Biological Control Agents*, ed. R.U. Ehlers, 175–189. Dordrecht: Springer.

Berninger, T., Ó. González López, A. Bejarano, C. Preininger, and A. Sessitsch. 2018. Maintenance and assessment of cell viability in formulation of non-sporulating bacterial inoculants. *Microbial Biotechnology* 11, no. 2: 277–301.

Bharadwaj, G., R. Shah, B. Joshi, and P. Patel. 2017. *Klebsiella pneumoniae* VRE36 as a PGPR Isolated from *Saccharum officinarum* Cultivar Co 99004. *Journal of Applied Biology and Biotechnology* 5, no. 1: 47–52.

Bharti, N., and M. Suryavanshi. 2021. Quality control and regulations of biofertilizers: Current scenario and future prospects. In *Biofertilizers*, eds. A. Rakshit, V.S. Meena, M. Parihar, H.B. Singh, and A.K. Singh, 133–141. Woodhead Publishing. https://www.sciencedirect.com/science/article/pii/B978012821667500018X.

Bhat, M.A. 2019. Plant growth promoting rhizobacteria (PGPR) for sustainable and eco-friendly agriculture. *Acta Scientific Agriculture* 3: 23–25.

Blanco, M. 2011. *Supply of and Access to Key Nutrients NPK for Fertilizers for Feeding the World in 2050.* Madrid: Universidad Politécnica de Madrid (UPM). https://www.researchgate.net/profile/Maria-Blanco-33/publication/236272173_Supply_of_and_access_to_key_nutrients_NPK_for_fertilizers_for_feeding_the_world_in_2050/links/02bfe511e5f39575f7000000/Supply-of-and-access-to-key-nutrients-NPK-for-fertilizers-for-feeding-the-world-in-2050.pdf.

Bottone, E.J. 2010. *Bacillus cereus*, a volatile human pathogen. *Clinical Microbiology Reviews* 23, no. 2: 382–398.

Brisk Insights. 2016. Biofertilizer Market by Type (Nitrogen Fixing, Phosphate Solubilizing, Potash-Mobilizing), by Application (Seed and Soil Treatment), by Crop Type (Cereals and Grains, Pulses and Oil Seeds, Fruits and Vegetables, Plantations), by Microorganisms (*Azotobacter, Rhizobium, Azospirillum* and *Cyanobacteria*), Industry Size, Growth, Share and Forecast to 2022. http://www.briskinsights.com/report/biofertilizermarket.

Cakmakci, R., M. Turan, M. Gulluce, and F. Sahin. 2014. Rhizobacteria for reduced fertilizer inputs in wheat (*Triticum aestivum* spp. *Vulgare*) and barley (Hordeum Vulgare) on aridisols in Turkey. *International Journal of Plant Production* 8, no. 2: 163–182.

Caradonia, F., V. Battaglia, L. Righi, G. Pascali, and A. La Torre. 2019. Plant biostimulant regulatory framework: Prospects in Europe and current situation at international level. *Journal of Plant Growth Regulation* 38, no. 2: 438–448.

Carletti, P., A.C. García, C.A. Silva, and A. Merchant. 2021. Towards a functional characterization of plant biostimulants. *Frontiers in Plant Science* 12: 677772.

Celador-Lera, L., A. Jiménez-Gómez, E. Menéndez, and R. Rivas. 2018. Biofertilizers based on bacterial endophytes isolated from cereals: Potential solution to enhance these crops. In *Stress Management and Agricultural Sustainability*, ed. V.S. Meena, 1:175–203. Role of Rhizospheric Microbes in Soil. Singapore: Springer. https://doi.org/10.1007/978-981-10-8402-7_7.

Chen, C., Z. Yue, C. Chu, K. Ma, L. Li, and Z. Sun. 2020. Complete genome sequence of *Bacillus* sp. strain WR11, an endophyte isolated from wheat root providing genomic insights into its plant growth-promoting effects. *Molecular Plant-Microbe Interactions* 33, no. 7: 876–879.

Cheng, X., D.W. Etalo, J.E. van de Mortel, E. Dekkers, L. Nguyen, M.H. Medema, and J.M. Raaijmakers. 2017. Genome-wide analysis of bacterial determinants of plant growth promotion and induced systemic resistance by *Pseudomonas fluorescens. Environmental Microbiology* 19, no. 11: 4638–4656.

Clemente, J.M., C.R. Cardoso, B.S. Vieira, I. da Mata Flor, and R.L. Costa. 2016. Use of bacillus Spp. as growth promoter in carrot crop. *African Journal of Agricultural Research* 11, no. 35: 3355–3359.

Cole, B.J., M.E. Feltcher, R.J. Waters, K.M. Wetmore, T.S. Mucyn, E.M. Ryan, G. Wang, S. Ul-Hasan, M. McDonald, and Y. Yoshikuni. 2017. Genome-wide identification of bacterial plant colonization genes. *PLoS Biology* 15, no. 9: e2002860.

Compant, S., C. Clement, and A. Sessitsch. 2010. Plant growth-promoting bacteria in the rhizo and endosphere of plants their role, organization, mechanisms involved and prospects for utilization. *Soil Biology and Biochemistry* 42: 669–678.

Dal Cortivo, C., M. Ferrari, G. Visioli, M. Lauro, F. Fornasier, G. Barion, A. Panozzo, and T. Vamerali. 2020. Effects of seed-applied biofertilizers on rhizosphere biodiversity and growth of common wheat (*Triticum aestivum* L.) in the field. *Frontiers in Plant Science* 11: 72.

Deng, Y., Y. Zhu, P. Wang, L. Zhu, J. Zheng, R. Li, L. Ruan, D. Peng, and M. Sun. 2011. Complete genome sequence of *Bacillus subtilis* BSn5, an endophytic bacterium of *Amorphophallus konjac* with antimicrobial activity for the plant pathogen *Erwinia carotovora* subsp. *carotovora*. *Journal of Bacteriology* 193, no. 8: 2070.

Dhaked, B.S., S. Triveni, R.S. Reddy, and G. Padmaja. 2017. Isolation and screening of potassium and zinc solubilizing bacteria from different rhizosphere soil. *International Journal of Current Microbiology and Applied Sciences* 6, no. 8: 1271–1281.

Dipak, K.H., and P. Aloke. 2020. Role of biostimulant formulations in crop production: An overview. *International Journal of Applied Research in Veterinary Medicine* 8: 38–46.

Dixon, G.R., and U.F. Walsh. 2002. Suppressing *Pythium ultimum* induced damping-off in cabbage seedlings by biostimulation with proprietary liquid seaweed extracts. *Acta Horticulturae* 635:103–106. https://doi.org/10.17660/ActaHortic.2004.635.13.

du Jardin, P. 2012. *The Science of Plant Biostimulants–A Bibliographic Analysis, Ad Hoc Study Report*. Ad hoc study report. European Commission. https://orbi.uliege.be/bitstream/2268/169257/1/Plant_Biostimulants_final_report_bio_2012_en.pdf.

du Jardin, P. 2015. Plant biostimulants: Definition, concept, main categories and regulation. *Scientia Horticulturae* 196: 3–14.

du Jardin, P., L. Xu, and D. Geelen. 2020. Agricultural functions and action mechanisms of plant biostimulants (PBs). In *The Chemical Biology of Plant Biostimulants*, eds. D. Geelen and L. Xu, 1–30. John Wiley & Sons Ltd. https://doi.org/10.1002/9781119357254.ch1.

Duan, J., W. Jiang, Z. Cheng, J.J. Heikkila, and B.R. Glick. 2013. The complete genome sequence of the plant growth-promoting bacterium *Pseudomonas* Sp. UW4. *PLoS One* 8, no. 3: e58640.

Eberl, L., and P. Vandamme. 2016. Members of the genus *Burkholderia*: Good and bad guys. *F1000Research* 5. doi: 10.12688/f1000research.8221.1.

EBIC. 2012. http://www.biostimulants.eu/.

Eickhout, B., A. van Bouwman, and H. Van Zeijts. 2006. The role of nitrogen in world food production and environmental sustainability. *Agriculture, Ecosystems & Environment* 116, no. 1–2: 4–14.

El-Nahrawy, S., and A.E.-D. Omara. 2017. Effectiveness of co-inoculation with pseudomonas koreensis and rhizobia on growth, nodulation and yield of common bean (*Phaseolus vulgaris* L.). *Microbiology Research Journal International* 21, no. 6: 1–15.

FAO. 2011. *Changes in Global Potato Production*. Rome: Food and Agriculture Organization of the United Nations. http://www.slideshare.net/rtbcgiar/from-a-poverty-lens-to-a-food-security-lenspotatoes-to-improve-global-food-security-and-sustainability.

FAO. 2013. *FAOSTAT Statistics Database*. Rome: Food and Agriculture Organization of the United Nations.

FAO. 2018. *The Future of Food and Agriculture – Alternative Pathways to 2050*. Rome: Food and Agriculture Organisation of the United Nations. http://www.fao.org/3/CA1553EN/ca1553en.pdf.

Fasusi, O.A., and O.O. Babalola. 2021. The multifaceted plant-beneficial rhizobacteria toward agricultural sustainability. *Plant Protection Science* 57, no. 2: 95–111.

García-Fraile, P., E. Menéndez, L. Celador-Lera, A. Díez-Méndez, A. Jiménez-Gómez, M. Marcos-García, X.A. Cruz-González, P. Martínez-Hidalgo, P.F. Mateos, and R. Rivas. 2017. Bacterial probiotics: A truly green revolution. In *Probiotics and Plant Health*, eds. V. Kumar, M. Kumar, S. Sharma, and R. Prasad, 131–162. Singapore: Springer.

García-Fraile, P., E. Menéndez, and R. Rivas. 2015. Role of bacterial biofertilizers in agriculture and forestry. *AIMS Bioengineering* 2, no. 3: 108–205.

Gholami, A., S. Shahsavani, and S. Nezarat. 2009. The effect of plant growth promoting rhizobacteria (PGPR) on germination, seedling growth and yield of maize. *International Journal of Agricultural and Biosystems Engineering* 3, no. 1: 9–14.

Gouda, S., R.G. Kerry, G. Das, S. Paramithiotis, and J.K. Patra. 2018. Revitalization of plant growth promoting rhizobacteria for sustainable development in agriculture. *Microbiology Research* 206: 131–140.

Goulding, K. 2016. Soil acidification and the importance of liming agricultural soils with particular reference to the United Kingdom. *Soil Use and Management* 32, no. 3: 390–399.

Gupta, R., A. Kumari, A. Noureldeen, and H. Darwish. 2021. Rhizosphere mediated growth enhancement using phosphate solubilizing rhizobacteria and their tri-calcium phosphate solubilization activity under pot culture assays in rice (*Oryza sativa.*). *Saudi Journal of Biological Sciences* 28: 3692–3700.

Halpern, M., A. Bar-Tal, M. Ofek, D. Minz, T. Muller, and U. Yermiyahu. 2015. The use of biostimulants for enhancing nutrient uptake. *Advances in Agronomy* 130: 141–174.

Hamidi, B., M. Zaman, S. Farooq, S. Fatima, R.Z. Sayyed, Z.A. Baba, T.A. Sheikh, et al. 2021. Bacterial plant biostimulants: A sustainable way towards improving growth, productivity, and health of crops. *Sustainability* 13: 2856.

Han, H.S., and K.D. Lee. 2006. Effect of co-inoculation with phosphate and potassium solubilizing bacteria on mineral uptake and growth of pepper and cucumber. *Plant, Soil and Environment* 52, no. 3: 130–131.

Han, J., J. Shi, L. Zeng, J. Xu, and L. Wu. 2015. Effects of nitrogen fertilization on the acidity and salinity of greenhouse soils. *Environmental Science and Pollution Research* 22, no. 4 (February 1): 2976–2986.

Hanan, Y.M., and T. El Sebai. 2019. Effect of bio-stimulant (Phosphate solubilizing microorganisms) on yield and quality of some sugar beet varieties. *Egyptian Journal of Applied Sciences* 34, no. 7: 114–129.

Herrmann, L., and D. Lesueur. 2013. Challenges of formulation and quality of biofertilizers for successful inoculation. *Applied Microbiology and Biotechnology* 97: 8859–8873.

Holden, N., L. Pritchard, and I. Toth. 2009. Colonization outwith the colon: Plants as an alternative environmental reservoir for human pathogenic enterobacteria. *FEMS Microbiology Reviews* 33, no. 4: 689–703.

Htwe, A.Z., S.M. Moh, K. Moe, and T. Yamakawa. 2019. Biofertilizer production for agronomic application and evaluation of its symbiotic effectiveness in soybeans. *Agronomy* 9, no. 4: 162.

Htwe, A.Z., S.M. Moh, K.M. Soe, K. Moe, and T. Yamakawa. 2019. Effects of biofertilizer produced from *Bradyrhizobium* and *Streptomyces griseoflavus* on plant growth, nodulation, nitrogen fixation, nutrient uptake, and seed yield of mung bean, cowpea, and soybean. *Agronomy* 9, no. 2: 77.

Hussein, K.A., and J.H. Joo. 2015. Isolation and characterization of rhizomicrobial isolates for phosphate solubilization and indole acetic acid production. *Journal of the Korean Society for Applied Biological Chemistry* 58, no. 6: 847–855.

Islam, F., T. Yasmeen, Q. Ali, S. Ali, M.S. Arif, S. Hussain, and H. Rizvi. 2014. Influence of pseudomonas aeruginosa as PGPR on oxidative stress tolerance in wheat under Zn stress. *Ecotoxicology and Environmental Safety* 104: 285–293.

Kecskés, M.L., A.T.M.A. Choudhury, A.V. Casteriano, R. Deaker, R.J. Roughley, L. Lewin, R. Ford, and I.R. Kennedy. 2016. Effects of bacterial inoculant biofertilizers on growth, yield and nutrition of rice in Australia. *Journal of Plant Nutrition* 39, no. 3: 377–388.

Khatoon, Z., S. Huang, M. Rafique, A. Fakhar, M.A. Kamran, and G. Santoyo. 2020. Unlocking the potential of plant growth-promoting rhizobacteria on soil health and the sustainability of agricultural systems. *Journal of Environmental Management* 273: 111118.

Kumari, B., M. Mallick, M.K. Solanki, A.C. Solanki, A. Hora, and W. Guo. 2019. Plant growth promoting rhizobacteria (PGPR): Modern prospects for sustainable agriculture. In *Plant Health under Biotic Stress*, eds. R. Ansari and I. Mahmood, 109–127. Singapore: Springer. https://doi.org/10.1007/978-981-13-6040-4_6.

Laditi, M., O. Nwoke, M. Jemo, R. Abaidoo, and A. Ogunjobi. 2012. Evaluation of microbial inoculants as biofertilizers for the improvement of growth and yield of soybean and maize crops in savanna soils. *African Journal of Agricultural Research* 7, no. 3: 405–413.

Le Mire, G., M.L. Nguyen, B. Fassotte, P. du Jardin, F. Verheggen, P. Delaplace, and M.H. Jijaki. 2016. Implementing Plant biostimulants and biocontrol strategies in the agroecological management of cultivated ecosystems: A review. *Biotechnology, Agronomy, Society and Environment* 20, no. 1: 299–313.

Lovatt, C.J. 2015. Use of a Natural Metabolite to Increase Crop Production. USA: Google Patents. http://www.freepatentsonline.com/y2016/0088842.html.

Mącik, M., A. Gryta, and M. Frąc. 2020. Biofertilizers in agriculture: An overview on concepts, strategies and effects on soil microorganisms. *Advances in Agronomy* 162: 31–87.

Mahajan, A., and R. Gupta. 2009. Bio-fertilizers: Their kinds and requirement in India. In *Integrated Nutrient Management (INM) in a Sustainable Rice-Wheat Cropping System*, eds. A. Mahajan, R.D. Gupta, 75–100. Dordrecht: Springer.

Markets and Markets. 2017. *Biostimulants Market by Active Ingredient (Humic Substances, Seaweed, Microbials, Trace Minerals, Vitamins & Amino Acids), Crop Type (Row Crops, Fruits & Vegetables, Turf & Ornamentals), Formulation, Application Method, and Region—Global Forecast to 2022*. USA: MarketsandMarkets™ INC. http://www.marketsandmarkets.com.

Martínez-Hidalgo, P., M. Maymon, F. Pule-Meulenberg, and A.M. Hirsch. 2019. Engineering root microbiomes for healthier crops and soils using beneficial, environmentally safe bacteria. *Canadian Journal of Microbiology* 65, no. 2: 91–104.

Mehnaz, S., N.K. Arora, S. Mehnaz, and R. Balestrini. 2016. An overview of globally available bioformulations. In *Bioformulations for Sustainable Agriculture*, 267–281. New Delhi: Springer. https://doi.org/10.1007/978-81-322-2779-3_15.

Mendes, R., P. Garbeva, and J.M. Raaijmakers. 2013. The rhizosphere microbiome: Significance of plant beneficial, plant pathogenic, and human pathogenic microorganisms. *FEMS Microbiology Reviews* 37: 634–663.

Mendis, H.C., V.P. Thomas, P. Schwientek, R. Salamzade, J.T. Chien, P. Waidyarathne, J.W. Kloepper, and L. De La Fuente. 2018. Strain-specific quantification of root colonization by plant growth promoting rhizobacteria *Bacillus firmus* I-1582 and *Bacillus amyloliquefaciens* QST713 in non-sterile soil and field conditions. *PLoS One* 13, no. 2: e0193119.

Mishra, J., and N.K. Arora. 2016. Bioformulations for plant growth promotion and combating phytopathogens: A sustainable approach. In *Bioformulations: For Sustainable Agriculture*, eds. N.K. Arora, S. Mehnaz, and R. Balestrini, 3–33. New Delhi: Springer.

Mitter, E.K., M. Tosi, D. Obregón, K.E. Dunfield, and J.J. Germida. 2021. Rethinking crop nutrition in times of modern microbiology: Innovative biofertilizer technologies. *Frontiers in Sustainable Food Systems* 5: 29.

Mohammadi, K., and Y. Sohrabi. 2012. Bacterial biofertilizers for sustainable crop production: A review. *ARPN Journal of Agricultural and Biological Science* 7, no. 5: 307–316.

Monteiro, G., G. Nogueira, C. Neto, V. Nascimento, and J. Freitas. 2021. Promotion of Nitrogen Assimilation by Plant Growth-Promoting Rhizobacteria. In *Nitrogen in Agriculture-Physiological, Agricultural and Ecological Aspects*. IntechOpen.

Mushtaq, Z., S. Faizan, and A. Hussain. 2021. Role of microorganisms as biofertilizers. In *Microbiota and Biofertilizers: A Sustainable Continuum for Plant and Soil Health*, eds. K.R. Hakeem, G.H. Dar, M.A. Mehmood, and R.A. Bhat, 83–98. Cham: Springer International Publishing. https://doi.org/10.1007/978-3-030-48771-3_6.

Nguyen, M.L. 2018. Biostimulant Effects of Rhizobacteria on Wheat Growth and Nutrient Uptake under Contrasted N Supplies. PhD Dissertation, Belgium: Université De Liège - Gembloux Agro-Bio Tech.

Niazi, A. 2014. Genome-Wide Analyses of *Bacillus amyloliquefaciens* Strains Provide Insights into Their Beneficial Role on Plants. Doctoral Thesis, Uppsala: Acta Universitatis Agriculturae Sueciae. https://pub.epsilon.slu.se/11421/.

Niazi, A., S. Manzoor, S. Asari, S. Bejai, J. Meijer, and E. Bongcam-Rudloff. 2014. Genome analysis of *Bacillus amyloliquefaciens* subsp. *plantarum* UCMB5113: A rhizobacterium that improves plant growth and stress management. *PLoS One* 9, no. 8: e104651.

Nicot, P.C., A. Stewart, M. Bardin, and Y. Elad. 2016. Biological control and biopesticide suppression of *Botrytis*-incited diseases. In *Botrytis – the Fungus, the Pathogen and Its Management in Agricultural Systems*, eds. S. Fillinger and Y. Elad, 165–187. Cham: Springer International Publishing. https://doi.org/10.1007/978-3-319-23371-0_9.

Nori, S.S., S. Kumar, S. Khandelwal, and S. Suryanarayan. 2019. Biostimulant Formulation for Improving Plant Growth and Uses Thereof. Google Patents. https://patents.google.com/patent/US10358391B2/en.

Novello, G., P. Cesaro, E. Bona, N. Massa, F. Gosetti, A. Scarafoni, V. Todeschini, G. Berta, G. Lingua, and E. Gamalero. 2021. The effects of plant growth-promoting bacteria with biostimulant features on the growth of a local onion cultivar and a commercial zucchini variety. *Agronomy* 11, no. 5: 888.

Oancea, F., I. Răut, and V. Zamfiropol-Cristea. 2017. Influence of soil treatment with microbial plant biostimulant on tomato yield and quality. *Journal of International Scientific Publications* 5: 156–165.

Odoh, C.K., C.N. Eze, U.K. Akpi, and V.U. Unah. 2019. Plant growth promoting rhizobacteria (PGPR): A novel agent for sustainable food production. *American Journal of Agricultural and Biological Sciences* 14: 35–54.

Orozco-Mosqueda, M.D., A. Flores, B. Rojas-Sánchez, C.A. Urtis-Flores, L.R. Morales-Cedeño, M.F. Valencia-Marin, S. Chávez-Avila, D. Rojas-Solis, and G. Santoyo. 2021. Plant growth-promoting bacteria as bioinoculants: Attributes and challenges for sustainable crop improvement. *Agronomy* 11, no. 6: 1167.

Paungfoo-Lonhienne, C., M. Redding, C. Pratt, and W. Wang. 2019. Plant growth promoting rhizobacteria increase the efficiency of fertilisers while reducing nitrogen loss. *Journal of Environmental Management* 233: 337–341.

Peter, A.J., E.L.D. Amalraj, and V.R. Talluri. 2020. Commercial aspects of biofertilizers and biostimulants development utilizing rhizosphere microbes: Global and Indian scenario. In *Rhizosphere Microbes: Soil and Plant Functions*, eds. S.K. Sharma, U.B. Singh, P.K. Sahu, H.V. Singh, and P.K. Sharma, 655–682. Singapore: Springer. https://doi.org/10.1007/978-981-15-9154-9_27.

Raimi, A., R. Adeleke, and A. Roopnarain. 2017. Soil fertility challenges and biofertiliser as a viable alternative for increasing smallholder farmer crop productivity in Sub-Saharan Africa. *Cogent Food & Agriculture* 3, no. 1: 1400933.

Ray, P., V. Lakshmanan, J.L. Labbé, and K.D. Craven. 2020. Microbe to microbiome: A paradigm shift in the application of microorganisms for sustainable agriculture. *Frontiers in Microbiology* 11: 3323.

Ribbeck-Busch, K., A. Roder, D. Hasse, W. De Boer, J.L. Martínez, M. Hagemann, and G. Berg. 2005. A molecular biological protocol to distinguish potentially human pathogenic *Stenotrophomonas maltophilia* from plant-associated *Stenotrophomonas rhizophila*. *Environmental Microbiology* 7, no. 11: 1853–1858.

Rivett, D.W., and T. Bell. 2018. Abundance determines the functional role of bacterial phylotypes in complex communities. *Nature Microbiology* 3, no. 7: 767–772.

Romero-Perdomo, F., J. Abril, M. Camelo, A. Moreno-Galván, I. Pastrana, D. Rojas-Tapias, and R. Bonilla. 2017. Azotobacter chroococcum as a potentially useful bacterial biofertilizer for cotton (*Gossypium hirsutum*): Effect in reducing N fertilization. *Revista Argentina de Microbiología* 49, no. 4: 377–383.

Saad, M.M., A.A. Eida, and H. Hirt. 2020. Tailoring plant-associated microbial inoculants in agriculture: A roadmap for successful application. *Journal of Experimental Botany* 71, no. 13: 3878–3901.

Schmidt, R., E. Ervin, and X. Zhang. 2003. Questions and answers about biostimulants. *Golf Course Management* 71, no. 6: 91–94.

Sekar, J., R. Raj, and V.R. Prabavathy. 2016. microbial consortial products for sustainable agriculture: Commercialization and regulatory issues in India. In *Agriculturally Important Microorganisms: Commercialization and Regulatory Requirements in Asia*, eds. H.B. Singh, B.K. Sarma, and C. Keswani, 107–132. Singapore: Springer. https://doi.org/10.1007/978-981-10-2576-1_7.

Shaikh, S., and R.Z. Sayyed. 2015. Role of plant growth-promoting rhizobacteria and their formulation in biocontrol of plant diseases. In *Plant Microbes Symbiosis: Applied Facets*, ed. N.K. Arora, 37–351. New Delhi: Springer. 10.1007/978-81-322-2068-8_18.

Sharma, H.S.S., C. Fleming, C. Selby, J. Rao, and T. Martin. 2014. Plant biostimulants: A review on the processing of macroalgae and use of extracts for crop management to reduce abiotic and biotic stresses. *Journal of Applied Phycology* 26, no. 1: 465–490.

Sukul, P., J. Kumar, A. Rani, A.M. Abdillahi, R.B. Rakesh, and M.H. Kumar. 2021. Functioning of plant growth promoting rhizobacteria (PGPR) and their mode of actions: An overview from chemistry point of view. *Plant Archives* 21, no. 1: 628–634.

Tabassum, B., A. Khan, R. Tariq, M. Razman, S. Muhammad, I. Khan, N. Shahid, and K. Aaliya. 2017. Review bottlenecks in commercialisation and future prospects of PGPR. *Applied Soil Ecology* 121: 107–117.

Takors, R. 2012. Scale-up of microbial processes: Impacts, tools and open questions. *Journal of Biotechnology* 160, no. 1–2: 3–9.

Traon, D., L. Amat, F. Zotz, and P. du Jardin. 2014. *A Legal Framework for Plant Biostimulants and Agronomic Fertiliser Additives in the EU-Report to the European Commission, DG Enterprise & Industry*. European Commission Enterprise & Industry Directorate. https://orbi.uliege.be/bitstream/2268/169265/1/A%20legal%20framework%20for%20plant%20biostimulants%20and%20agronomic%20fertilisers%20additives.pdf.

Trivedi, P., C. Mattupalli, K. Eversole, and J.E. Leach. 2021. Enabling sustainable agriculture through understanding and enhancement of microbiomes. *New Phytologist* 230, no. 6 (June 1): 2129–2147.

Trivedi, P., P.M. Schenk, M.D. Wallenstein, and B.K. Singh. 2017. Tiny microbes, big yields: Enhancing food crop production with biological solutions. *Microbial Biotechnology* 10, no. 5: 999–1003.

United Nations, Department of Economic and Social Affairs, Population Division. 2015. *World Population Prospects: The 2015 Revision, Key Findings and Advance Tables*. New York.

Upadhyay, S., and D. Singh. 2015. Effect of salt-tolerant plant growth-promoting rhizobacteria on wheat plants and soil health in a saline environment. *Plant Biology* 17, no. 1: 288–293.

Uribe, D., J. Sanchez-Nieves, and J. Vanegas. 2010. Role of microbial bio-fertilizers in the development of a sustainable agriculture in the tropics. In *Soil Biology and Agriculture in the Tropics*, ed. P. Dion, 21:235–250, Soil Biology. Berlin: Springer.

Uzcátegui-Negrón, M., J. Serrano, P. Boiron, V. Rodriguez-Nava, A. Couble, D. Moniée, K.S. Herrera, H. Sandoval, V. Reviákina, and M. Panizo. 2011. Reclassification by Molecular Methods of Actinobacteria Strains Isolated from Clinical Cases in Venezuela. *Journal de Mycologie Médicale* 21, no. 2: 100–105.

Vasseur-Coronado, M., H.D. du Boulois, I. Pertot, and G. Puopolo. 2021. Selection of Plant Growth Promoting Rhizobacteria Sharing Suitable Features to Be Commercially Developed as Biostimulant Products. *Microbiological Research* 245: 126672.

Vejan, P., R. Abdullah, T. Khadiran, S. Ismail, and A.N. Boyce. 2016. Role of plant growth promoting rhizobacteria in agricultural sustainability-A review. *Molecules* 21: 5–17.

Verma, G., and D.D. Yadav. 2019. Effect of fertility levels and biofertilizers on the productivity and profitability of chickpea (*Cicer arietinum*). *Indian Journal of Agronomy* 64, no. 1: 138–141.

Vessey, J.K. 2003. Plant growth promoting rhizobacteria as biofertilizers. *Plant and Soil* 255: 571–586.

Vijay, K.K.K., S.K. Raju, M. Reddy, J. Kloepper, K. Lawrence, D. Groth, M. Miller, H. Sudini, and D. Binghai. 2009. Evaluation of commercially available PGPR for control of rice sheath blight caused by *Rhizoctonia solani*. *Journal of Pure and Applied Microbiology* 3, no. 2: 485–488.

Wan, B., Y. Liu, Y. Wu, S. Liu, G. Wang, D. Zhang, and Y. Jiang. 2016. Screening, identification of phosphate- and potassium-solubilizing PGPR and its promoting effect on Tobacco. *Journal of Henan Agricultural Sciences* 45, no. 9: 46–51.

Yazdani, M., M.A. Bahmanyar, H. Pirdashti, and M.A. Esmaili. 2009. Effect of phosphate solubilization microorganisms (PSM) and plant growth promoting rhizobacteria (PGPR) on yield and yield components of corn (*Zea Mays* L.). *World Academy of Science, Engineering and Technology* 49: 90–92.

Zhu, Q., X. Liu, T. Hao, M. Zeng, J. Shen, F. Zhang, and W. De Vries. 2018. Modeling soil acidification in typical Chinese cropping systems. *Science of the Total Environment* 613: 1339–1348.

14 Microbial Biostimulants for Plant Protection against Phyllosphere Pathogens

Ziyu Shao, Peter Dart, and Peer M. Schenk
The University of Queensland

CONTENTS

14.1 INTRODUCTION

The plant's phyllosphere is considered a special microbial environment due to its biotic and abiotic factors (e.g., availability of nutrients at the leaf surface). Consequently, interactions between microbial biocontrol species, the plant host, and leaf pathogens are very complex. Resistance against leaf pathogens can be achieved through direct antimicrobial activities or by priming and stimulating the plant's defense activities. Antagonistic activities of beneficial microbial species to control leaf pathogen are highly diverse and are affected by several ecological factors.

This chapter summarizes the recent examples and knowledge gaps of phyllosphere biocontrol that focuses on beneficial bacterial biostimulants and biocontrol agents against common filamentous fungal and bacterial leaf pathogens.

Microbiome composition analyses demonstrate that the plant host's genotype and phenotype are the main factors that shape phyllosphere microbial diversity and richness (Figure 14.1) (Laforest-Lapointe et al., 2017). As shown in Figure 14.1, leaf bacteria that coexist with any introduced

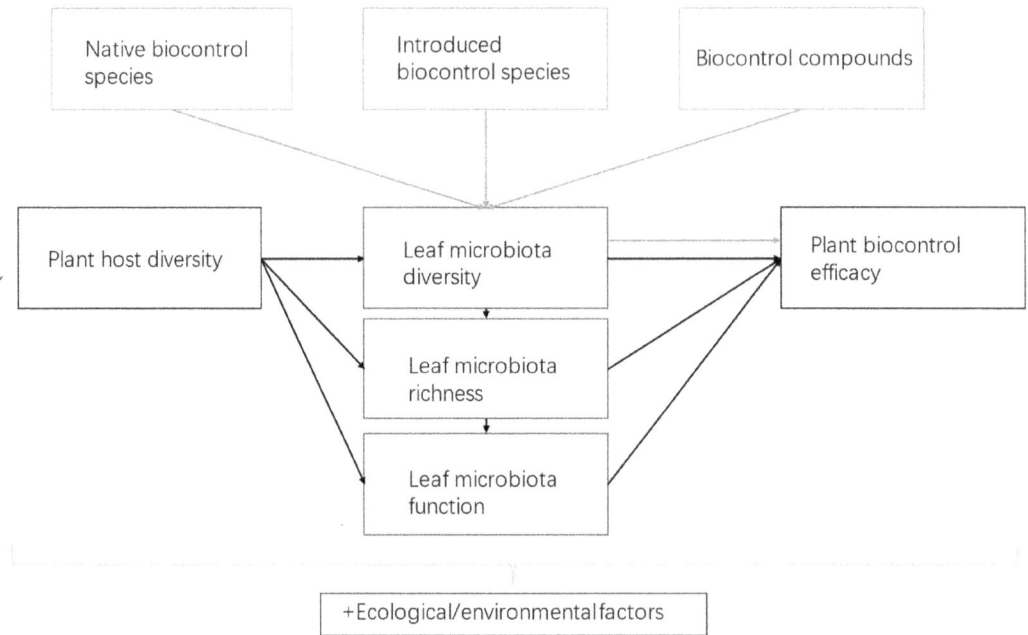

FIGURE 14.1   Relationships/interactions between phyllosphere beneficial strain features and plant protection efficiency.

biocontrol species and/or biocontrol compounds determine their ability to suppress plant disease through different modes of action (Laforest-Lapointe et al., 2017; Legein et al., 2020). Other environmental or ecological factors (e.g., temperature, weather and humidity) are also contributing to the system.

The aims of this chapter are (i) to increase our understanding of the factors that determine the effectiveness of phyllosphere biocontrol, (ii) to list current knowledge gaps and (iii) to make suggestions to further optimize leaf pathogen control through the use of microbial agents.

14.2   ROLE OF THE HOST SPECIES

Regardless of whether in natural or controlled ecosystems, foliar bacterial communities are strongly dependent on the host species and genotype (Purahong et al., 2018; Rastogi et al., 2012; Vorholt, 2012). Here are some examples that explain the specificity of host-dependent (and cultivar-dependent) biocontrol outcomes in the phyllosphere.

14.2.1   Specificity

Phyllosphere interactions between plant hosts and bacterial strains are very complicated. In most cases, such interactions are very specific and vary between different plant hosts, bacterial species, pathogen types, modes of interaction, site of interaction (endophyte or epiphyte), field conditions and other biotic and abiotic factors.

For example, lettuce leaf bacteria composition and diversity are sensitive to the weather and their dynamics are related to geographical and seasonal factors (Rastogi et al., 2012). Kiwifruit plant phyllosphere diversity is related to plant species; for example, one species' (*Actinidia deliciosa*) leaves contained *Pseudomonas* genus strains (both harmless and pathogenic) that were 13 times more abundant than on another species (*Actinidia chinensis*) (Purahong et al., 2018). The species specificity is related to the host's pathogen defense abilities (e.g., kiwifruit species against *Pseudomonas syringae* pv. *actinidiae*) and its ability to support microbial biocontrol activities (Purahong et al., 2018; Thapa & Prasanna, 2018).

Biocontrol cases can also be microbe species specific. Foliar biocontrol can be narrow or broad spectrum, individual strains can act as biocontrol agents on their source-of-origin host crops as well as on other hosts, while the same species can have different biocontrol outcomes from different experimental field trials (Janakiev et al., 2020; Kim et al., 2011). That means that plant–microbe interactions in the phyllosphere require a systems biology approach to be understood as well as comprehensive analyses in order to develop reliable leaf pathogen biocontrol strategies.

14.2.2 Cultivar Differences

The plant's genotype can influence the plant's capability of harboring biocontrol species and pathogens. Some biocontrol cases have shown strong cultivar effects where different plant cultivars affect the disease suppression capacity of beneficial strains. Some *Pseudomonas* strains can reduce severities of potato diseases (e.g., late blight caused by *Phytophthora infestans*), but disease-susceptible potato cultivars (e.g., Bintje) gained higher protection efficiency than that of resistant biovars (e.g., Lady Claire) (De Vrieze, Germanier, Vuille, & Weisskopf, 2018). Similar effects have also been shown in plum (*Prunus domestica*) experiments, where a native *Bacillus thuringiensis* isolate was the only putative biocontrol strain which actively antagonized the growth of a wide range of both native and reference strains of plum pathogens. Four plum varieties had differing populations of resident phyllosphere communities, which also varied between seasons (Janakiev et al., 2020).

The cultivar difference can be explained by microbiota diversity in some cases, rather than the genetic differences between the beneficial biocontrol bacteria. For example, studies found that under the same pathogen attack (e.g., *Pseudomonas syringae* pv. *actinidiae*), different kiwifruit species have different leaf bacterial community responses following the attack, providing different opportunities to screen for biocontrol candidates (Purahong et al., 2018).

14.3 ROLE OF BIOCONTROL BACTERIA

There is a strong correlation between the foliar bacterial diversity (even at family/genus level) and leaf metabolites (mainly nitrogen- and carbohydrate-containing compounds) that are exuded and shape the phyllosphere microbiomes (Ke et al., 2021). Just as for the health response in animals to variations in their microbiomes, the microbial taxa composition in the phyllosphere is critical for plant health (Singh et al., 2019). An interaction of the plant with a "healthy" microbiome that harbors biocontrol species/compounds ensures that it is also well protected against pathogens (Ke et al., 2021).

14.3.1 Taxonomic Analyses of Phyllosphere Microbes

Managing the taxa dynamics of phyllosphere microbiomes may be one of the most important challenges to optimize biocontrol efficiency against leaf pathogens. In the past, bacterial fungicides/biopesticides mainly focused on soil bacterial isolates (plant growth-promoting rhizobacteria, PGPR), e.g., *Pseudomonas* and *Rhizobium* spp. (Pliego et al., 2011), which have promoted pathogen suppression *in vitro* and/or in the rhizosphere. However, rhizosphere microbial isolates may or may not be suitable for phyllosphere applications, for example, because they cannot resist ultraviolet (UV) light exposure effects on their viability.

How plants can select/stimulate and control phyllosphere species and the microbial community composition to maintain plant health is still unclear. Leaves are colonized by soil microbes as seedlings emerge from the soil and are further inoculated by wind and rain, and then these microbial networks become correlated with plant disease immunity (Chen et al., 2020). The most abundant phyllosphere bacterial phyla include *Actinobacteria*, *Proteobacteria* (mainly Alphaproteobacteria and Gammaproteobacteria, e.g., *Pseudomonas fluorescens*) (Rastogi et al., 2012), *Firmicutes* and *Bacteroidetes* (e.g., *Bacillus* spp.) (Thapa & Prasanna, 2018), and these have been screened *in vitro* and strains are selected from several plant hosts to provide biocontrol activities in the phyllosphere. The predominant species constituted a large portion (e.g., >70%) of the phyllosphere community in plants (Chen et al., 2020; Vorholt, 2012). Changes in the *Arabidopsis* genome that affects fitness also resulted in changes in the phyllosphere microbiome, a dysbiosis akin to that in the human inflammatory bowel diseases (Chen et al 2020). There are several commercial biocontrol products in the market now that use these dominant rhizosphere bacteria, especially *Bacillus* spp., as agents against pathogens (Christopher, 2019).

The commonly found, dominant bacterial species in the plant's phyllosphere may not be the dominant species in every plant host. For example, Actinobacteria is one of the most abundant groups that occupied up to 40% of the phyllosphere bacterial community in rice, whereas soybean, white clover and *Arabidopsis* harbored significantly fewer Actinobacteria species (<10%) (Vorholt, 2012). In some cases, dominant species on a particular crop have been selected as biocontrol agents, rather than the predominant species across a range of common crops. Common dominant genera, especially *Methylobacterium*, *Sphingomonas* and *Hymenobacter*, are characterized through their important functions (plant defense, etc.) in several crops (e.g., lettuce) (Janakiev et al., 2020; Singh et al., 2019). New members of dominant species from different crops have been found, including *Achromobacter*, *Comamonas*, *Curtobacterium*, *Enterobacter*, *Leclercia*, *Microbacterium*, *Pantoea*, *Sphingobacterium* and *Stenotrophomonas* (Sahu et al., 2021). These species target specific pathogens with distinct inhibition mechanisms; for instance, *Microbacterium* and *Stenotrophomonas* isolates can inhibit the plant pathogen *Magnaporthe oryzae* on rice leaves by upregulating the jasmonic acid (JA) defense gene *OsCERK1* through interaction between bacterial peptidoglycan and plant receptors (Sahu et al., 2021). Such interactions are specific to the plant species and biocontrol species, as discussed in the following sections. The various roles of dominant phyllosphere species are still unclear at present. It has been suggested that each plant species has its own core and rare phyllosphere microbial groups (Rastogi et al., 2012).

Interestingly, in some cases, leaf microbiota strain genetic modifications or response to chemicals addition changes only the rare genera or non-dominant species composition that occupy a small portion (e.g., <20%) of the phyllosphere microbiome. In those cases, the core phyllosphere microbiota composition can only be slightly changed by external signals (Perazzolli et al., 2020). Gammaproteobacteria, for example, which is the most abundant phyllosphere phylum that contains *Enterobacteriaceae*, *Orbaceae*, *Pseudomonadaceae* and others, were quite consistent across plant species (Massoni et al., 2020) or only slightly changed following addition of a sugar (Perazzolli et al., 2020). Modification of the conserved microbiota might need introduction of external bacterial species and/or compounds that modify the rare communities first before affecting the core groups indirectly (Perazzolli et al., 2020).

Hence, the current biocontrol theory has moved to a strategy to manage the native microbial communities by modifying the balance between the core species and the rare species (Chen et al., 2020; Legein et al., 2020). The most common way is to increase/reintroduce a single or combined native species (two or more strains) and/or compounds to affect the existing microbial communities (Janakiev et al., 2020; Xu & Jeger, 2013). The introduced species can belong to core or rare groups, but, due to the leaf biocontrol specificity of these introduced species, and differences in responses between plant and microbe species tested, it is very hard to compare the results from different experiments.

In summary, there are different strategies to manage taxa in the phyllosphere environment that are useful tools in the selection of biocontrol strains: (i) use of rhizosphere isolates, (ii) use of other dominant species from a particular crop, (iii) use of predominant foliar species across plant species, and (iv) methods for rebalancing or maintaining community structure of native core and rare microbial species.

14.4 BACTERIAL BIODIVERSITY

Bacterial communities in the phyllosphere are highly variable both between plant species, stage of development and the environmental conditions (e.g., seasons). Also, the phyllosphere microbiome-"carrying" capacity is limited and the microbiome constituents do not occupy the whole leaf surface and are distributed often in clumps and what determines this is not yet well understood (Leveau, 2019; Meyer & Leveau, 2012). It may be directly linked to the local nutrient availability (Remus-Emsermann et al., 2012). In this harsh environment with high UV radiation and high microbial diversity and competition between cells, foliar bacterial systems have formed a dynamic diversity to balance the dominant species versus the rare groups and the native species versus the introduced organisms (Vorholt, 2012).

14.4.1 NATIVE PHYLLOSPHERE BACTERIA

Most phyllosphere bacterial species usually become dominant or native because of their functions. For instance, *Methylobacterium*, the dominant phyllosphere genus in a range of plant species such as *Vitis* (grape) and strawberry, promotes the host's growth by converting methanol from stomata into growth biostimulants (Abanda-Nkpwatt et al., 2006).

Using native species and/or compounds produced by native species to gain consistent biocontrol effects is a kind of biocontrol method. To avoid disrupting the original phyllosphere microbial diversity away from an equilibrium, the emphasis has moved to managing the composition of the rarer groups of species to achieve biocontrol. For example, treatment of grapevines with tagatose, which is a sugar-like sweetener compound from *Sterculia setigera* gum or derived from milk lactose with wide-scale use in foods, can suppress infections by downy mildew (*Plasmopara viticola*) and powdery mildew (*Erysiphe necator*) in the field by increasing the relative abundance of the native beneficial genera *Methylobacterium* on the host leaves. This shift in the phyllosphere population dynamics is thus directly related to a nutritional/antinutritional role, which is also dependent on the location and composition of the existing (native) microbiome. This is sufficient to shift its response toward controlling the pathogen (Perazzolli et al., 2020). Tagatose phyllosphere sprays are thus likely to have a plant prebiotic effect.

Some dual combinations of nine different native rhizosphere and phyllosphere beneficial strains further enhanced the disease suppression of potato late blight phytophthora over that obtained with the single strains. The mixture of two native phyllosphere *Pseudomonas* spp. (e.g., *P. frederiksbergensis* S19 and *P. fluorescens* S49) used to treat *Phytophthora* spp. infection in potato *in vitro* (leaf disk; infection bioassays) provided the most disease control across three cultivars with different sensitivities to phytophthora from sensitive to resistant. Only the phyllosphere strain S35 was active as a single strain against pathogens on all three cultivars, while the activity of other strains varied across the cultivars (De Vrieze et al., 2018; Hunziker et al., 2015). The use of combinations of native strains as inoculants now needs field evaluation.

The variation in pathogenic species already resident in the phyllosphere of different crop microbial communities is also worth investigating, e.g., *Pseudomonas syringae* and *Pseudomonas graminis* in plums which *Bacillus thuringiensis* was able to control (Janakiev et al., 2020). Those naturally occurring pathogens come in combinations, which requires specific defense mechanisms/biocontrol methods against all the pathogens (Purahong et al., 2018). On the contrary, native foliar species evolve with common pathogens often present in the phyllosphere microbiome; they also evolve specific interactions controlling further pathogen development, while introduced species/strains may not be as effective (Janakiev et al., 2020). The management of introduced inoculant species and the effect of augmenting a native phyllosphere species, on this microbe–pathogen interaction needs further study.

14.4.2 Introduced Biocontrol Bacteria

Maintaining or modifying the delicate balance between the native species and the introduced species can affect the plant protection outcomes. In a lettuce study, the introduced pathogen *Rhizoctonia solani* increased the diversity of native phyllosphere microbes, but further introduction of beneficial strains (*Bacillus amyloliquefaciens*) onto the same hosts resulted in a reduced diversity of endemic bacteria as well as fungal pathogens (Erlacher et al., 2014; Rosier et al., 2016).

The introduced species selected for its biocontrol potential may even attack native microorganisms rather than the target pathogen. For example, a *Trichoderma harzianum* inoculum did not suppress the pathogen *Botrytis cinerea* on strawberry leaves, but changed the native fungal diversities (Sylla et al., 2013). The communication occurring between components of the phyllosphere microbiome is partly determined by the community structure of dominant and rare species, and the mode of action of the beneficial biocontrol microbes will have impacts on the biocontrol outcomes.

14.5   BACTERIAL RICHNESS

The overall microbial species richness in foliage communities is relatively high (Vorholt, 2012). Externally applied biocontrol species and native phyllosphere species affect each other via signal transduction and exchanges of molecules (Chen et al., 2020). Such interactions can also affect the biocontrol outcomes. Recent insights into interactions in the phyllosphere microbiome networks obtained through quorum sensing and quenching are helping to understand the community-level interactions involved in biocontrol activity (Ma et al., 2013).

14.5.1 Quorum Sensing in the Phyllosphere

Quorum sensing is a way that Gram-negative bacterial populations can communicate with each other via production of small molecular inducers which affect gene expression, when they reach a threshold concentration, and thereby limit population numbers and density. Quorum sensing is a typical population density-dependent system, which is also related to the bacterial diversity in the phyllosphere. Quorum sensing and quenching (the degradation of the quorum signaling compound) pathways are employed by both beneficial species and pathogenic species, providing the signal connection between, and within, both populations of microbes. For example, production of secondary metabolites difficidin and bacilysin by beneficial strains of *Bacillus amyloliquefaciens* is an outcome of quorum sensing, and these compounds suppress the virulence of the bacterial pathogen (*Xanthomonas oryzae* pv. *oryzae*) on rice seedling leaves via modification of the pathogen gene expression (Wu et al., 2015).

Quorum sensing molecules can have some bacteria species-related specificity; for example, *Pseudomonas* spp. mainly have DNA-binding transcriptional activators (LuxR), while *Vibrio* spp. have membrane-binding receptors (LuxPQ) (Papenfort & Bassler, 2016). The quorum dynamics are complex, and the plant host's responses are also involved in the feedback loop, which can trigger biocontrol responses (Legein et al., 2020; Papenfort & Bassler, 2016). Plants can benefit from the bacterial quorum sensing molecules, for example by upregulating the salicylic acid defense pathway when exposed to the quorum sensing activators, *N*-acyl homoserine lactones (AHLs), and this priming of the salicylic pathway leads to control of the bacterial pathogen multiplication and density and its disease-forming potential (Legein et al., 2020; Schenk & Schikora, 2014).

Quorum sensing represents a pathway for moderating interactions between the phyllosphere microbiome bacteria that lead to biocontrol, and this needs further elucidation. A question that remains is as follows: How can the phyllosphere microbiome population density be increased to induce or produce quorum sensing molecules? Is there a limiting nutrient whose addition could be the inducer? Perhaps it could operate through a prebiotic additive in the inoculum that promotes the growth of strains that produce the quorum sensing compound, for example the use of milk powder as the inoculum carrier. Is the biocontrol response to tagatose addition to grapevines mediated in this manner (Perazzolli et al., 2020)?

14.5.2 DYNAMICS OF LEAF MICROBIAL COMMUNITIES

Other signal exchanges within phyllosphere microbial communities also affect their dynamics. For example, the relationship between leaf microbial community richness and pathogen virulence is crucial to leaf pathogen biocontrol outcomes. In some cases, native bacterial richness is directly related to the susceptibility to pathogen invasion. For example, when treating tobacco wildfire disease in the field, some beneficial strains can suppress the pathogens indirectly via increasing the abundance of *Sphingomonas* and *Pantoea* spp. in the host leaf microbiome (Qin et al., 2019). The underlying communication networks include quorum sensing and other pathways such as nitrogen fixation and auxin (indole acetic acid) production (Qin et al., 2019).

The kiwifruit pathogen *Pseudomonas syringae* pv. *actinidiae* (*Psa*) regulates its population and virulence via AHLs that require the presence of special receptors/translators (LuxI/R) for quorum sensing to take place (Ma et al., 2013; Papenfort & Bassler, 2016). However, *Psa* does not have AHL or LuxI/R genes, but only has three putative LuxR solos instead (Patel et al., 2014). The absent molecules have to be obtained from exogenous sources, and one of the LuxR solos' role is most likely to interact with plant host epiphytic bacteria (e.g., *Lactobacillus plantarum*) (Patel et al., 2014; Purahong et al., 2018). As mentioned above, AHLs can also trigger plant host immune responses (Hughes & Sperandio, 2008; Rosier et al., 2016). Thus, controlling the native or introduced foliar species richness in a manner that can interfere with this pathway can reduce the pathogenicity of *Psa* or inhibit its invasion (Ma et al., 2013; Purahong et al., 2018).

Further studies of phyllosphere microbiome dynamics are required to explore the inter-species communications between the native bacteria including pathogens and the introduced pathogens and/or biocontrol species.

14.6 OPTIMIZATION OF PATHOGEN BIOCONTROL EFFICACY

14.6.1 MODE OF ACTION

According to previous studies, phyllosphere biocontrol modes of action can be divided into direct and indirect groups, which involve several different mechanisms, including antibiosis, parasitism and/or predation, competition for space and nutrients, induced plant resistance and plant hormone interference (Figure 14.2) (Legein et al., 2020; Pliego et al., 2011). Different modes of action are related to bacterial functions in the phyllosphere, which also link to communication between microorganisms. Environmental factors also play an important role in the activity of biocontrol microbes in the phyllosphere.

14.6.1.1 Colonization

Foliar bacteria need to form aggregates with extracellular polymeric substances to colonize plant leaves along with other specific features such as the production of biosurfactants to enhance leaching abilities and increase leaf membrane permeability (Vorholt, 2012).

How different beneficial bacteria can colonize the phylloplane region of the phyllosphere is poorly understood, but this can have big consequences for plant health. Su et al. (2019) evaluated the effects of different phases of colonization of a *Rhodopseudomonas palustris* strain Gj-22 on tobacco mosaic virus control of *Nicotiana benthamiana* and showed that only specific stages of colonization and development/dissolution of bacterial aggregates formed will result in the priming of the defense system of the host plants. The site of interaction on the leaf and the concentration of bacterial aggregates also contributed to the signaling (Su et al., 2019).

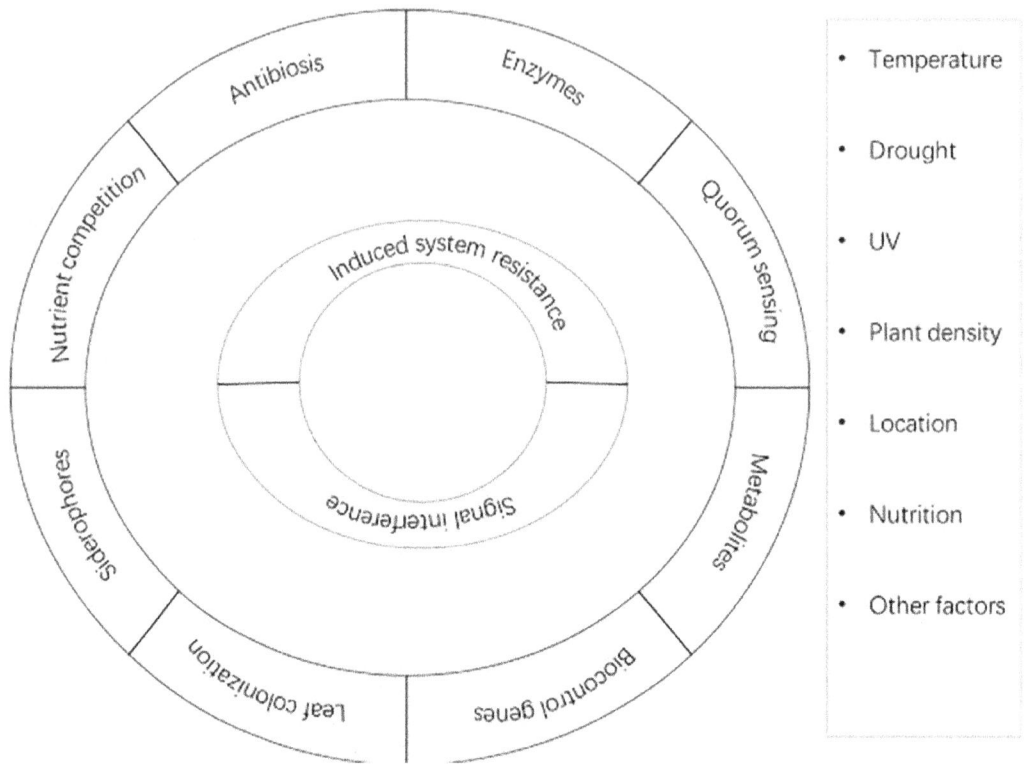

FIGURE 14.2   Modes of action of phyllosphere biocontrol species.

Priming a plant's defense system for action occurs when the processes by which the natural pathogen infects the host plant are mimicked, resulting in the activation of specific defense responses. For example, *Pseudomonas protegens* (strain CS1) can colonize lemon leaves with lemon pathogen *Xanthomonas citri* subsp. *citri*, and it will then release the siderophore pyochelin to help the plant and other bacteria to generate reactive oxygen species (ROS) that act against the pathogen (Michavila et al., 2017).

14.6.1.2   Secondary Metabolites

The secondary metabolites of phyllosphere microbes include infusible compounds (mainly volatiles), enzymes (mainly hydrolytic enzymes) and regulators (e.g., of community population density) (Legein et al., 2020; Vogel et al., 2012). Spraying these metabolites on foliage can trigger plant immune gene responses (Sanchez et al., 2012). Interestingly, phyllosphere species can also produce phenazines, similar to rhizosphere species (Legein et al., 2020; Miquel Guennoc et al., 2018). Phenazines can inhibit fungal mycelial growth by affecting fungal histone acetylation processes (Miquel Guennoc et al., 2018). How and whether phenazines from foliar species can act as biocontrol compounds is still unknown.

The bacterial surfactant production by biocontrol strains such as *Bacillus amyloliquefaciens* leads to biofilm formation on the leaf, and this enhances the biocontrol effects. The interference of biofilm synthesis by some ascomycetes fungi (Miquel Guennoc et al., 2018) may also influence the biocontrol effect of a biofilm produced on the leaf by biocontrol strains (Bais et al., 2004). The hypothesis needs field experiments for validation of practical applications.

14.6.1.3   Biocontrol Genes

Identifying the mode of actions of beneficial strains by sequence analysis of their genomes is critical when screening for potential biocontrol candidates. The traditional method to find plant protection genes in foliar bacterial species was via a forward mutant/genetic screen. For example, Vogel et al. (2012) introduced more than 5,000 Tn5 mutant strains of *Sphingomonas* sp. strain Fr1, which has biocontrol activity against the model pathogen *Pseudomonas syringae* pv. tomato DC3000, to Arabidopsis plants to find the mutant, pathogen suppression strains and thereby the active genes in the parent strain. Several transposon insertion sites revealed the presence of hydrolytic enzyme genes (e.g., some histidine kinase and transglycosylase genes) as being involved in the biocontrol activity (Vogel et al., 2012). This may enable the location of the same gene sequences in other strains of *Sphingomonas* sp. for testing as potential biocontrol candidates.

Now, studies are using next-generation technologies to perform directed selection of candidate biocontrol microbes based on the particular biocontrol gene(s). For example, rice phyllosphere actinomycetes such as some strains from *Streptomyces* genera (Actinobacteria), *Saccharothrix* genera (Actinomycetales) and Bacillaceae can have genetic domains for nonribosomal peptide synthetase (NRPS) genes and/or type I polyketide synthase (PKS) genes (Abderrahmani et al., 2011; Harsonowati, et al., 2017). The enzymes NRPS and PKS can catalyze several biological functions in fungi, including iron acquisition, drug resistance, antimicrobial activity (Schrettl et al., 2007; Süssmuth et al., 2011). These compounds can suppress rice blast disease *Pyricularia oryzae* by up to 88% in Indonesia (Harsonowati et al., 2017).

14.6.2 Monitoring of Biocontrol Microbes

In order to optimize the efficiency of biocontrol agents, the temporal tracking of the surviving inoculant population in the host's phyllosphere is important. Options are selection of strains with rifampicin resistance and use of plate counting, or PCR, but it does not readily discriminate between the DNA of living and dead cells, or the use of a green fluorescent protein construct. Tut et al. (2021) used propidium monoazide (PMA)-based quantitative PCR to monitor viable populations of two commercial phyllosphere biocontrol strains in lettuce, *Bacillus subtilis* and *Gliocladium catenulatum* (Tut et al., 2021). The PMA tool can aggregate the DNA of dead cells and block its amplification. There was a considerable amount of non-viable DNA (Fittipaldi et al., 2012). A readily usable tool for quantitative monitoring of the populations of inoculants in the phyllosphere will be invaluable support for progressing putative inoculants into commercialization.

Other techniques can also detect single cells to separate lysate cells from the population, such as MLTreeMap web, which require software frameworks (Stark et al., 2010). But the method needs professional technical support. Having a readily usable method for monitoring both biocontrol inoculant strain and pathogen populations in the phyllosphere will greatly help in the progress to commercialization of many strains with disease control potential, which have been characterized in the last decade.

14.6.3 Other Biocontrol Species

Biocontrol microbes other than phyllosphere bacteria can also be used in crop protection. Yeasts, for example, may be used for biocontrol as well. Apple phyllosphere yeasts, including *Metschnikowia pulcherrima*, *Candida subhashii* and *Aureobasidium pullulans*, were able to inhibit the growth of filamentous fungi (e.g., *Gibberella fujikuroi*) *in vitro* by synthesizing antifungal enzymes to digest fungal cell walls (Hilber-Bodmer et al., 2017). Some of these phyllosphere yeasts can also colonize on plant root exudates so that they can potentially control soilborne pathogens. However, soil yeasts have a more potent activity against soilborne pathogens (Hilber-Bodmer et al., 2017). Therefore, phyllosphere yeasts are also potential biocontrol candidates for phyllosphere and soilborne pathogens.

The complexity of the interactions in the biocontrol processes, influenced by plant species, plant genotype, resident native phyllosphere bacterial species community diversity and richness of species within it, microbe-to-microbe communication, mode of interaction between microbial strains and pathogens, has made it difficult to select inoculants that consistently control disease and is why so few are on the market at present (Dart, Shao, & Schenk, 2021).

REFERENCES

Abanda-Nkpwatt, D., Musch, M., Tschiersch, J., Boettner, M., & Schwab, W. (2006). Molecular interaction between *Methylobacterium extorquens* and seedlings: Growth promotion, methanol consumption, and localization of the methanol emission site. *Journal of Experimental Botany, 57*(15), 4025–4032. doi:10.1093/jxb/erl173

Abderrahmani, A., Tapi, A., Nateche, F., Chollet, M., Leclere, V., Wathelet, B., … Jacques, P. (2011). Bioinformatics and molecular approaches to detect NRPS genes involved in the biosynthesis of kurstakin from Bacillus thuringiensis. *Applied Microbiology and Biotechnology, 92*(3), 571–581. doi:10.1007/s00253-011-3453-6.

Bais, H. P., Fall, R., & Vivanco, J. M. (2004). Biocontrol of *Bacillus subtilis* against infection of *Arabidopsis* roots by *Pseudomonas syringae* is facilitated by biofilm formation and surfactin production. *Plant Physiology, 134*(1), 307–319. doi:10.1104/pp.103.028712.

Chen, T., Nomura, K., Wang, X., Sohrabi, R., Xu, J., Yao, L., … He, S. Y. (2020). A plant genetic network for preventing dysbiosis in the phyllosphere. *Nature, 580*(7805), 653–657. doi:10.1038/s41586-020-2185-0.

Christopher, A. D. (2019). Taxonomy of registered *Bacillus* spp. strains used as plant pathogen antagonists. *Biological Control, 134*, 82–86. doi:10.1016/j.biocontrol.2019.04.011.

Dart, P., Shao, Z., & Schenk, P. M. (2021). Biopesticide commercialisation in Australia: Potential and challenges. In O. Koul (Ed.), *Developement and Commercialization of Biopesticides: Costs and Benefits*. Melbourne: Elsevier.

De Vrieze, M., Germanier, F., Vuille, N., & Weisskopf, L. (2018). Combining different potato-associated *Pseudomonas* strains for improved biocontrol of *Phytophthora infestans*. *Frontiers in Microbiology, 9*, 2573. doi:10.3389/fmicb.2018.02573.

Erlacher, A., Cardinale, M., Grosch, R., Grube, M., & Berg, G. (2014). The impact of the pathogen Rhizoctonia solani and its beneficial counterpart *Bacillus amyloliquefaciens* on the indigenous lettuce microbiome. *Frontiers in Microbiology, 5*, 175. doi:10.3389/fmicb.2014.00175.

Fittipaldi, M., Nocker, A., & Codony, F. (2012). Progress in understanding preferential detection of live cells using viability dyes in combination with DNA amplification. *Journal of Microbiological Methods, 91*(2), 276–289. doi:10.1016/j.mimet.2012.08.007.

Harsonowati, W., Astuti, R. I., & Wahyudi, A. T. (2017). Leaf blast disease reduction by rice-phyllosphere actinomycetes producing bioactive compounds. Journal of General Plant Pathology, 83(2), 98–108.

Hilber-Bodmer, M., Schmid, M., Ahrens, C. H., & Freimoser, F. M. (2017). Competition assays and physiological experiments of soil and phyllosphere yeasts identify *Candida subhashii* as a novel antagonist of filamentous fungi. *BMC Microbiology, 17*(1), 4. doi:10.1186/s12866-016-0908-z.

Hughes, D. T., & Sperandio, V. (2008). Inter-kingdom signalling: Communication between bacteria and their hosts. *Nature Reviews Microbiology, 6*(2), 111–120. doi:10.1038/nrmicro1836.

Hunziker, L., Bonisch, D., Groenhagen, U., Bailly, A., Schulz, S., & Weisskopf, L. (2015). *Pseudomonas* strains naturally associated with potato plants produce volatiles with high potential for inhibition of *Phytophthora infestans*. *Applied and Environmental Microbiology, 81*(3), 821–830. doi:10.1128/AEM.02999-14.

Janakiev, T., Dimkic, I., Bojic, S., Fira, D., Stankovic, S., & Beric, T. (2020). Bacterial communities of plum phyllosphere and characterization of indigenous antagonistic *Bacillus thuringiensis* R3/3 isolate. *Journal of Applied Microbiology, 128*(2), 528–543. doi:10.1111/jam.14488.

Ke, M., Ye, Y., Li, Y., Zhou, Z., Xu, N., Feng, L., … Qian, H. (2021). Leaf metabolic influence of glyphosate and nanotubes on the *Arabidopsis thaliana* phyllosphere. *Journal of Environmental Sciences (China), 106*, 66–75. doi:10.1016/j.jes.2021.01.002.

Kim, Y. C., Leveau, J., McSpadden Gardener, B. B., Pierson, E. A., Pierson 3rd, L. S., & Ryu, C. M. (2011). The multifactorial basis for plant health promotion by plant-associated bacteria. *Applied and Environmental Microbiology, 77*(5), 1548–1555. doi:10.1128/AEM.01867-10.

Laforest-Lapointe, I., Paquette, A., Messier, C., & Kembel, S. W. (2017). Leaf bacterial diversity mediates plant diversity and ecosystem function relationships. *Nature, 546*(7656), 145–147. doi:10.1038/nature22399.

Legein, M., Smets, W., Vandenheuvel, D., Eilers, T., Muyshondt, B., Prinsen, E., … Lebeer, S. (2020). Modes of action of microbial biocontrol in the phyllosphere. *Frontiers in Microbiology, 11*, 1619. doi:10.3389/fmicb.2020.01619.

Leveau, J. H. (2019). A brief from the leaf: Latest research to inform our understanding of the phyllosphere microbiome. *Current Opinion in Microbiology, 49*, 41–49. doi:10.1016/j.mib.2019.10.002.

Ma, A., Lv, D., Zhuang, X., & Zhuang, G. (2013). Quorum quenching in culturable phyllosphere bacteria from tobacco. *International Journal of Molecular Sciences, 14*(7), 14607–14619. doi:10.3390/ijms140714607.

Massoni, J., Bortfeld-Miller, M., Jardillier, L., Salazar, G., Sunagawa, S., & Vorholt, J. A. (2020). Consistent host and organ occupancy of phyllosphere bacteria in a community of wild herbaceous plant species. *The ISME Journal, 14*(1), 245–258. doi:10.1038/s41396-019-0531-8.

Meyer, K. M., & Leveau, J. H. (2012). Microbiology of the phyllosphere: A playground for testing ecological concepts. *Oecologia, 168*(3), 621–629. doi:10.1007/s00442-011-2138-2.

Michavila, G., Adler, C., De Gregorio, P. R., Lami, M. J., Caram Di Santo, M. C., Zenoff, A. M., … Vincent, P. A. (2017). *Pseudomonas protegens* CS1 from the lemon phyllosphere as a candidate for citrus canker biocontrol agent. *Plant Biology (Stuttg), 19*(4), 608–617. doi:10.1111/plb.12556.

Miquel Guennoc, C., Rose, C., Labbe, J., & Deveau, A. (2018). Bacterial biofilm formation on the hyphae of ectomycorrhizal fungi: A widespread ability under controls? *FEMS Microbiology Ecology, 94*(7). doi:10.1093/femsec/fiy093.

Papenfort, K., & Bassler, B. L. (2016). Quorum sensing signal-response systems in Gram-negative bacteria. *Nature Reviews Microbiology, 14*(9), 576–588. doi:10.1038/nrmicro.2016.89.

Patel, H. K., Ferrante, P., Covaceuszach, S., Lamba, D., Scortichini, M., & Venturi, V. (2014). The kiwifruit emerging pathogen *Pseudomonas syringae* pv. *actinidiae* does not produce AHLs but possesses three luxR solos. *PLoS One, 9*(1), e87862. doi:10.1371/journal.pone.0087862.

Perazzolli, M., Nesler, A., Giovannini, O., Antonielli, L., Puopolo, G., & Pertot, I. (2020). Ecological impact of a rare sugar on grapevine phyllosphere microbial communities. *Microbiological Research, 232*, 126387. doi:10.1016/j.micres.2019.126387.

Pliego, C., Kamilova, F., & Lugtenberg, B. (2011). Plant growth-promoting bacteria: Fundamentals and exploitation. In *Bacteria in Agrobiology: Crop Ecosystems*, 295–343. doi:10.1007/978-3-642-18357-7_11.

Purahong, W., Orru, L., Donati, I., Perpetuini, G., Cellini, A., Lamontanara, A., … Spinelli, F. (2018). Plant microbiome and its link to plant health: Host species, organs and *Pseudomonas syringae* pv. *actinidiae* infection shaping bacterial phyllosphere communities of kiwifruit plants. *Frontiers in Plant Science, 9*, 1563. doi:10.3389/fpls.2018.01563.

Qin, C., Tao, J., Liu, T., Liu, Y., Xiao, N., Li, T., … Meng, D. (2019). Responses of phyllosphere microbiota and plant health to application of two different biocontrol agents. *AMB Express, 9*(1), 42. doi:10.1186/s13568-019-0765-x.

Rastogi, G., Sbodio, A., Tech, J. J., Suslow, T. V., Coaker, G. L., & Leveau, J. H. (2012). Leaf microbiota in an agroecosystem: Spatiotemporal variation in bacterial community composition on field-grown lettuce. *The ISME Journal, 6*(10), 1812–1822. doi:10.1038/ismej.2012.32.

Remus-Emsermann, M. N., Tecon, R., Kowalchuk, G. A., & Leveau, J. H. (2012). Variation in local carrying capacity and the individual fate of bacterial colonizers in the phyllosphere. *The ISME Journal, 6*(4), 756–765. doi:10.1038/ismej.2011.209.

Rosier, A., Bishnoi, U., Lakshmanan, V., Sherrier, D. J., & Bais, H. P. (2016). A perspective on inter-kingdom signaling in plant-beneficial microbe interactions. *Plant Molecular Biology, 90*(6), 537–548. doi:10.1007/s11103-016-0433-3.

Sahu, K. P., Kumar, A., Patel, A., Kumar, M., Gopalakrishnan, S., Prakash, G., … Gogoi, R. (2021). Rice blast lesions: An unexplored phyllosphere microhabitat for novel antagonistic bacterial species against *Magnaporthe oryzae*. *Microbial Ecology, 81*(3), 731–745. doi:10.1007/s00248-020-0161-3.

Sanchez, L., Courteaux, B., Hubert, J., Kauffmann, S., Renault, J. H., Clement, C., … Dorey, S. (2012). Rhamnolipids elicit defense responses and induce disease resistance against biotrophic, hemibiotrophic, and necrotrophic pathogens that require different signaling pathways in *Arabidopsis* and highlight a central role for salicylic acid. *Plant Physiology, 160*(3), 1630–1641. doi:10.1104/pp.112.201913.

Schenk, S. T., & Schikora, A. (2014). AHL-priming functions via oxylipin and salicylic acid. *Frontiers in Plant Science, 5*, 784. doi:10.3389/fpls.2014.00784.

Schrettl, M., Bignell, E., Kragl, C., Sabiha, Y., Loss, O., Eisendle, M., … Haas, H. (2007). Distinct roles for intra- and extracellular siderophores during *Aspergillus fumigatus* infection. *PLoS Pathogens, 3*(9), 1195–1207. doi:10.1371/journal.ppat.0030128.

Singh, P., Santoni, S., Weber, A., This, P., & Peros, J. P. (2019). Understanding the phyllosphere microbiome assemblage in grape species (Vitaceae) with amplicon sequence data structures. *Scientific Reports, 9*(1), 14294. doi:10.1038/s41598-019-50839-0.

Stark, M., Berger, S. A., Stamatakis, A., & von Mering, C. (2010). MLTreeMap–accurate maximum likelihood placement of environmental DNA sequences into taxonomic and functional reference phylogenies. *BMC Genomics, 11*, 461. doi:10.1186/1471-2164-11-461.

Su, P., Zhang, D., Zhang, Z., Chen, A., Hamid, M. R., Li, C., … Liu, Y. (2019). Characterization of *Rhodopseudomonas palustris* population dynamics on tobacco phyllosphere and induction of plant resistance to Tobacco mosaic virus. *Microbial Biotechnology, 12*(6), 1453–1463. doi:10.1111/1751-7915.13486.

Sussmuth, R., Muller, J., von Dohren, H., & Molnar, I. (2011). Fungal cyclooligomer depsipeptides: From classical biochemistry to combinatorial biosynthesis. *Natural Product Reports, 28*(1), 99–124. doi:10.1039/c001463j.

Sylla, J., Alsanius, B. W., Kruger, E., Reineke, A., Strohmeier, S., & Wohanka, W. (2013). Leaf microbiota of strawberries as affected by biological control agents. *Phytopathology, 103*(10), 1001–1011. doi:10.1094/Phyto-01-13-0014-R.

Thapa, S., & Prasanna, R. (2018). Prospecting the characteristics and significance of the phyllosphere microbiome. *Annals of Microbiology, 68*(5), 229–245. doi:10.1007/s13213-018-1331-5.

Tut, G., Magan, N., Brain, P., & Xu, X. (2021). Molecular assay development to monitor the kinetics of viable populations of two biocontrol agents, *Bacillus subtilis* QST 713 and *Gliocladium catenulatum* J1446, in the phyllosphere of lettuce leaves. *Biology (Basel), 10*(3). doi:10.3390/biology10030224.

Vogel, C., Innerebner, G., Zingg, J., Guder, J., & Vorholt, J. A. (2012). Forward genetic in planta screen for identification of plant-protective traits of *Sphingomonas* sp. strain Fr1 against *Pseudomonas syringae* DC3000. *Applied and Environmental Microbiology, 78*(16), 5529–5535. doi:10.1128/AEM.00639-12.

Vorholt, J. A. (2012). Microbial life in the phyllosphere. *Nature Reviews Microbiology, 10*(12), 828–840. doi:10.1038/nrmicro2910.

Harsonowati, W., Astuti, R., & Wahyudi, A.T. (2017). Leaf blast disease reduction by rice-phyllosphere actinomycetes producing bioactive compounds. *The Phytopathological Society of Japan and Springer Japan, 83*, 98–108. doi:10.1007/s10327-017-0700-4.

Wu, L., Wu, H., Chen, L., Yu, X., Borriss, R., & Gao, X. (2015). Difficidin and bacilysin from *Bacillus amyloliquefaciens* FZB42 have antibacterial activity against *Xanthomonas oryzae* rice pathogens. *Scientific Reports, 5*, 12975. doi:10.1038/srep12975.

Xu, X. M., & Jeger, M. J. (2013). Combined use of two biocontrol agents with different biocontrol mechanisms most likely results in less than expected efficacy in controlling foliar pathogens under fluctuating conditions: A modeling study. *Phytopathology, 103*(2), 108–116. doi:10.1094/PHYTO-07-12-0167-R.

Index

Note: **Bold** page numbers refer to tables and *italic* page numbers refer to figures.